TVA
AND THE
TELLICO DAM
1936-1979

TVA
AND THE
TELLICO DAM
1936-1979

A BUREAUCRATIC CRISIS
IN POST-INDUSTRIAL AMERICA

BY WILLIAM BRUCE WHEELER AND MICHAEL J. McDONALD

THE UNIVERSITY OF TENNESSEE PRESS / KNOXVILLE

Frontispiece: The Tellico Dam, nearly completed, 1978. Courtesy Tennessee Valley Authority.

∞ The paper used in this book meets the minimum requirements of the American National Standard for Permanence of Paper for Printed Library Materials, Z39.48–1984. Binding materials have been chosen for durability.

LIBRARY OF CONGRESS CATALOGING-IN-PUBLICATION DATA

Wheeler, William Bruce, 1939–
  TVA and the Tellico Dam, 1936–1979.

  Bibliography: p.
  Includes index.
  1. Tennessee Valley Authority. 2. Tellico Dam (Tenn.)
I. McDonald, Michael J., 1934–   . II. Title.
III. Title: Tennessee Valley Authority and the Tellico
Dam, 1936–1979.
HD9685.U7T387   1986   353.0082'3'09768   85-22224
ISBN 0-87049-492-9 (alk. paper)

CONTENTS

Preface   *page ix*

1. TVA and the Search for a New Mission   *3*

2. The Decision to Build the Tellico Dam   *23*

3. The Little Tennessee Valley and the Process of
    Modernization   *46*

4. Storm Clouds on the Horizon:
    The Rise of Opposition   *64*

5. The Search for Justification:
    The Benefit-Cost Problem   *87*

6. Attacks and Counterattacks:
    The Tellico Debate Goes National   *124*

7. The Model City as Microcosm:
    The Timberlake Imbroglio   *158*

8. The Snail Darter, the Courts, and the Changing of the
    Guard at TVA   *184*

    Conclusion: TVA and the Tellico Dam   *214*

    Appendix: Land Purchases by TVA at Tellico,
    1967–1976   *221*

    Notes   *226*

    A Note on Sources   *271*

    Acknowledgments   *276*

    Index   *278*

ILLUSTRATIONS

PHOTOS   *following page 110*
The Little Tennessee River
Aerial View of Fort Loudoun
The Little Tennessee River
Gordon Clapp
Arthur Jandrey
General Herbert Vogel
Aubrey "Red" Wagner
Gabriel O. Wessenauer
Louis Van Mol
Aerial View of the Tellico
   Project Site
J. Porter Taylor
George Palo
Reed Elliot
Robert Howes
Paul Evans
Richard "Dick" Kilbourne
E. P. "Phil" Ericson
Don McBride
Minnard "Mike" Foster

Justice William O. Douglas
Directors Frank Smith, "Red"
   Wagner, and Don McBride
   Conferring with Alfred
   "Ted" Guthe
Aerial View of an Archaeologi-
   cal Dig
Worth Greene and Two Chero-
   kee Visitors
A. J. "Flash" Gray
James Gober
F. B. "Red" Williams
TVA Model of the Planned City
   of Timberlake
Lynn Seeber
Wagner Cheered by TVA
   Employees on his Last Day
TVA Model of the Tellico Dam
Richard M. Freeman, Charles
   H. Dean, Jr., and S. David
   Freeman

MAPS   1. TVA Projects in the Tennessee Valley Region   *page 20*
2. Little Tennessee River Watershed   *25*
3. Tellico Dam Site and Vicinity   *26*

TABLES   1. Crude Birth Rates, 1930 and 1960   *page 59*
2. Average Family Size, 1900–1960   *60*
3. Acreage Purchased by TVA for the Tellico Project   *222*

CHARTS   1. Organization of TVA, 1969   page 5
          2. Tellico Land Purchases, Mean Farm Size Per Year, 1966–76
             page 223
          3. Tellico Land Purchases, Acres Purchased by Year, 1967–76
             page 224

PREFACE

On November 29, 1979, the Tennessee Valley Authority (TVA) closed the gates on the Tellico Dam, impounding the last freeflowing stretch of the Little Tennessee River and creating Tellico Lake, the last in a series of manmade lakes extending upriver to Fontana Dam in North Carolina. The agency had planned the Tellico Project for over forty years, and now it was finally finished. Its completion brought TVA to the end of one of the bitterest controversies the valley had experienced in modern times, one whose skirmishes stretched in impassioned debates from the quiet lawns of county courthouses to the corridors of power in Washington.

Whatever the significance of this controversy at the local, regional, and national levels, its greatest importance lies at the institutional level. Above all, a history of the Tellico Project is a history of TVA in the post-industrial era. In a sense, Tellico and its anguished controversy signify the end of an American era.

In 1913, Senator George Norris, one of TVA's "founding fathers", had defended the "conservationist" views of Gifford Pinchot, who supported San Francisco's attempts to create a reservoir in California's Hetch-Hetchy Valley in Yosemite National Park to improve the city's water supply. The conservationist ethic espoused a centralizing "efficient" use of resources for public benefit. Samuel P. Hays described the "conservationists" of the Progressive era in a lucid fashion—and in a way that seems perfectly to describe TVA in modern times:

> Conservationists were led by people who promoted the "rational" use of resources, with a focus on efficiency, planning for future use, and the application of expertise to broad national problems. But they also promoted a system of decision-making... a process by which the expert would decide in terms of the most efficient dovetailing of all

competing resource uses according to criteria which were considered to be objective, rational, and above the give and take of political conflict.[1]

In contrast, John Muir had opposed the "efficient" use of the Hetch-Hetchy Valley and espoused the "preservationist" view reflected in the Sierra Club's stance against the commercial exploitation of a wilderness area. And if Hays's description of "conservationists" is an apt description of TVA's attitudes at the time of the Tellico controversy, so too is Hays's description of the "preservationists" a fairly good portrait of many of the Tellico opponents: "A host of political impulses, often separate and conflicting, diffuse and struggling against each other within the larger political order."

Fifty years had passed between Norris's defense of Hetch-Hetchy and TVA's decision to proceed with Tellico; it is the implication of these fifty years that needs to be considered. Tellico's completion marked the death of the agency as it had been known. The dam became a battleground for TVA's conservationism, which sought to use a water project for the economic benefit of the region's public; and TVA's opponents, who wanted to preserve the last free-running stretch of a "wild" scenic river for a different public good, as an oasis in a land of manmade lakes. The preservationist-engendered Environmental Defense Funds, Environmental Impact Statements, the National Environmental Protection Act, and the Endangered Species Act were to bring TVA to its knees. In the half-century since the Hetch-Hetchy controversy, preservationism had gained a new lease on life. It had challenged technological modernization and the technocratic dream of progress through the application of scientific, bureaucratic, and managerial skills. What is ironic is that though the tactical victory at Tellico was in the end won by TVA and the conservationists, they did not win through the application of their skills or their ethic. The victory rather constituted a fitting, if limited, end to the American obsession with technocratic power: Tellico would be approved through a blatant piece of political chicanery—a Veblenian dream turned nightmare in the hands of politicians.

Conservationists and preservationists are not embodiments of good and evil, however, and to relate the Tellico story properly, we have tried to "get inside" TVA and its opponents. Our story is ultimately a history of TVA, and we have tried to plumb the agency's institutional will and its collective mentality. In the process we have come to know the men and women at TVA who were at one time or another involved in the Tellico Project, not as stereotypical "TVA employees" but as individuals with particular strengths and frailities, accomplishments and disappointments. We have attempted, too, to understand the Little Tennessee River Valley as it was both before and during the Tellico episode. Economic and census reports, demographic data, published memoirs, and old histories all helped. In addition, we visited and interviewed people who were intimately affected by the Tellico Project in a different way from most TVA employees. As we came to understand the valley's intricate class structure, its religious and non-religious ways of thinking, its changing collective mood, here too we came to see behind the cliches to its inhabitants' complexities, desires, and fears—ironically, many of the same ones that their counterparts at TVA possessed.

Finally we have tried to understand those who opposed the project (and many of these were people of the valley). This has been a more difficult task, for many are still bitter, and memories of being removed from their land, signing petitions, going to rallies, making speeches, and writing letters still rankle. Our conversations vividly reminded them of their "failure." But they were determined that their side of the story should be told. In time, we came to see the complex nature of the opposition and the chasms—sometimes vast—between opponents themselves. For some the Tellico battle was their "one great fight," their personal Pickett's Charge against what they considered to be the forces of darkness.

There are, then, no heroes or villains in this book. Ours is a story of people swept into the maelstrom of controversy, controversy which changed some of their lives forever. But this is a book about more than people, whether in the agency or in opposition to it. It is a history of TVA in the modern era, seen through the prism of the Tellico Project. An interpretive history of the Tellico Dam controversy it surely is, and a regional history of an important episode in American life as well. But primarily it is a history of the Tennessee Valley Authority in the post-industrial era—not a definitive history, to be sure, but a history which consciously and analytically sets out to explain a major American (and regional) institution in changing times.

It could be argued that it is too soon to evaluate TVA. After all, the cliche has it that a historian never writes about anything later than her or his birthdate. But TVA occupies a strange place in American historiography. The seminal works on it have been written by people who could remember the agency's founding and who could still interview in person most of their subjects. We desired to capture the participants' thoughts and feelings as soon after the actual events as possible, and to bring to the subject a freshness and intensity which many historical works do not possess.

If contemporary history is to allow readers to understand better themselves and their own era, they must read about themselves and their own times, rather than translating the "lessons" of, say, the Civil War or the American Revolution into presentist terms. We have chosen to deal with the near past because so much of the aura of the living institution remains to be captured.

The growing acceptance of oral history has changed many of the ground rules concerning what history is and what it is not. Significant collections of oral data on TVA are being compiled (at Memphis State University and within TVA) which will be extremely valuable to later historians. But we had our own questions, which might be neglected in this process. We especially wanted to capture a sense of immediacy before it was lost. Ironically, however, one real loss of immediacy may have occurred in the written record.

One of the authors had collaborated on an earlier book about TVA's population removal at Norris Dam. In the course of that project's exhaustive research in TVA records, no question concerning its purpose had ever been raised by TVA. But in connection with this new history, TVA's Office of General Counsel insisted on examining all TVA files before they were used by us. Paralegals were regularly assigned to remove certain material which, as they put it, might pertain to pending or possible litigation. We will never know what has been removed. We were told informally that "a good deal" of material had been kept from us, and we heard through another source that one TVA person had commented that "we have to be careful what they see." We are now familiar enough with the records, however, to believe that if the missing material might alter a detail or two, it would not change the principal story we have told.

TVA
AND THE
TELLICO DAM
1936-1979

1

TVA AND THE SEARCH FOR A NEW MISSION

On February 5, 1959, General Manager Aubrey "Red" Wagner sent a memorandum to twenty-six carefully selected individuals within the vast Tennessee Valley Authority (TVA) system. The memorandum invited them to a "brain-storming" session to be held in the Watts Bar hydroelectric plant conference room on Friday, February 13. In his memorandum, Wagner made clear what the agenda would be:

> A few weeks ago the staff met with the Board and summarized our investigations of possible future dam and reservoir projects. It seemed clear from this discussion that if past methods for justifying and financing such projects are continued in the future, few, if any, more dams will be built in the Tennessee Valley. At the same time, it may be possible to demonstrate that added projects would contribute enough to further regional development to amply justify their construction. This could well be a field in which TVA could plow new ground. It may only depend on how ingenious and resourceful we can be in finding a basis for evaluating a project's usefulness and for financing its construction. . . . Come if you can and bring all the optimism you have.[1]

In spite of its sterile atmosphere, the Watts Bar conference room was a favorite meeting place for TVA officials. For one thing, an invitation to a meeting at Watts Bar implied a recognition of its recipient's power and status in the huge agency. Perhaps more important, however, was that from the conference room windows TVA executives could look at an example of what their skills had wrought in the past: the Watts Bar Dam and Reservoir, the most recent in a series of nine dams that had tamed over 600 miles of the Tennessee River from Knoxville, Tennessee, to Paducah, Kentucky. The executives could not help being filled with a sense of pride and accomplishment.

A little before 10:00 A.M. the conference room started to fill.

Although they might not have been well known outside TVA, the group members were among the most talented and powerful ever to gather in the Tennessee Valley. Chief of Power Gabriel O. Wessenauer and his aide, R. A. Kampmeier, had come from Chattanooga. Tough, honest, and thoroughly committed to the agency, Wessenauer was rumored to be the most powerful single individual in TVA. In fact, in 1954 he had been offered the position of general manager but had turned it down, possibly because, as some joked, he was already more powerful than the general manager. "As long as you agreed with 'Wes,' you'd have gotten along with him fine," said one former TVA man. The agriculturists, whose influence was declining within the organization, were represented by Leland Allbaugh. Deliberate, judicious, professorial, he had come to TVA from Iowa State University, where he had been a professor of agriculture. J. Porter Taylor, one of the few native southerners to reach an elevated position in TVA, was there from Navigation; Richard "Dick" Kilbourne, a veteran forester, from the Tributary Area Committee; Dr. O. M. Derryberry from Health; Paul Evans from the Office of Information; Lawrence L. Durisch from Government Relations; and Reed Elliot from Project Planning. In all, the group represented a kind of Who's Who of the Tennessee Valley Authority (see Chart 1). Most had been with TVA since the 1930s and were committed to its projects with a fervor almost religious in intensity.[2]

Wagner's opening remarks set the tone for the meeting. The general manager wanted to build more dams, create more reservoirs. With the exception of Melton Hill, Wagner mourned, TVA had not proposed a single multipurpose project since 1951, since the costs of such projects had risen beyond their traditional benefits of power, navigation and flood control. "We would like to find the ways and means," Wagner declared, "for pushing a water-resource development program over the next big hump. Assuming that additional dam and reservoir projects will make a valuable addition to the . . . region," he went on, the problem would be how to justify and finance them so that TVA's benefit-cost ratio would be more acceptable.[3]

The February 13 Watts Bar Conference was a major event in TVA's history. Since the end of World War II, the agency had searched for a new mission, one which would be as dramatic as its earlier construction of high dams on the main Tennessee River and as important as its wartime mission to provide electrical power for the Aluminum Company of America (ALCOA) and the Manhattan Project at Oak Ridge. The changing national mood, outside criticism, and internal divisions had made the search for a new mission a tortuous one. But by 1959 Wagner had accumulated enough power to imprint

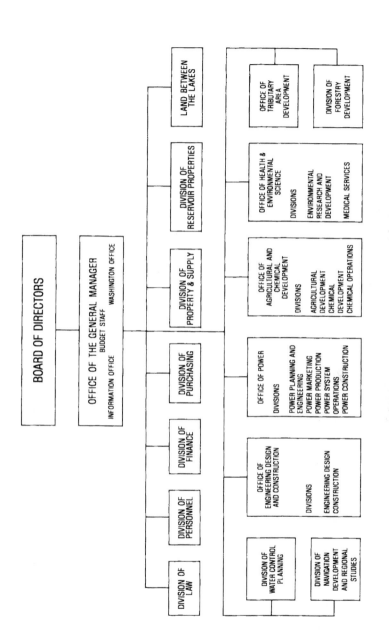

Chart 1. Organization of TVA, 1969.

his philosophy, ideas, and personality upon a TVA looking for decisive leadership. The Watts Bar Conference clearly suggested that Wagner's "new mission" was to be the industrial development of the valley and the Tennessee River's tributaries through the construction of more dams and reservoirs. The old Fort Loudoun Extension Project was soon to be revived and rechristened Tellico. To Wagner and TVA, the Tellico Project became indispensable as the showcase of the authority's new mission and as the logical outcome of historical events and forces which had moved both inside and outside TVA during its long history. In short, in order for TVA to survive and continue to serve the valley, Tellico must be completed.

As the end of World War II neared, people within TVA recognized that the agency was at a crossroads. The removal of Arthur E. Morgan from the TVA Board of Directors in 1937 signaled the triumph of those within TVA whose conception of the agency's mission was narrower than Morgan's. These men saw the organization principally as a provider of cheap electric power and fertilizer and as a government agency which would work in cooperation with the region's farmers and the extension service to assist agriculture, in particular through test demonstration farm units (5,500 in Tennessee alone). Morgan's notion had been much broader, encompassing a complete "general socioeconomic plan for the valley." When fellow director Harcourt Morgan labeled A. E. Morgan's ideas "impracticable and highly visionary," the third director, David E. Lilienthal, agreed, and Arthur Morgan's hands were effectively tied and his days numbered.[4]

Gradually Lilienthal became the dominant force within TVA, due to his forceful personality and his control of all functions relating to electric power. An articulate champion of the agency's role as an uplifter of the valley through the production of cheap power, Lilienthal effectively snuffed out plans within TVA for a larger mission. Such ideas, Lilienthal reasoned, interpreted too broadly the 1933 Act of Congress which had created the agency, and they were sure to get TVA in deeper trouble with conservatives already screaming about its "socialistic" inclinations. Besides that, a broadly construed mission smacked too much of the policies of Arthur Morgan.

As early as 1934, Lilienthal had been able to suppress demands within TVA for a comprehensive regional planning study to examine all aspects of the Tennessee Valley and undertake multipurpose projects to meet the region's socioeconomic needs. Internally known as R.P.1, the proposed regional planning project aimed at analyses of population movements, classification and location of resources, studies of improved facilities for transportation and communication, the development of a basic land use plan, and a full plan for economic development which would, it was believed, affect future population patterns in the valley. The push for R.P.1 was led by the talented Earle S. Draper, head of the Division of Land Planning and Housing, who clearly shared Arthur Morgan's vision of TVA as a comprehensive agency. In March 1934 advocates of R.P.1 asked the board for permission to do a comprehensive regional plan.[5]

But Lilienthal would have none of it. In a memorandum to his fellow directors, dated the same day that Draper submitted his request to the board, Lilienthal launched an indirect salvo against R.P.1. Using the rather curious—and inaccurate—argument that the board would be abrogating its own authority if planning was done by any group except the board, Lilienthal then attacked the whole notion of comprehensive planning itself, concluding with what was probably his real concern: "The economic implications of allocation of land use and the drawing of the physical plan for land involved in this project have infinite ramifications."[6]

Lilienthal was not able to kill R.P.1 outright in 1934, but he was able to cast doubt on the whole notion of TVA's comprehensive planning, while requiring Draper to make frequent detailed reports to the board where Lilienthal could further impede the project. Draper responded as best he could: "It is not my intention to get a large number of men and attempt to produce extreme activity." But Lilienthal's mind was already made up. As Arthur Morgan's power waned and Lilienthal's simultaneously increased, R.P.1 gradually was strangled, until finally, in 1937–38, with Arthur Morgan gone, it was unceremoniously dropped.[7]

Later, in *TVA: Democracy on the March* (1944), Lilienthal defended his scuttling of TVA's larger mission as a planning agency by emphasizing democratic planning by the people themselves, a people economically uplifted by cheap electric power, jobs and prosperity:

> If this conception of planning is sound, as I believe, then it is plain that in a democracy we always must rest our plans upon "here and now," upon "things as they are." How many are the bloody casualties of liberal efforts to improve the lot of man, how bitter the lost ground and disillusionment because of failure to understand so simple and yet so vital an issue of human strategy. . . .
>
> Everyone has heard the story of the man who was asked by a stranger how he could get to Jonesville; after long thought and unsuccessful attempts to explain the several turns that must be made, he said, so the anecdote runs: "My friend, I tell you; if I were you, I wouldn't start from here." Some planning is just like that; it does not start from here; it assumes a "clean slate" that never has and never can exist.[8]

Hence, for whatever personal, ideological, or political reasons, Lilienthal had effectively bound TVA to his conception of it as essentially a power company. Under his watchful eye, the Office of Power became the most influential division within the agency, with an annual budget dwarfing those of all other offices and divisions combined. Nonpower functions received scant attention.

However, the "power people" who dominated TVA were not without internal opposition. By the end of World War II, a fissure had developed between them and the "water people" (more accurately designated by some as "nonpower resource development people"). Although this division had existed almost from the agency's creation and had continued in the running battles between Arthur Morgan and Lilienthal, by 1945 it had become even more pronounced, creating a climate of conflict from which ultimately would come TVA's new mission.

As much as anything, the difference between "power people" and "water people" was one of attitude. "The power and water people tended to look at things differently," remembers former TVA Chief Engineer (1959–69) George Palo. The "power people," with offices in Chattanooga, concentrated their attention and skills on the production of low-cost electric power. Like their chief mentor David Lilienthal, they tended to be strict constructionists when thinking of the powers delegated to the agency by Congress. Many were also suspicious of regional planning, development, and working with states, local communities, and citizens. These activities they believed were outside TVA's proper functions. "The power people didn't know much of what was going on in the rest of TVA," recalls former Head of Navigation J. Porter Taylor. Nor, it appears, did they care.[9]

The "water people" had a philosophy much broader in scope, which in one sense made them heirs of Arthur Morgan's larger vision for TVA. They also seem to have been considerably more aware of the shifting public mood and external politics than the more aloof "power people." "Water people" were concentrated in areas like Navigation, Regional Studies, Economics, Recreation, Forestry, and Water Control Planning. They envisioned a broad and multifaceted role for the agency. Looking upon the reservoirs which the hydroelectric dams created as excellent sites for industrial development, local and regional planning, and experimentation, they saw the reservoirs as tools to effect a comprehensive approach to regional uplift. Most of the "water people" were nonsoutherners, the plurality if not majority of them coming from the Midwest. They brought to TVA a midwestern view of agriculture, a mental picture of vast, flat farmlands with black, fertile soil, mechanization, huge silos and storage facilities, and high yields. By that standard, most of the Tennessee Valley was submarginal farmland. The "water people" were appalled at the proud and grizzled mountain farmer of the 1930s, scratching a bare living out of poor soil with insufficient land, fertilizer, equipment, and money. In their eyes, the salvation of the region lay not in agriculture but in industry, reforestation, regional and community planning. "TVA is more than a power company," was their rallying cry. Battles with the "power people" tended to be fought over the agency's future direction and, symbolic of that, its budget allocation. As former Director of Information Paul Evans put it, "There was always an effort to make sure you got your share."[10]

The domination of Lilienthal (he left the agency in 1946 to become chairman of the new Atomic Energy Commission) and the outbreak of World War II, during which TVA had to concentrate almost all of its efforts on producing power for defense industries, postponed any open confrontation between the "power" and the "water" people within TVA. But as the war neared its end, many within the agency began to ask openly, "What does TVA do now?" Some within the agency even expected that TVA, with much of its main work done, would fold. In fact, in late 1947 General Manager George F. Gant sent a memorandum to division heads ordering them to plan for turning over their functions to other agencies.

In 1942, however, farsighted General Manager Gordon Clapp (a career TVA employee with definite "water" sympathies) had directed the Department of Regional Studies to report on postwar plans for the agency. By May 1944, Steps I and II of the department's working papers had been completed (Step III would be finished in late 1945). "You will note," cautioned Regional Studies Director Howard K. Menhinick, "that the reports cover numerous activities not contemplated as TVA programs." In TVA parlance, that meant that the agency's postwar future lay in being more than merely a provider of power. In language to thrill every "water" person, Step II of the working papers announced, "Further industrial development of the Tennessee Valley region is necessary to accomplish the best use of its human and physical resources."[11]

The reports, however, surfaced in a climate, both inside and outside TVA, which was definitely hostile to a "new," more extensive mission for the agency. Although Clapp's appointment as chairman of the TVA board in 1946 gave him the authority to push for a larger mission, he was in fact almost powerless to achieve his goal. Republican victories in 1946 had left Congress controlled by that party for the first time in TVA's history. While President Truman himself was friendly to TVA, there was a good deal of loose talk in congressional cloakrooms about cutting the agency's appropriations and even of dismantling the giant. The move by TVA's Office of Power away from hydroelectric dams, which Wessenauer had come to believe were too costly, and toward producing electricity through coal-fired steam plants had revived the opposition of private utilities; they feared that TVA was moving out of the river valleys with a view toward taking over all power production in the Southeast. In 1947, Republican majorities in Congress voted down appropriations for a TVA coal-fired steam plant to be built at New Johnsonville, Tennessee. Hence, while Clapp later declared that "TVA is controversial because it is consequential," in the late 1940s the chairman was too busy mollifying Congress to give much thought or attention to a more comprehensive "new mission" for the embattled agency.[12]

Nor was the atmosphere inside TVA more receptive toward a "new mission." Wessenauer, as chief of power, was determined to move the agency into the construction of coal-fired steam plants (later he was one of the pioneer advocates of TVA nuclear power activity). The best water sites had already been dammed, Wessenauer reasoned, and the Watts Bar steam plant (TVA's first, begun in August 1940 and delivering power by February 1942) had been comparatively quick and inexpensive to build. Including New Johnsonville, which Congress turned back in 1947 but ultimately—under a new Congress—approved in 1949, the Office of Power had plans for seven such steam plants. Once the congressional logjam was broken in 1949, all seven projects were approved within less than three years. Of course, this gave the "power people" the lion's share of appropriations, manpower, and attention in those years, forcing calls for a "new mission" even further into the background.[13]

Symptomatic of the triumph of a narrower approach to TVA's mission was the agency's shifting policy on land purchase. At Norris in 1934, TVA had purchased some 153,000 acres, eighty percent of which would be above the normal summer pool elevation of the reservoir. Subsequent projects at Chickamauga, Wheeler, Guntersville, and Pickwick had continued this "heavy purchase" policy. This policy was welcomed by the "broad constructionists" within the agency, people who saw TVA's role as one of comprehensive regional development. Yet, just as clearly, the policy was opposed by the agency's agriculture people, led by Director Harcourt Morgan, and by the power people. During one conversation, in which was broached the subject of aiding the region through the development of tourism on the reservoirs and shorelines, Harcourt Morgan said sourly, "I'd rather have pellagra than tourists."[14]

Lilienthal's eventual domination of TVA presaged the abandonment of the "heavy purchase" policy. In 1944, probably prompted by Lilienthal, General Manager Gordon Clapp ordered Assistant General

Manager Arthur S. Jandrey to review the agency's land acquisition policies with a view toward selling unneeded property. The report, submitted in July 1944 and informally titled the "Jandrey Report," became the basis of the "small purchase" policy. At the outset Jandrey flatly stated that the agency had acquired more land than it needed, that it ought to assess all acquired land to designate which was surplus property, and that it should sell all unneeded land. Thus, prodded by the dominant "strict constructionists" and fearful that, if the agency did not dispose of surplus property rapidly, Congress would force it to do so, TVA in 1946 began divesting itself of the thousands of acres it had acquired on the agency-made reservoirs. By 1960 it had sold roughly 147,000 acres of shoreland to private owners and developers and had transferred an additional 171,000 acres to other public agencies. Also, beginning with the Kentucky Project (1939), TVA stopped purchasing land above its reservoirs' high-water marks.[15]

TVA's land purchase policy can be used as an eerily accurate barometer of the power struggles within the agency itself. The shift from a "heavy" to a "small" purchase policy, like the defeat of efforts for comprehensive planning, a strict construction of the TVA Act, and the budgetary predominance of the Office of Power, signaled the hegemony within TVA of the "power people" and their narrow notion of the agency's mission. Hence those within TVA who called for a more comprehensive and ambitious "new mission" were forced to bide their time and wait for a more receptive climate of opinion both inside and outside the troubled agency. In the early 1950s, a frequent TVA consultant, Syracuse University's Professor Roscoe Martin, could still say that TVA's chief problem was one of "program definition"— a polite way of saying that the agency had yet to discover a "new mission" for itself in the postwar valley.[16]

Dwight Eisenhower's election to the presidency in 1953 appeared to spell even more difficulties for the agency. Although Eisenhower had tried to portray himself as being above the savage and tawdry world of politics, some of his campaign statements, his cabinet appointments (especially George Humphrey at Treasury, Ezra Taft Benson at Agriculture, and Charles Wilson at Defense), and his almost theological commitment to balancing the federal budget made it clear what the general tone of his administration would be. An internationalist in foreign policy, the new president showed a strong conservative bent in domestic affairs.

TVA could not fail to sense Eisenhower's unfriendliness. The general's much-quoted post-inauguration statement—"If I do anything, it's going to be less government than more government"—was correctly perceived as a slap at TVA and other federal agencies which were, in the view of conservatives, unwisely and unlawfully impinging on the private sector. Those at TVA knew that difficult times lay immediately ahead.[17]

The attack began almost immediately. In mid-May 1953 the president recommended to Congress deep cuts in TVA's appropriations, including funding approved by the preceding Truman administration. The next month, in a speech in South Dakota, Eisenhower spoke darkly of "creeping socialism" (a phrase conveniently borrowed from Herbert Hoover) and, when asked to offer some examples of "creeping socialism," he replied that TVA certainly fit the bill. Nor was the president merely posturing for the press and the conservatives in his own party; in a cabinet meeting held on July 31, 1953, the subject of TVA arose, and the president exploded, "I'd like to see us sell the whole thing, but I suppose we can't go that far."[18]

Even worse, from TVA's point of view, was to follow. The Dixon-Yates imbroglio, which was on the front pages sporadically from late 1953 through 1955, posed a clear threat to the agency. The essential issue in the complex Dixon-Yates controversy was whether the Eisenhower administration would encourage private power systems (in this case a combination of Middle South Utilities [Dixon] and the Southern Company [Yates]) to provide power to the Atomic Energy Commission and private suppliers while simultaneously preventing TVA from increasing its generating capacity. Although the scheme ultimately died amidst charges of conflicts of interest, special privilege, and political favoritism, no one could doubt that the Eisenhower administration intended to shackle TVA or at least to limit its scope.[19]

With a new chairman of the TVA Board of Directors, Eisenhower would have an even better opportunity to affect the fortunes of TVA. The administration made it clear that Clapp, whose term was due to expire in May 1954, would not be recommended to the Senate for reappointment. Indeed, some close to Eisenhower toyed with the idea—which Clapp himself quashed—of persuading Clapp to resign early. In other words, Clapp would have to go, and the quicker the better.[20]

In a news conference in May 1954, President Eisenhower stated that the name of Clapp's replacement would be submitted to the Senate "as soon as he [Eisenhower] found a man who was completely nonpolitical in his position and status, who was in his opinion a professionally well-qualified man, whose general philosophical approach agreed with his and whose integrity and probity was above reproach." At the same time, Sherman Adams wrote privately that the administration was casting about for a "very strong business person" and, by implication, a person who would bring the giant agency (22,000 employees in 1954) to heel.[21]

The first name considered by the administration was that of Harry C. Carbaugh, a Chattanooga businessman with strong personal ties to key Tennessee Republicans (most notably to Congressman Howard Baker, Sr.), a man described as "a sound conservative individual without any New Deal tendencies." But Carbaugh's nomination would have been so tainted by charges of political cronyism that his name was grudgingly shelved; while Carbaugh undoubtedly was able, his political connections would have embarrassed the "nonpolitical" Eisenhower.[22]

In the meantime, factions in the Tennessee Valley began lobbying for Clapp's retention. Beginning in the late 1940s, TVA, in response to external threats, had helped organize various private interest groups to argue the agency's case before state and federal government bodies. In 1947 TVA assisted in the foundation of the Tennessee Valley Public Power Association (TVPPA). By 1950 the agency was helping to fund some TVPPA publicity projects and providing research and information to the private pressure group. In 1953 Citizens for TVA (CTVA) was formed, composed largely of valley area mayors, judges, political figures, editors, power distributors, and labor leaders. By 1954 CTVA claimed some 35,000 members, boasted the active support of Tennessee Governor Frank Clement, and was once referred to by General Herbert Vogel as a "political front organization" for TVA.[23]

Citizens for TVA, pressuring the Eisenhower administration to reappoint Clapp, presented the president with a petition bearing some 60,000 signatures. Of course Eisenhower was unmoved. Instead, the administration nominated General Herbert Vogel to head TVA. Vogel formerly had been with the Corps of Engineers, a group traditionally hostile to TVA. In spite of his moderate tone before the Senate Public Works Committee, Vogel was widely believed to have been charged by the president with dismantling TVA. He was easily confirmed in August 1954, and for those within TVA, things could not have appeared darker.[24]

Ironically, the Eisenhower administration's moves to hobble TVA aided those who advocated a broader and more comprehensive mission for the beleaguered agency. Responding to the political winds, TVA made two momentous decisions which would affect the agency for the next two decades. Though these were short-range decisions in response to immediate problems, they would chart TVA's course for years to come.

The first decision was intended to free TVA from yearly congressional reviews, with their regular marshaling of the private utilities lobby. TVA proposed to the president in 1955 that all money needed for power (construction, maintenance, and delivery) be self-financed through TVA bonds to be paid off by power revenues. If the Office of Power did not have to come to Congress every year for appropriations, the arguments of the private utility interests would be weakened. Hence TVA recommended to the Bureau of the Budget that the TVA statute "be amended to permit the Agency to raise funds for capital expansion by issuing revenue bonds."[25]

The move for self-financing of TVA power was opposed by private power interests outside the Tennessee Valley; those within the valley were pretty much in the agency's pocket. But in 1957 the parsimonious Eisenhower became convinced of the value of self-financing, and Congress belatedly approved the plan in August 1959. Freed from annual controversies in Congress, the "power people" felt secure enough not to oppose those other TVA interests that called for a "new mission." Obviously those new projects would be the subject of congressional scrutiny. In other words, the "power people" could continue to operate and grow, no matter what the agency's "new mission" might be.[26]

The first issuing of revenue bonds to finance additions to the power system brought forth a flurry of controversy, principally from private power interests outside the valley and from the national Chamber of Commerce, an old foe of TVA. Yet the first bond issue in November 1960, for $50 million, sold briskly, and a second issue was scheduled for 1961. By then the Office of Power, still under Wessenauer, was toying with the idea of nuclear power, but such power would be so expensive that it was doubtful that revenue bonds alone could cover the staggering costs. Hence the Office of Power, already self-financing, was willing to embrace a "new mission" for TVA as a way to lure additional funds from Congress through appropriations for non-power projects. "Power wanted a shield to get more money beyond self-financing," remembered former TVA staffer A. J. "Flash" Gray. In short, the "power people's" opposition to a "new mission" had been virtually eliminated.[27]

An analyst critical of TVA might say that the agency seems to have a knack for making decisions which solve short-range problems but, in the long run, prove injurious. By putting the power program on a self-financing basis, TVA was able to avoid annual head-on fights with private power lobbyists and with Congress over appropriations. With those yearly battles gone, the private power interests outside the valley lost a rallying cry, an issue, and considerable national support. Too, congressional scrutiny of power projects was virtually eliminated. However, by separating the power program from congressional appropriations, TVA rendered its nonpower projects even more vulnerable to congressional investigations. Clad in the mantle of power production, most TVA budget requests eventually had gone through. Without that "shield," they would now encounter more difficulty.

TVA's second momentous decision came in response to Eisenhower administration wishes. In his message to Congress in July 1953, the president made it clear that government agencies (such as TVA) were to cooperate with state, local, and private interests in development and program implementation. This "partnership principle," with the private sector taking the initiative, was viewed seriously by the Eisenhower administration, as evidenced by the bungled Dixon-Yates incident and the 1954 passage of the small watersheds program. This latter program called for private initiative to develop these watersheds and for government agencies merely to offer technical assistance. TVA could read the handwriting on the wall.[28]

The "partnership principle" introduced a new dimension in TVA's relationship with state and local agencies and the private sector. Although under Lilienthal TVA had prided itself on its "grass roots" orientation, in fact the agency had maintained this posture by guiding state and local agencies to do its will. "Partnership" threatened to be an intrusion of a new egalitarianism, and this TVA resisted.

By the 1950s, an "engineering mentality" had come to dominate the agency. This attitude was founded on the dual assumptions that (1) most, if not all, of the valley's problems could be solved by technology; and (2) most politicians were ignoramuses, back-slappering pawns of greedy "robber barons." Although TVA employees were active in local churches, Boy Scout and Girl Scout troops, and neighborhood Parent-Teacher Associations, they traditionally shied away from joining business groups like Rotary, Kiwanis, or Chambers of Commerce. In short, many within TVA viewed cooperation with state and local interests with as much enthusiasm as a prisoner shows for walking the last mile. Yet the potential of a "partnership" was enough to give hope to those who called for a "new mission" for the agency. As with self-financing of the power program, the "partnership principle" gave the broad constructionists within TVA new opportunities, new life.[29]

Hence a combination of fortuitous events and forces aided those who advocated a "new mission" for TVA. The Office of Power was no longer hostile. Indeed, Wessenauer even took to the stump in what were for him rare public appearances to support a more comprehensive role for TVA, perhaps reasoning that the new role might foster increased power consumption, which would naturally make new power projects more attractive. Too, Wessenauer probably had begun to appreciate the fact that multipurpose projects could secure congressional appropriations which might well be used to keep rates low.

At the same time that Wessenauer and the "power people" were beginning to embrace the notion of a "new mission," others within TVA were as well. Economists and planners in the Office of Regional Planning, formerly the Department of Regional Studies, had long hoped that the old idea of a comprehensive regional plan, which had been killed by Lilienthal and Harcourt Morgan, could be revived. They became more vocal. In addition, the Navigation and Natural Resources people realized that a new, broader mission would further their interests. And the once-important Project Planning Branch within the Division of Water Control recognized that unless it embraced the "new mission," its tasks and influence would seriously diminish. Only Agriculture and Fisheries and Wildlife people remained in opposition. But their influence had so deteriorated within the agency that their protests were virtually ignored. After having wandered so long in the wilderness, the prophets of a "new mission" suddenly had become the "saviors" of the valley and of TVA.[30]

The person most responsible for guiding TVA toward its "new mission" was Aubrey J. "Red" Wagner, a man whom later historians probably will assess as the most important individual in TVA in its first half-century. In his last twenty-four years with the agency (1954–78)—as general manager, board member, and chairman— Wagner, a career TVA man, had the opportunity to stamp his ideas and personality upon an organization much in need of strong leadership.

Like many of TVA's first generation, Wagner was a native midwesterner. Born in 1912 in Hillsboro, Wisconsin (pop. 804 in 1910), in 1927 "Red" Wagner moved with his family to Madison, where he enrolled in the University of Wisconsin in 1929. Graduating with a B.S. in civil engineering in 1933, Wagner arrived in Knoxville the next year with a wife, a two-week-old baby, a worn-out Dodge, and less than twenty-five dollars in his pocket. Before he left Wisconsin, he had gained TVA's promise that he would be employed for at least one year. He remained for forty-four.[31]

Wagner began his career with TVA as a field engineer, doing navigation surveys for the General Engineering and Geology Division. At the time, the Navigation Branch of that division contained but four people: Branch Chief C. T. Barker, Wagner, J. Porter Taylor, and Albert Newton. An avid hiker, camper, and fisherman, Wagner loved field work. But he also enjoyed times when the three neophyte

engineers would sit in the office swapping stories and talking over
various projects and goings-on within TVA. Taylor, the most effusive
of the three, had the southerner's gift for telling a good story and for
picturesque speech interspersed with mild cursing. Wagner tended
to be quieter, more pensive and private than the others, although he
had a knack for hitting the nail squarely on the head with a minimum
of verbiage. All were competent engineers, totally devoted to their
work and to the agency.

Apparently Barker recognized in Wagner, more than in the other
two, certain traits he appreciated: a talent for thoughtful and cogent
debate, a highly pleasing personality, and the diplomatic skill to win
others to his point of view without deprecating their own. Possibly
Barker's background (he was from Minnesota) made him look kindly
on Wagner the Wisconsinite. Whatever the reason, Barker took an
interest in Wagner's career and "brought him along" through Navigation,
until by 1943 Wagner was a full-fledged navigation engineer, by 1946
assistant chief of the River Transportation Division, and by 1948 chief
of the Navigation and Transportation Branch. From there he went up
like a rocket: assistant general manager of TVA in 1951, general
manager in 1954, member of the board in 1961, and chairman of the
board in 1962.

How was Wagner able to rise so rapidly in an organization
brimming with talent? In addition to his obvious skills, he had the
backing of several important figures at key times in his career, Barker
being but the first. Gordon Clapp knew Wagner and appreciated his
ideas and talents, and when General Manager John Oliver was casting about for an assistant general manager in 1951, undoubtedly he
consulted Clapp. It is probable that Clapp recommended Wagner for
the position. Then, in 1954, on his way out, Clapp very likely hoped to
entrench his views in the agency by making Wagner general manager.[32]

As general manager, Wagner had to work closely with Marguerite Owen in Washington. Owen, as director of TVA's Washington
office, was the agency's "chief lobbyist" and a woman with political
connections and political savvy. So powerful was she within TVA that
many former employees half-jokingly refer to her as the "fourth
member of the board." Owen acquired enormous respect for Wagner,
bordering on veneration, and when President John F. Kennedy had
the opportunity to place someone on the TVA board, Owen brought
Wagner's name to Kennedy's attention. Certainly, Wagner had all the
skills necessary to handle the jobs given him, but he also had powerful connections which were important in advancing his meteoric
career.

One other point must be made with respect to Wagner's rapid

rise in a TVA hierarchy so rigid and rank-conscious that TVA employees expected its perquisites long after they had retired. Although he was not from the Office of Power (Wagner was always known as a Navigation man), he was acceptable to the "power people." Early in his career, Wagner was a strict constructionist when it came to TVA's mission. Moreover, Wessenauer and Kampmeier were fellow Wisconsinites. Both Wessenauer and Wagner were extremely active in the Lutheran Church and often saw each other at synod meetings. Hence, as Wagner rapidly climbed the TVA ladder, the power people offered no opposition. If they had, he might well have ended his days as a division head.[33]

Wagner elicited fierce loyalty from the agency to which he was dedicated for so long. Although a man of enormous power, he almost never bullied any TVA employee, preferring to win people to his point of view by the force of reason. He had a warm temper, but years of practice had given him an almost unshakable self-control, and his emotions rarely escaped confinement. Unfailingly courteous himself, he was puzzled and nonplussed by rudeness in others. Coworkers even saw him laugh at himself from time to time. In sum, Wagner was reasonable but often stubborn. He personalized TVA activities but occasionally found humor in that, was temperamental but self-controlled, courteous but often inflexible—truly a paradoxical man guiding a paradoxical agency.

By the time Wagner reached the general manager's office, his ideas had changed considerably. In his mind, TVA needed a fusion of Lilienthal's ideas of regional uplift through the production of cheap power and those of the broad constructionists who called for more comprehensive planning and regional development. Wagner firmly believed that industrialization was the key to the valley's future, but that cheap electric power alone was no guarantee that such development would come after World War II. TVA, he reasoned, could bring industrialization to the valley through a combination of cheap power and a series of multipurpose planned development projects. Hence Wagner, at least in his own mind, had fused TVA's two factions into one, much as a teacher brings two fighting schoolboys together to shake hands. Using his oratorical talents and his powers of persuasion, he patiently guided TVA toward his middle-path fusion, between the visions of Arthur Morgan, whose ideas Wagner privately despised, and the hard-headed pragmatism of David Lilienthal.

Wagner drove the increasingly willing agency toward his own "new mission" ideal. To lessen external criticism and simultaneously assuage the "power people," he pushed hard for self-financing of power projects. Working within the "partnership principle" enunciated by the Eisenhower administration, he encouraged agency cooperation with state and local government agencies and private groups, even helping to create those groups where they did not exist. More important in the short run, Wagner was the key figure in the conversion of Vogel who, like a redeemed sinner, became an even stronger defender of the agency than some of its closest friends and long-time employees. Later it was widely believed that Wagner's influence on Vogel had "saved" TVA.[34]

Since the main river projects were all completed (see Map 1) and no attractive sites were left on the main river, Wagner reasoned that future multipurpose projects taking advantage of Eisenhower's "partnership principle" would have to be undertaken on the tributaries. Tributary area development had been an early concern of TVA, but the influence of Lilienthal and Harcourt Morgan and the emergence of World War II had submerged the whole notion. Now Wagner revived it. His idea was that each tributary area was a microcosm of the entire valley and that "we needed miniature TVAs in the tributaries." In the early 1950s, soon after he became assistant general manager, Wagner organized the Tributary Area Committee within TVA. Apparently concerned that "agriculture people" might capture control of the tributary area movement, Wagner skewed the committee's membership away from agriculture and toward his own preference for industrial development. Included on the committee were Reed Elliot, director, Division of Water Control Planning; J. Porter Taylor, director, Division of Navigation and Flood Control; James Van Mol, chief budget officer; and Richard "Dick" Kilbourne, director, Division of Forestry. Kilbourne, whose major concern up to that time had been trying to convince industries to use the valley's forest products, was named committee chairman, reporting to Wagner. When he became general manager in 1954, Wagner continued to take great interest in the committee's activities, and Kilbourne continued to report directly to him. As one former TVA employee remembered, "Wagner loved the tributary program. . . . The idea of factories in the fields was great for him."[35]

Before such projects could be started, however, Wagner had to reverse the minimal land purchase policy advocated by the Jandrey Report in 1944. In 1958 he ordered Ashford Todd, Jr., TVA director of property and supply, to prepare a new report on the agency's land disposal program. Completed on August 4, 1958, the report was received by Wagner two days later.[36]

It is clear from Todd's memorandum that Wagner was already convinced that the policy ought to be reversed. This was even more evident in Wagner's reply:

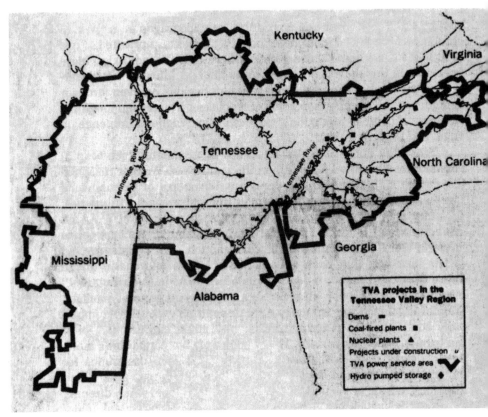

Map 1. TVA Projects in the Tennessee Valley Region.

> Present Board members, seeing the extensive recreational use of the reservoirs . . . have simply asked for assurances that in the years to come there will be fully adequate provision for the general public to have a complete opportunity for free access to the reservoirs and for their full use and enjoyment. I believe that this may justify some reexamination of our land allocations. . . .

The reexamination shocked both Wagner and the board, for neither had known the extent to which the agency had divested itself of valuable shoreline. Moreover, some of the private developments on former TVA land sickened many within the agency. By January 1960, the board confessed that TVA probably had sold too much and subsequently ordered all further sales to stop. Wagner had made his point, and his multipurpose projects could now proceed.[37]

    The examples cited provide a good number of clues to Wagner's philosophy and managerial style. Products of the Depression, he and his generation grew to maturity in the last stages of America's industrial age. To them, government initiative to save the private sector was to be welcomed rather than feared. In their view, most social problems had economic roots and could be cured by more industry and more jobs. Ironically, Wagner was called on to lead TVA into the post-industrial age, where his generation's philosophy would need serious modification. This ideology was to lead both Wagner and TVA into almost inevitable dilemmas.

In terms of managerial style, Wagner liked to bypass the agency's divisions by setting up special committees composed of people from numerous divisions and reporting directly to him. In theory, such committees would consist of experts from each of the concerned areas within TVA. But in practice this innovation did not always work. For one thing, such assignments were often made in addition to the normal work loads of the committee members. Too, since each committee member typically thought of himself (there were almost no women) as a spokesman for his own branch or division, committee meetings tended to be long and not always harmonious. Committee reports often expressed a divergence of views and few recommendations. Worse, Wagner usually had made up his mind about what the committee's final recommendations ought to be even before the committee was formed. He showed little patience with reports that did not concur with his preconceived conclusions, once refusing even to acknowledge receipt of a report on which a committee had worked for over a year.

Stubborn, occasionally inflexible, Wagner nevertheless provided the strong leadership TVA so desperately needed in those crucial years.[38] Wagner was extremely popular with the vast majority of TVA employees. He was one of their own, a career TVA man whose loyalty to the agency was fierce and unshakable. Unquestionably, his handling of Vogel had been deft and may well have saved the agency. In any case, the *perception* that it had done so surely increased Wagner's stature. And although the man was sometimes inflexible and occasionally irascible, he was also unfailingly courteous to almost everyone, always stopping to say "please" or "thank you" to people from division head to file clerk to janitor. In 1978 when Wagner walked across the TVA Plaza in Knoxville on the way to his last board meeting, employees stood at the windows of the TVA Twin Towers, many with tears in their eyes, and cheered him.

When Wagner called the meeting at Watts Bar for February 13, 1959, he was preparing to launch TVA on its new mission. After years of defensiveness, the agency would take back the initiative. Years of internal and external conflict had forced TVA into the narrow mission of power, navigation, and flood control. But changing times had forced the agency to reassess its role in the Tennessee Valley. It had concluded that only a series of multipurpose projects would bring the desired results of industrialization, recreation, tourism, and jobs, and that those multipurpose projects would have to come on the tributaries. The agency had been forced to accept pressure from Washington to cooperate with state and local governments and the private sector in constructing those projects. At Watts Bar, Wagner issued a clarion call not just for "more dams" but for a new era for TVA, an era into which he was prepared to lead the agency. And while the fusion which Wagner envisioned then existed largely in his mind, he had the power and talent to impose that vision on the diverse giant agency. Although he was not to be completely successful in fusing TVA's traditionally conflicting missions, he was at least putting the agency "on the march" once again.

Years later, people wondered why TVA had been so determined to build the Tellico Dam. But to Wagner and to many within the agency, Tellico was much more than a dam, and considerably more was at stake than its mere completion. In short, it was to be the showcase of the new mission. By February 13, 1959, the day of the Watts Bar Conference, the decision to build the Tellico Dam—and to return to a large purchase policy to make Tellico a multipurpose project—was, in the minds of many within TVA, almost irreversible.

2

THE DECISION TO BUILD THE TELLICO DAM

On July 7, 1959, Reed Elliot, head of TVA's Division of Water Control Planning, met with Chief Engineer George Palo. Both men belonged to the agency's first generation and were tough-minded and fiercely loyal to TVA. Both appreciated blunt, straight talk. And both realized that their meeting would be an important one.

Since the Watts Bar Conference the preceding February, where Wagner had made his momentous "more dams" speech, an air of excitement had stirred the spartan offices of TVA. In April 1959, Wagner had created the Future Dams and Reservoirs Committee (shortened to the unfortunate acronym of FUDAR) to investigate more imaginative methods of computing benefit-cost analyses for future projects. That same year, the general manager had formed the Future Hydro Projects Committee to set up a priority list for agency projects. Back in Elliot's own division, younger and more energetic engineers in the Project Planning Branch were pushing Elliot "to get stuff off the drawing board and into the pipeline." In short, Wagner's "more dams" speech had unleashed a torrent of energy long repressed in the Eisenhower years.[1]

At the July 7 meeting, Elliot came right to the point. "George, here are twenty-five hydro [dam] projects which we've been studying." Elliot placed the list before the chief engineer. "I've got a benefit-cost ratio on all these projects. All but one have a benefit-cost ratio of .25, the one exception being that of the Fort Loudoun Extension, which has a benefit-cost ratio of between .5 and .6. It looks like Fort Loudoun Extension is the best available project."[2]

In one sense, the Elliot-Palo meeting was to prove as momentous as Wagner's "more dams" speech. Gradually the rippling effects of the Elliot-Palo meeting began to turn the huge ship that was TVA toward what was to become the Tellico Project.

The day after the meeting, Palo recapitulated the conversation in a memo to Elliot, adding that he wanted to take a report on the Fort Loudoun Extension to the general manager by April 1, 1960, so that the project could be included in the 1962 fiscal year budget. Two days later, Palo was pressing Elliot for a commitment to the April 1, 1960 deadline. Quick to sense which way the wind was blowing, Elliot swung his division into action. In a memorandum to Robert Frierson and Don Mattern, the division head explained that "we had planned, in setting up the 1960 budget, to concentrate on Fort Loudoun Extension. I see no reason why a report [to Palo] could not be available by spring of 1960."[3]

Although TVA publicly denied its keen interest in the Fort Loudoun Extension Project, clearly various parts of the agency were gearing up for action. In December 1959, TVA officials inspected the area thoroughly. Early the next year, Wagner, who apparently had become deeply committed to the project, ordered a number of offices within the agency to conduct independent benefit-cost ratio studies, even moving the important FUDAR Committee from a general calculation of benefits to a specific analysis of Fort Loudoun Extension. In June, Director of Information Paul Evans and Wagner agreed to rechristen Fort Loudoun Extension as the Tellico Project (see Map 2). Although some within the agency hung back, many more saw Tellico as the vehicle of TVA's new mission.[4]

But there were problems from the very beginning. Most serious were the difficulties with the benefit-cost ratio. As Elliot subtly warned Palo in July 1959, the "traditional," easily computed benefits of power, navigation, and flood control would not yield a favorable ratio. Yet the quantification of other benefits proved difficult. FUDAR simply failed, and other offices and private consultants did little better. In frustration, Wagner wrote, "It is essential that we move ahead as rapidly as possible to apply to the Tellico project any new and different methods we can devise for justification and financing."[5]

Of course, such activity could not escape outside notice for long. Early in 1961 rumors began to surface that the agency was seriously studying the Tellico Project and had decided to go ahead with it. Opposition began to mount. That opposition, however, was for the most part localized, divided, and poorly organized. For this reason, TVA felt that it could push the project through quickly and easily, and on April 15, 1963, the board, led by Wagner, agreed to start the Tellico Project in fiscal year 1965. At last the project, which had been on the drawing boards since 1936, appeared to be underway. But more trouble lay ahead.

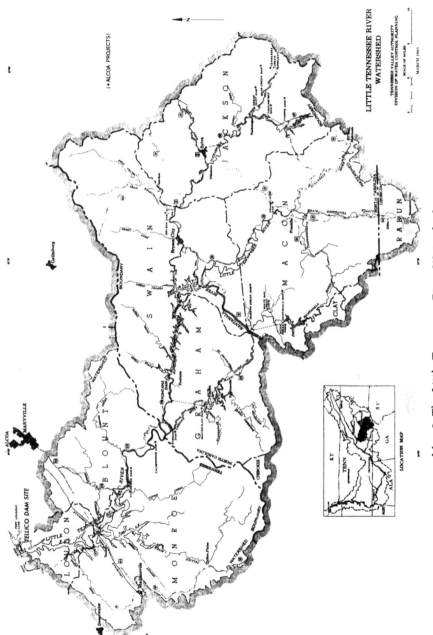

Map 2. The Little Tennessee River Watershed.

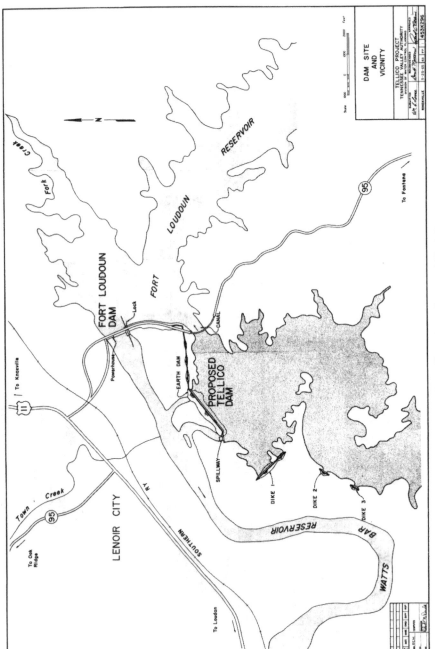

Map 3. The Tellico Dam Site and Vicinity.

If the Fort Loudoun Dam[6] could have been built far enough downstream, below the point where the Little Tennessee River flows into the main channel, there would never have been a Tellico Dam (see Map 3). Throughout spring and summer of 1936, TVA engineers and geologists had labored mightily to discover such a site. They made core drillings, economic analyses, and reservoir studies for a number of potential sites below the confluence of the Tennessee and Little Tennessee Rivers, and at least four places were marked for further examination. As J. Porter Taylor, who worked on at least one of the 1936 studies, recalls, "Hendon Johnston [supervisor of the core drilling teams] did his damnedest to find a site which would back up both rivers."[7]

But no acceptable site could be found. Even though studies and reports continued to be done into 1937, it had become clear rather early that Fort Loudoun Dam would have to be placed above the junction of the Tennessee and the Little Tennessee, at the Belle Canton Islands site. The dam would provide hydroelectric power, ease flooding problems downstream at Chattanooga, and extend the nine-foot navigation channel to Knoxville. At the conclusion of one report, River Transportation Division Chief C. T. Barker added, almost as an afterthought, "Plans for the Coulter Shoals [Fort Loudoun] Project provide ultimately for the placing of a dam across the mouth of the Little Tennessee River and diverting the flow of this river through a dredged canal into the Tennessee River above the proposed Coulter Shoals Dam." Hence the concept that was to become the Tellico Dam Project was born.[8]

Since the Fort Loudoun and Tellico Projects were intimately linked, decisions concerning one project ultimately affected the other. Two controversies arose at the outset which would have an important bearing on the history of Tellico Dam.

The first debate was over the land purchase policy for the Fort Loudoun Project. As has been noted above, there had been considerable feeling within TVA, particularly on the part of Harcourt Morgan and the agricultural interests, that the agency's large purchase policy was misguided in taking valuable agricultural land which never would be flooded. Of course, those within the agency who were pushing for a more comprehensive mission for TVA disagreed, countering that planned shoreline development was essential to lift the valley out of its economic woe.[9] The battle was never a close one. As we might expect, David Lilienthal backed Harcourt Morgan.

Equally important, local landowners began to protest even the modest policy adopted by the board on June 13, 1941, complaining about TVA's acquisition of some land for public access to the proposed Fort Loudoun reservoir. John Snyder, director of TVA's land acquisition program, complained to General Manager Gordon Clapp, "Our Fort Loudoun program will have to bear the brunt of unfortunate publicity for which this department was not responsible. An organization has been formed which is soliciting the owners on both sides of the river to sign a petition, which will be to the effect that either we change our plan of taking or they will refuse to sell." Snyder also reminded Clapp of similar protests at the Guntersville Project in Alabama in the mid-1930s. With the construction of Fort Loudoun Dam already underway, and with the directors determined to pursue a limited land purchase policy, the agency, fearing bad publicity that it could ill afford, ultimately decided to follow a modest course that was nearly a foregone conclusion.[10]

The land purchase policy implemented at Fort Loudoun would have an important effect on the later Tellico Dam controversy. As the years passed, many landowners along the Little Tennessee River came to take it for granted that TVA's modest land purchase policy was immutable, that what had been would always be. Apparently it was not well known that the agency's *original* policy, begun at Norris, had been a major purchase policy. Nor did it occur to the residents that if those who sought a more comprehensive mission for TVA ever gained the upper hand within the agency, that original policy might well be restored. Hence, most landowners assumed that the land purchase policy employed at nearby Fort Loudoun would be followed for the Tellico Project, and even though they were aware of TVA's desire to build what was to become the Tellico Dam, they believed that all was well.[11]

The second important controversy was over the power capacity at Fort Loudoun. The original plans had called for the installation of three generating units with a total capacity of 96,000 kilowatts. However, the possibility of a dam across the Little Tennessee and a diversionary canal meant that more power might be produced at the Fort Loudoun Dam if a fourth generating unit were added. Even though construction had begun in July 1940, studies by design engineers continued to see whether a fourth generating unit, necessitating a fourth spillway, were feasible. Although they concluded that the additional generating unit would not be needed until the Little Tennessee River was diverted into the Fort Loudoun pool, they recommended that space be left for the fourth unit, which could be added later. Ultimately that scheme was approved by the board on March 13, 1941, and architectural plans for the dam were altered even as it was under construction.[12]

In a sense, the decision to leave space for a fourth generating unit committed the agency to a sister dam on the Little Tennessee, which was called "Fort Loudoun Extension" from about 1940 to 1960. Indeed, roughly six weeks prior to the board's March 13, 1941, approval of the revised plans for Fort Loudoun, Clapp ordered Chief Engineer T. B. Parker "to proceed with foundation exploratory work for the possible Fort Loudoun extension across the Little Tennessee River." At the same time, TVA was planning to ask Congress for authorization to initiate the project. Even so, while the building of the dam and the installation of a fourth generating unit at Fort Loudoun would increase power production, the increase would be so modest that later critics questioned whether the additional power generated by the Tellico Project would be worthwhile. Hence, in later determination of a benefit-cost ratio for Tellico, increased power production would play a small role. Again, a decision made in the 1940s in connection with the Fort Loudoun Dam would return years later to plague the agency.[13]

As World War II approached, increased defense requirements prompted the agency to rush through a number of projects. In summer 1941 a controversy over the Douglas Dam pitted TVA against Senator Kenneth McKellar. Apparently the senior senator from Tennessee was piqued because TVA had not consulted him on the order in which it would build future dams. Lilienthal had only contempt for McKellar, and each snub made the senator more intemperate. "Your discourteous letter of September 30th has been received," began one frigid letter from McKellar to the TVA director. The outbreak of war, however, strengthened TVA's hand, and the Douglas Project was approved in time for construction to begin in February 1942. The dam was completed in just over twelve months.[14]

Undoubtedly TVA was an important contributor to the war effort. And, the Douglas controversy notwithstanding, the war facilitated numerous TVA projects. Kentucky, Watts Bar, Fort Loudoun, Appalachia, Chatuge, Ocoee Number 3, Nottely, and Cherokee Dams were all under construction when the war began and continued to receive appropriations. Fontana and Douglas Dams were started and completed during the war. Indeed, World War II lent many of these projects a sense of urgency and offered TVA both a mission and an opportunity.

By April 1942, W. L. Voorduin of the Water Control Planning Department had completed a comprehensive study of the project, an activity authorization had been drawn up, and a recommendation had been made to General Manager Gordon Clapp. Interestingly, Voorduin's report listed as justifications for the project merely the three traditional justifications of flood control, navigation, and power.

On July 3 the board approved the project, and an application for a preference rating was filed with the War Production Board (WPB) on July 30.[15]

The WPB, however, had some strong reservations. In its effort to use the war as an excuse to start as many construction projects as possible, TVA had overstepped its bounds. Perhaps the most serious difficulty was that TVA, trying to get as many projects authorized and underway as possible, had not installed the third and fourth generating units at Fort Loudoun Dam. Because of the critical materials needed for that equipment, the WPB had refused to grant preference ratings for the third and fourth generating units. Without that additional power capacity, the WPB reasoned, a dam across the Little Tennessee River was unnecessary. Hence on August 3, 1942, the TVA board instructed the general manager to withdraw the agency's application for a project rating for the Fort Loudoun Extension. The board, however, made it clear that TVA intended to submit a similar application "at a later date" and said that engineering and field surveys would continue. Clearly, the agency had not abandoned its plans for a dam near the mouth of the Little Tennessee River.[16]

At war's end, a host of uncompleted projects, the movement of the Office of Power toward coal-fired steam plants, an unfriendly Congress, and a general conservative shift in the nation forced the Fort Loudoun Extension Project almost to the bottom of TVA's 1945 report on postwar construction programs. After thirteen "urgent" projects, the Fort Loudoun Extension rated twenty-first of twenty-two "non-urgent" ones. But the agency never abandoned its conviction that the Fort Loudoun Extension ultimately should be built. As one TVA official who refused to be identified explained, "There was always talk about *when,* not *whether,* the project would be undertaken." In April 1946 Gordon Clapp wrote to a Mr. Glover of the Milk Producers Association that "we believed the project would be built some day; we had no plans for immediate construction and no request pending for appropriation with which to build it." In a somewhat different vein, General Manager John Oliver responded to a query from Senator Estes Kefauver in 1951 by explaining that "costs have so increased that the project is not now economically attractive." Hence, while TVA made the Fort Loudoun Extension a low priority, it was never abandoned entirely.[17]

Until the 1959 Watts Bar Conference and Wagner's "more dams" speech, the project was kept on the back burner. The climate of opinion in the Eisenhower administration, in addition to the move toward coal-fired steam plants and the attacks by private power interests and other business groups, diverted the agency's attention.

As general manager, Wagner continued to issue TVA's official position that the Fort Loudoun Extension project was not as economically feasible as other projects. Far more attractive in the 1950s were projects on the tributaries favored by the Eisenhower administration because they were cooperative efforts merging the private and public sectors.[18]

Early in his career at TVA, Wagner became convinced that widespread industrial development of the tributary areas was preferable to the concentration of industries in the region's few large cities. "Wagner was rural-oriented," remembers one former TVA employee. His was a rurality, however, tinged with industrial modernization. Indeed, as late as the mid-1960s, when the "factories-in-the-fields" concept was being seriously questioned by economists and regional planners, Wagner stubbornly held his ground. In a 1966 interview with the Knoxville Journal, Wagner maintained, "I think it is clear the future of East Tennessee and the whole Tennessee Valley is in further industrialization. . . . It is an area that never can support great numbers of people with agriculture."[19]

During the Eisenhower years, tributary area development offered a number of possibilities attractive to TVA. In 1953 President Eisenhower had stressed cooperation between public and private interests in developing the nation's watersheds. Congress responded to the president's message with the Small Watershed Program (Public Law 566), which stated that such projects must be initiated by private landowners' or citizens' groups and must secure the cooperation of appropriate state and local agencies. Almost immediately tributary area development became fashionable.[20]

TVA quickly recognized how it could benefit from the national trend toward tributary area development. Such projects might neutralize Eisenhower administration criticisms and hostility toward the agency. Too, the opportunity to tap private as well as state and local government funding offered an alluring prospect in the lean years of the 1950s. Hence, although tributary area development was hardly a new idea at TVA (the notion existed as far back as 1936), economic and political forces in the 1950s aided in its revival.[21]

Soon after becoming general manager, Wagner began pushing tributary programs. The first test of this approach came in connection with the Chestuee watershed, a project so modest that many former TVA people deny that it was the beginning of the agency's tributary program. The Chestuee watershed near Athens, Tennessee had been identified by TVA in 1938 as one of its one hundred "problem areas." The following year local residents had requested

TVA's aid in arresting the periodic flooding of valuable farmland. Until 1951 the agency's response had been limited to doing studies and making reports, many of which went far beyond the residents' requests for flood control assistance. Ultimately TVA did little except to direct local residents in the removal of debris from the creeks and make some plans for the future development of the area.[22]

Although the Chestuee Project was a modest one, TVA learned lessons it would apply later to other projects. Although private landowners' or citizens' groups might initiate the process by coming to TVA to request assistance, it was possible for the agency to bend these groups to its will. For example, a group might petition TVA to help in a flood control project and wind up with a comprehensive plan for area development. And since the agency had no intention of allowing citizen groups to have much of a real voice in area planning, the whole process was a most ticklish one. As Richard Kilbourne later recalled, "We didn't like these local groups telling us what we ought to do. . . . We had to be diplomatic about the whole business."[23]

If TVA could dominate citizen groups which came to the agency for technical assistance, why could it not create such groups to apply for help, thus giving the agency access to an area while still conforming to the letter, if not the spirit, of the Eisenhower administration's decrees? And since those citizen groups that did come to TVA were notoriously tightfisted when it came to money, why couldn't such groups be used to increase the agency's congressional appropriations for "citizen-initiated" tributary projects? "They dangled the bait of dams, power, jobs, and money," commented one former TVA employee who requested that his name be withheld, "and the locals went for it almost every time. Of course, the first people on the bandwagon were local political leaders, who then could claim credit for all the TVA money coming in[to the area]." In the early 1960s, Ralph Dimmick of the University of Tennessee agriculture faculty innocently asked a TVA employee whether local citizen groups were really self-formed or TVA-initiated. Years later Dimmick shook his head and smiled. "I was told the word had gone out to 'put a red flag on Ralph Dimmick.' Until then, I really didn't know the answer."[24]

Another lesson TVA learned from the Chestuee Project was that many within the agency were unenthusiastic about tributary area development. As Kilbourne recalls, "The 'power people' didn't have much of a role in all this. They sort of stood on the sidelines and pooh-poohed the whole thing. They thought they were above all this nonsense." If a tributary project did not involve construction of some kind, still others were uninterested. Moreover, TVA's regional planners were openly critical, maintaining that a project-by-project approach on the tributaries entailed an exceedingly limited view of regional development and sacrificed a more comprehensive approach to regional planning. "They wanted to start up little TVA's all over the valley," one former TVA employee explained, "and that was taking the easy way out by trying to repeat TVA's successes, only on a much smaller scale."[25]

Wagner, however, was determined to push tributary area development. Indeed, one can see certain similarities between Wagner's behavior in the tributary area development program and in the later Tellico controversy. Although many TVA people maintain that they could argue with Wagner and always receive a fair hearing, there was that inflexibility about the man. While he was sometimes able to change his mind, once he was committed to an idea he saw it through with a tenacity normally reserved for converts at a revival meeting. Moreover, he tended to personalize particular projects, entwining his own ego with a project's completion and ultimate success, and this made it difficult for him to maintain perspective. It was as if he had to prove himself with each TVA project or accomplishment. Success required determination, and failure was inconceivable.[26]

Wagner was able to initiate a number of important tributary projects, the most important of which perhaps was the Beech River Project, the first major multipurpose tributary undertaking. Southwest of Nashville in Henderson and Decatur Counties, the Beech River area suffered from flooding and erosion, lack of drinkable water for the city of Lexington, and massive outmigration of its younger population due to lack of employment opportunities.[27]

"Beech River is where we really took off," recalled Kilbourne. Ironically, the project originated in complaints from local citizens that TVA had been responsible for flooding their land. The agency responded with a comprehensive effort which included flood control, reforestation, and planning for industrial development.

"Beech River became a *total* project," Wagner later boasted. "Anything needed for the development of the area, we could do." In 1953 the agency projected that its involvement in the area would produce improved land use, greater nonfarm employment and income, and better community services and facilities. TVA "encouraged" the creation of the Beech River Watershed Development Association (BRWDA) in 1955, undertook construction of dams and planning projects, and in general focused its enormous talent on a watershed area badly in need of help. The results were truly impressive.[28]

Years later, Kilbourne spoke of the Beech River Project with a mixture of exhilaration and wistfulness. "There was the idea floating around that TVA could come in and completely remake the community."

Indeed, TVA's success at Beech River prompted a number of communities, usually acting through their local chambers of commerce, to seek TVA assistance and, predictably, TVA money. "Other groups heard about Beech River," said Kilbourne, "and they wanted in." But, having provided the model at Beech River, the agency was financially unable to take on many tributary projects. When the newly-formed Elk River Association appealed to TVA for a dam, planning, and area development, Chairman Herbert Vogel replied coldly that the benefit-cost ratio would make it infeasible unless state and local agencies could assume part of the cost or unless factors could be added into the benefit-cost ratio so as to make the project more defensible before Congress. The same message was communicated to the Duck River Development Association. TVA encouraged both groups to lobby Congress for increased appropriations for TVA (both did), but the agency itself kept a low profile. By that time, Wagner had more pressing concerns, principally the major dam construction program called for in his 1959 "more dams" speech.[29]

This is not to say, however, that Wagner lost interest in tributary area development. After all, it was "his" program. When Wagner moved onto the board in 1961, he created the Office of Tributary Area Development (OTAD) with Kilbourne as head. And throughout his career he continued to nurture TVA's tributary program. And OTAD's accomplishments, noted in its yearly newsletters, were indeed impressive. In roughly a dozen tributary areas, OTAD set up local committees, identified needs, and moved to meet those needs. Projects were as diverse as the valley itself. Educational programs from practical nursing to auto mechanics, school dropout counseling, "townlift" projects, encouragement of tourism and recreation, collecting abandoned automobiles from roadsides, and preparation of historical brochures all were undertaken. Industrial site surveys were made; solid waste disposal problems were solved; reforestation and forest industry planning were completed; dams were planned and built. Indeed, as Kilbourne later stated, there was nothing OTAD would not do.[30]

More relevant to this study, however, is the clear relationship between TVA's tributary program and the Tellico Project. Many of the people involved on the tributaries (Elliot, Taylor, Mattern and Kilbourne are but a few examples) became intimately involved with the controversial later project. As evidenced by Vogel's 1959 letter to Joe Sir of the Elk River Association, TVA had begun to toy with the idea of going beyond the traditional factors of power, navigation, and flood control in calculating benefit-cost analyses. As will be seen, those recalculations would become crucial to the Tellico Project. Also, many people involved in the tributary program came to realize that the small purchase policy followed since Fort Loudoun Dam was not compatible with their ambitious schemes for multipurpose projects. Multiple uses of the shoreline might be cranked into the benefit-cost ratio, too, if TVA were to acquire shore land. Simultaneously, many in the agency came to understand that a *major* multipurpose construction project would do more to unify TVA than the admittedly modest tributary projects. Many there had scoffed at the tributary program as "piddling," not equal in stature to the mighty dam projects of a few years before. Last, the tributary projects taught many TVA people how to work with and manipulate local citizens' groups, political wheelhorses, and development associations. While constantly insisting that it was nonpolitical, TVA became politically deft at the local level. "They drank a lot of coffee out there," remarked one TVA employee, "but they got people thinking TVA's way." Some TVA people had mixed feelings, believing that what TVA needed was "an honest project" and not "pork-barrel" programs. Others, however, went along. In short, the tributary area development program became TVA's training ground for Tellico. Therefore, when Wagner issued his 1959 call for more dams, TVA was ready to go.

Six weeks after the Watts Bar Conference, Wagner created the Future Dams and Reservoirs Committee (FUDAR). He charged the committee with identifying economic by-products arising from dam and reservoir construction and devising methods for quantifying in dollars those incremental values so they could be added to a project's benefit-cost ratio, thus making the project more appealing to Congress. To underline the importance of the work, Wagner staffed FUDAR with some of TVA's best people. Elliot, Taylor, Kilbourne, and Lowry were old hands who had worked in the tributary program. To that group were added Newton Dicks (Finance), J. E. Gilleland (Power Resources), Robert Marquis (Office of General Counsel), and F. E. Gartrell (Health and Safety). Although the committee's personnel would change somewhat, FUDAR continued to be a collection of some of the agency's most talented people. For chairman, the general manager picked Robert Howes from the Division of Reservoir Properties.[31]

FUDAR functioned poorly from the very beginning. To begin with, each committee member had taken on the assignment in addition to his regular duties, and so could not devote full time to the project. Indeed, even scheduling meetings was difficult; not until two months after its creation did FUDAR hold its first meeting.[32]

When the group finally did assemble, other difficulties surfaced. In theory, the committee's composition should have insured a melding of good ideas from throughout the agency. It didn't work out that way.

Instead, each person attacked the problem using his own background and expertise and often was unwilling to give in to other points of view. Some wanted a "real" project to work on, believing that each reservoir project had to be approached differently. Taylor and Elliot were engineers used to pat formulas for quantifying benefits; they had little patience with benefits that were difficult to quantify. Chairman Howes, who had come to the agency in 1934, was a landscape architect whose TVA specialty was recreation. Years of labor on recreation benefits had taught him to be less rigid in determining formulas, and without much success he urged the committee to be more flexible. "I was seen by some as an interloper, a usurper, because I wasn't in water resources." Years later, Howes said in frustration, "Nothing went right about that . . . committee's performance."[33]

Wagner had hoped to receive some kind of progress report from FUDAR by July 1, 1959. But one year after that deadline the committee still had little to report, and J. R. Perry acknowledged to Howes that FUDAR was in great difficulty. Perry suggested that the committee needed more expertise (a land economist, for example). Of course, that was probably what FUDAR *least* needed.[34]

Undoubtedly Wagner appreciated the difficulty of the assignment and deadline he had given FUDAR. Yet it is clear that he had a definite purpose in pushing Howes for the report, a purpose of which neither Howes nor the other FUDAR members, with the possible exception of Elliott, were aware. Wagner knew of the Elliot-Palo meeting on July 7, 1959, in which the Fort Loudoun Extension (Tellico) Project had been discussed. Moreover, Wagner was familiar with the site, having done navigation studies in the Little Tennessee River area as early as 1938. Wagner wanted to use the FUDAR report to push the Tellico Project at the 1960 board meeting which would approve projects for 1962. Obviously the general manager wanted to keep momentum going after the Watts Bar Conference and had selected the Tellico Project as a means of doing that.[35]

FUDAR's lack of progress by mid-1960 clearly irritated Wagner. In August of that year, he was forced to go before the board to withdraw the Tellico Project because of its poor benefit-cost ratio. That same month Wagner directed FUDAR to concentrate all its efforts on justifying Tellico. He instructed the committee to circulate a progress report for comments by October 15, 1960, and to submit a final report no later than January 1, 1961. Chief Engineer George Palo was to keep the pressure on Howes and FUDAR to meet the assigned deadlines.[36]

Actually, if he had had it his own way, Wagner would not have bothered with economic justifications or benefit-cost ratios. Wagner,

Howes believed, "would have preferred to dump benefit-cost ratios and just go ahead with projects." Palo, too, had little patience with "fiddling around with figures." But Wagner and Palo were not to have things their way. The general manager was informed in no uncertain terms by TVA's Washington representative Marguerite Owen that Congress expected benefit-cost ratios to accompany requests for appropriations and probably would not approve a project without one.[37]

Gradually the FUDAR Committee identified three areas where significant benefits might be calculated and added to a benefit-cost ratio: land enhancement, recreation, and "general economic benefits." Land enhancement meant simply that TVA would make major land purchases above the proposed reservoir, develop that land, and eventually sell it at a profit. But how could a formula be devised which would measure future increases in land values? Would Congress accept land enhancement as a benefit in a benefit-cost ratio? And would TVA exceed its powers when it got into what amounted to the real estate development business? In comparison, computing recreation benefits should have been easy. But no one seems to have been able to agree on *how* such benefits could be measured. Average dollars per day spent by tourists, or some other formula? "General economic benefits" proved more difficult still. Could TVA justifiably count increased industrialization (which, it was assumed, would lead to better paying jobs, increased consumerism, higher sales tax collections, etc.)? Would this lead to "double-counting" with other categories? And would Congress allow it?

Howes and his committee were intelligent, dedicated men, but the task was simply beyond them. If they were to fail, however, it would not be from lack of effort. Since September 1959 they had surveyed past reservoir projects, made corrections, and applied various formulas to Tellico. They had studied tourism, migration, employment, and unemployment. On his own, Kilbourne had developed a method for measuring the value of water supply for domestic, industrial, and municipal purposes. Other government agencies, like the U.S. Fish and Wildlife Service, were requested to supply material. Yet in the end, by January 1961, FUDAR had failed.[38]

By January 6, Howes had finished a draft of the report to the general manager. The draft was to be circulated within FUDAR, which was then to meet on January 20 to compose the final version. In the draft, Howes effectively admitted that the committee had failed. By page three, he was wondering whether TVA should not simply abandon benefit-cost ratios as bases for justifying future dam and reservoir projects. The draft concluded with the wishful thought that perhaps if they kept working at it, the FUDAR members could actually come up with the formulas which for so long had eluded them.[39]

The draft was mocked and savagely criticized within TVA. Elliot had shown his copy to Robert Frierson, who told Elliot that the report was "indicative of [a] lack of real study, committee discussion and consideration, and has developed little if any useful ideas or suggestions which would be helpful." In another memo to Elliot, E. J. Rutter called the report "over-temperate" and said, "It seems obvious from the summarized results of nearly two years of thinking . . . that there has been no significant breakthrough toward new concepts for justifying such projects. . . . The FUDAR Committee appears to have not allowed itself to be overly swayed by wishful thinking."[40]

As Wagner had been elevated to the board, the final report was submitted to L. J. Van Mol, the new general manager, on April 7, 1961. It was essentially a cosmetic revision of the January draft. "Neither Wagner nor Van Mol ever acknowledged receiving the report," recalled Howes with just a touch of bitterness. "I guess they didn't like it." Some time after submitting the report, Howes ran into E. A. Shelley, assistant general manager. The subject of the report came up. Howes recalls that Shelly stated flatly, "Don't expect Van Mol to say anything to you about the FUDAR report." There was little to say after that.[41]

But while many looked on the FUDAR report as an unmitigated disaster, in some ways it was a success. Although unable to quantify benefits, FUDAR had identified two areas where such benefits might well accrue: land enhancement and recreation. Both offered real promise for making TVA's benefit-cost ratios more attractive. Moreover, the simple identification of these potential benefits influenced the shape of the Tellico Project, which Wagner was so eager to get started. Tellico would require massive land purchases above the proposed reservoir if the project was to be economically justified. And recreation, along with industrialization, would have to be a major factor in the land use scheme.

Probably realizing all this, Wagner did not immediately dissolve the FUDAR Committee but instead ordered it to keep working, especially on land enhancement. More people were brought in. Kerry Schell, an economist in the Division of Reservoir Properties, came up with a study showing "modest" land enhancement at Tellico. "It was too modest, and they shelved it," remembers Schell, now an agricultural economist with the University of Tennessee. Jack Knetsch, a former TVA employee who had left the agency to join Resources for the Future, was brought back as a private consultant, but his reports were disappointing. Captain Beverly McGruder, on loan from the Corps of Engineers, was given a shot at the problem by Elliot and

Mattern, without significant results. Finally, after seven years, thousands of dollars, and untold hours of work, E. A. Shelley admitted in the margin of a document, "This was a good idea and a noble effort has apparently been made, but I'm afraid the results are not especially significant. I think we recognized from the beginning this might be the case." Later, in another deft piece of marginalia, Shelley told Van Mol, "I don't think it's necessary to report to [the] board on this. Only AJW [Wagner] is aware that it has been in process. Suggest you just mention to him sometime the results of the study." Finally, in 1966, FUDAR was quietly dissolved.[42]

There is little doubt that by early 1960 Wagner was determined, favorable benefit-cost ratio or not, to push for the Tellico Project. In late 1959 he had ordered the Project Planning Branch to coordinate a number of studies of the area. In January 1960 those studies began trickling in: reservoir area inspection, mineral analysis, highway relocation reports, navigation potential, land cost estimates. By April, TVA design engineers and architects had completed drawings of elevations and sections of the dam. Nor did Wagner abandon the almost quixotic search for a more favorable benefit-cost ratio. In March 1960, over four months before Wagner withdrew the Tellico Project from board consideration, Mattern reported to Elliot that the best benefit-cost ratio the Project Planning Branch could come up with was 1:1, clearly insufficient to win congressional approval. That was when Wagner urged the FUDAR Committee to work harder to create defensible formulas which would convert land enhancement and recreation benefits into dollar terms. But the general manager also pressed a number of others into the quest for a better benefit-cost ratio for Tellico. All studies and proposals were to be filtered through Palo's office. "No later than January 1, 1961, I will expect to have a coordinated report from the Chief Engineer's office on the economic justification of the Tellico project. . . . [coming up with] as many new ideas and methods of justification and financing as can be developed by that time to complement conventional methods of economic justification." Clearly Wagner meant business.[43]

The narrow victory of John F. Kennedy in the 1960 presidential race added even more urgency. It was common knowledge that Kennedy supported public works projects to create jobs and stimulate regional economies. Moreover, as a neo-New Dealer, it was expected that the president would look with favor on TVA. Indeed, in the 1963 celebration of the agency's thirtieth anniversary, Kennedy attended festivities at Muscle Shoals and said, "The work of TVA will never be over. There will always be new frontiers for it to conquer." Because of the change in administrations, TVA felt that it should push immediately for approval of several projects, including Tellico. As one TVA planner later recalled, "When Kennedy became President, Tellico, Melton Hill and Elk River were dusted off as hurry-up projects. We looked at them very briefly and had to go back to the President in very fast order with our proposals, and that is how they got into the funding bill. Of course, we suffered the consequences of that forever after."[44]

TVA's almost frantic search in late 1960 for a better benefit-cost ratio for Tellico yielded few substantial results. Taylor, Howes, Kilbourne, and G. W. Hamilton, head of the Sanitation Section, were among those who made suggestions to Palo, but little in their reports was new. From the Office of Power came Wessenauer's grumpy response that preliminary figures had put power benefits too high and power costs too low. "It is my present opinion," Wessenauer wrote, "that construction of the project depends *entirely* on finding means of justification which go beyond navigation, flood control, and power programs." That, of course, was hardly news. Palo reasoned that if TVA could not increase the benefits figure, perhaps the ratio could be improved by lowering costs. By October 1960, roughly $5.5 million had been shaved from the project costs. In February of the following year Palo confessed to Wagner that all efforts so far had proved fruitless. "Though sound reasons can be advanced for constructing a hydro power project with a ratio just exceeding 1.0, the . . . inclusion of other benefits are [sic] needed to make Tellico an unquestionably justified project." That too was hardly news.[45]

Of course, such frenetic activity could not escape notice for long. Nor, apparently, did Wagner want it to. In the agency's annual year-end report, it was noted that the Tellico Project was "under consideration" and that its economic feasibility was being studied. Not mentioned was the fact that in four months little progress had been made.[46]

The *Knoxville News-Sentinel* broke the story in its 1961 New Year's Day edition. In reply to a host of queries, TVA Chairman Herbert Vogel denied that any final decision had been made, that benefits approximately equaled costs, and that TVA might have to ask state and local governments and "local individuals" to help foot the bill. Technically, Vogel was correct in all his denials, although it is quite clear that TVA itself was committed to the Tellico Project and had been for at least a year.[47]

Vogel's newspaper comments that day and later made it fairly clear to discerning readers that TVA was having difficulty coming up with adequate justification for undertaking the Tellico Project. On January 15, the *News-Sentinel* reported that "TVA is looking for a way to justify spending Federal money on the Tellico Dam." That, of course, was something of an understatement.[48]

Virtually every project TVA had undertaken had met with some opposition: disgruntled farmers who did not want to sell their land, private power and business interests, conservative groups on the national level such as the national Chamber of Commerce, and, in the case of Norris, church groups unhappy over the agency's initial reinterment policy (this was changed to meet the objections). But the opposition, never terribly strong, usually had been disorganized and easily mollified by TVA. Hence TVA had little reason to expect formidable opposition to the Tellico Project. Moreover, experience on earlier tributary projects had taught TVA how to create "locally-initiated" citizens groups, work with local political figures and the press, and in general manage public opinion. Soon after news of the Tellico Project's status broke on January 1, 1961, local business groups such as the Madisonville Jaycees and the Lenoir City Regional Planning Commission endorsed the project. Although it is difficult to prove, it is not at all unlikely that TVA itself drafted such endorsements and placed them in the proper hands for passage and distribution. For one thing, the wording of virtually all the endorsements was identical. And later, in 1963, TVA did try to induce the Greater Knoxville Chamber of Commerce to endorse the Tellico Dam.[49]

TVA moved to head off potential opposition in three ways. First, in response to inquiries, TVA officials stubbornly maintained that no decision had been made to build the Tellico Dam. In early 1961 local landowner Philip Appleby was complaining to anyone who would listen about TVA's intrusion into the lower end of the Little Tennessee Valley. All of Appleby's letters (to President Kennedy and Senators Kefauver and Gore) were referred to TVA for reply and information. In his letter to Gore on the subject, Director A. R. Jones wrote, "We have no immediate plans for construction of the project." That posture was retained even after the board's April 15, 1963 decision to request funds. On April 19 Van Mol wrote to the Loudon Regional Planning Commission, "Although this project is only tentative at present. . . . " While technically correct, such a public position tended to give the incorrect impression that TVA was doubtful about whether or not to go ahead. Therefore, until 1963, landowners in the area remained uncertain, confused, and generally silent.[50]

The second way in which TVA initially was able to minimize opposition to the Tellico Project was by being less than candid about how much land was going to be needed for the entire project. Since 1960 it had been fairly clear that TVA would have to buy considerably more land than that which would be flooded so as to capture the value of land enhancement for a favorable benefit-cost ratio. Indeed, without the value of land enhancement, many within TVA feared that an acceptable benefit-cost ratio could never be reached. Years after his retirement, former Chief Engineer George Palo admitted in a letter to Chairman David Freeman:

> At the hearing on January 8 [1979] Zygmund [sic] Plater[51] contended that Tellico was not a "hydro" project but a "land" project. In part he was right. Tellico was the first project ever proposed by TVA because of economic benefits expected on the shore of the reservoir. . . . Industry had, of course, come to Decatur from the Wheeler project, at Guntersville from that project, and at other points, but had not been considered as a part of project approval.

Yet TVA was careful not to make public the fact that it was strongly considering purchasing land up to one-half mile from the projected shoreline, or that it intended to wait approximately ten years for land values to increase, then to sell the land and count the profit as a "benefit" in the crucial benefit-cost ratio. In a February 1961 letter to H. S. Gangwer, manager of the Fort Loudoun Electric Co-Op, TVA Hydro Projects Section head William Loose cautioned Gangwer, "Please understand that this project has not been authorized, and the information shown on these maps [which sketched out a tentative taking line] is for your official use and not for public information." Such maps would prove important later. Moreover, many local residents assumed that the benchmarks (concrete pillars with brass plates on top, designed to show reservoir elevations) created by TVA in the early 1940s were actually the upper limits of the "taking line." Many later claimed that TVA had done nothing to correct that impression. Hence, confusion about how much land actually would be purchased also made it difficult for worried landowners to rally others to opposition.[52]

The third method TVA used to avoid serious opposition was encouraging the formation of semi-private groups sympathetic to the project. The agency continued to urge local chambers of commerce and municipal and regional planning commissions to publicly endorse the Tellico Project. TVA representatives met with officials of the Bowaters Southern Paper Company, which had a paper mill, a tree nursery, and sizable landholdings in the area, and of the Fort Loudoun Association, a group devoted to the preservation of the old French and Indian War garrison as a historic site. At the meeting with Bowaters, Vogel tried to defuse the potentially dangerous situation by assuring the company that the benefit-cost ratio had not been determined and that TVA would make no move until it had a solidly defensible one. As for the Fort Loudoun Association, the agency had repeatedly—and sincerely—promised that TVA would, in Gordon Clapp's words, "do anything that we can within reason" to preserve the historic structure. Both Bowaters and the Fort Loudoun Association apparently agreed to wait for further developments.[53]

In order to marshal pro-Tellico public opinion, TVA, working through OTAD (which knew how to do that sort of thing), assisted in the formation of the Little Tennessee River Valley Development Association (LTRVDA), a "voluntary and non-profit association" which on one hand would plan for the area's future development and on the other hold pro-Tellico meetings and lobby in Congress. Some meetings were held, some lobbying was done, and some study groups were formed. But, partly because of subsequent internal division, the association really never amounted to much. Ultimately it was dissolved.[54]

Meanwhile the agency was working furiously to firm up Tellico's benefit-cost ratio. Past performances of other TVA reservoirs were used to tighten land enhancement and recreation figures. On land enhancement, TVA began by assuming that without the project no inflation of land values would take place. Therefore, any rise in land values that did take place could be credited to the project. Given what happened to area shoreline property values in the 1960s and 1970s, that was at best a dubious assumption. In defense of those within TVA who labored over this problem, however, any other assumption would have made it nearly impossible to compute land enhancement benefits. Nevertheless, by October 1962 the agency realized it would have to purchase and hold approximately 16,500 acres of nonreservoir land to come up with a satisfactory benefit-cost ratio.[55]

TVA was never comfortable with the exceedingly speculative and fluid figures attached to land enhancement and recreation benefits. Indeed, a few within the agency even questioned whether TVA should claim such benefits at all. But they were ignored. In May 1962, the United States Senate, in Senate Document #97, had provided a rough guide to procedures to be used in evaluating and development water and related land resources. TVA immediately embraced these guidelines—indeed, the agency probably had been aware of the fact they were being formulated and had supported the effort. Later TVA asserted that its benefit-cost procedures were merely those outlined by the Senate, conveniently ignoring the fact that the agency had been using, or trying to use, similar procedures at least two years before the Senate guidelines came out. Even so, benefit-cost ratios varied significantly: in March 1962 Elliot claimed the ratio to be around 2:1, but Palo was more modest, asserting 1.53:1 (with "recreation" the big benefit at 2.41:1).[56]

To be sure, some within TVA still held back. At the Office of Power, Wessenauer was openly irritable. In a meeting early in 1963, the chief of power took the Tellico file, slammed it down on the table, and said, "Let's forget about this." Director A. R. Jones feared the repercussions of returning to a large land purchase policy, and made sarcastic comments both in meetings and in document margins. Chief Engineer Palo clearly was uncomfortable with the porous benefit-cost procedures. People in Agriculture as well as in Fisheries, Wildlife, and Forestry were openly unhappy. "But if anyone in Forestry had even whispered that," remembers one former TVA employee, "Ken Seigworth would have cut their heads off at dawn."

Yet Wagner was committed to Tellico, as were most at TVA, and the force of Wagner's personality, the majority within the agency, and the collective mentality of the early 1960s virtually assured its approval. The doubtful within TVA were neutralized (one former TVA employee claimed that Wagner appealed to Wessenauer as a personal friend not to stand in the way of the project), outside opposition to date had been minimal, and historic forces seemed to favor Tellico as the showcase of TVA's "new mission." On April 15, 1963, the board agreed to seek congressional appropriations to begin the Tellico Project in 1965. The board did say, however, that benefit-cost studies should continue.[57]

Even after board approval, TVA continued to maintain that the Tellico Project was still uncertain; Congress had not yet appropriated funds for the project. Over a month after the board's decision, A. J. Gray told the Tennessee State Planning Commission of "the uncertainty of the project," and the following September, almost five months after the board decision, TVA denied in a press release that the Tellico Project was getting "warmer."[58]

By mid-1963, then, TVA once again was ready to begin the project near the mouth of the Little Tennessee River. Much had changed between the board's first approval in 1942 and its recommitment to the project in April 1963. Shifts inside and outside the agency had strengthened those within TVA who advocated a comprehensive mission. The dearth of possible main river projects and the mood in Washington had forced the agency to consider more seriously projects on the tributaries. Moreover, increased congressional oversight of projects meant that proposals had to be more defensible than they perhaps had been before. Finally, Wagner's determination to unify TVA and get it moving again meant that a defensible major project had to be found, and quickly, so that TVA could regain its former momentum.

Most of the men involved in choosing and defending the Tellico Project had been with TVA from its inception: Wagner, Palo, Mattern, Elliot, Taylor, Howes, and others. Together these men had had a large role in reshaping the Tennessee Valley, accelerating its modernization, and raising the living standards of its citizens. If temporary disruptions of lives had occurred, that was, they and others reasoned, a small price to pay for the long-term benefits received. If opposition arose, they believed reason and results would cause it to evaporate. Decent, committed men, they had given as much or more to TVA and the valley as they had given to their own families.[59]

At the Watts Bar meeting in 1959, the Tellico Project, though it had existed in TVA's past under another name, had been merely an inspired hope in the mind of Aubrey Wagner. Four years later that hope had become a planning reality—an object of discussion and debate, a rallying point for a new mission, and a concern that would dominate the careers of many and absorb their energies during the decade ahead.

3

THE LITTLE TENNESSEE VALLEY
AND THE PROCESS OF MODERNIZATION

As TVA began to sketch out the scope and shape of the Tellico Project, many within the agency had developed a stereotyped image of the Tellico area and its people. Though awed by the natural beauty of Tellico, most TVA employees believed the area to be impoverished and in desperate need of modernization. This, they felt, only the agency could offer. At the same time, TVA generally viewed the people of the Tellico region much as it had the people of the Norris region in the early 1930s: as hardy, honest, and often courageous, but also comparatively poor, backward, unimaginative, and intensely opposed to change of any kind. In short, TVA saw the area as a static, unchanging pocket in a dynamic, technological, modern nation—a premodern deficiency that the Tellico Project would soon correct. Logically, therefore, there could be no opposition; even the more retrogressive of the area's citizens would come to appreciate and participate in the full benefits of the showcase of TVA's new mission.

Yet the region and its people were more complex and diverse than the agency gave them credit for. In many ways the area had traveled a long and uneven road from its frontier and Appalachian origins. Small enclaves of poverty and backwardness still existed, but to a large extent the Tellico region had crossed the chasm between the premodern and modern eras. The problem, as the people themselves saw it, was how to preserve many of their traditional institutions while simultaneously enjoying the benefits of modernity. Unfortunately, TVA would continue to hold its misconception of the area and its people even after a diverse local opposition to the project raised its head. Yet that misconception was born not of conspiracy or dishonesty, but rather of ignorance and a stubborn determination to repeat the prior successes of the agency. That flawed vision ultimately would prove tragic, both for TVA and for many of the region's people.

As in much of the American South, Sunday mornings traditionally have been quiet times in the valley of the Little Tennessee River.[1] Churches, most of them small, unpretentious frame structures, are filled. Inside, children fidget; parents in attractive, well-pressed clothes try simultaneously to maintain control and pay attention to the service; dating teenagers sit in pairs, apart from their respective parents. Well-used family Bibles are in nearly every hand. A songleader prods the congregation and a piano player of modest but sufficient skill through one of the many hymns sung throughout the service (rarely is there a choir). The singing, once started, is usually enthusiastic, and on the "old familiars" the harmonies are both inventive and pleasing, with a particular brand of southern female alto much in evidence. As the preacher sits aloof from all this, mentally and spiritually girding himself for the prayers and sermon to come, men, women, teenagers, and even children sing of trials, salvation, God's promise, and their own human frailties:

> The pains, the groans, and dying strife,
> Fright our approaching souls away;
> Still we shrink back again to life,
> Fond of our prison and our clay.[2]

Not everyone in the valley attends church, though most profess to be "believers." But, by unwritten rule, work beyond the minimum chores is rarely done on Sunday. Nonchurchgoing men often congregate at a gas station-general store known to be open or at a familiar spot by the side of the road. There they assume that distinctive squatting position which no "city boy" can hold for more than a minute, smoke (never stubbing out the butts but rather drowning them with spittle), and share experiences. Although "damns" or "hells" are sprinkled about liberally, sexual topics almost never come up, and, indeed, women are rarely mentioned at all. Jokes usually give way to stories—long, rambling discourses with many asides. While some of these tales—the story of one who came across a timorous bear while both were picking blackberries from the same bush, for example—are often told and heard, they never seem to age. As they once enjoyed well-worn punch lines on an old radio show, listeners anticipate the buildup, the climax, and the resolution with ill-disguised delight. When a local election, especially a race for sheriff or road commissioner, is brewing, all other subjects are swept aside as the candidates are deftly dissected ("that fella'd be as lost in that job as a dirty ball in high weeds"). Gradually these knots of men break up; Sunday afternoons are for families or for visiting. The men never ask each other whether they will gather again next week. Barring death, illness, or particularly bad weather, they will.

These Sunday morning rituals, the church services and the men's small conversation groups, go back as far as any participant can remember. Moreover, they are important in the valley, though not unique to it. They are two ways which people have of defining communities, maintaining community cohesion, and circulating community news. In a society which values privacy and minding one's own business, these institutions serve to tie together the multitude of communities that dot the valley landscape.

By the time TVA came to the Little Tennessee River Valley, some of the communities in that valley already had undergone considerable change and were well along in the modernization process. Others were not, communities on the river or valley floor generally moving faster than communities farther from that artery. Hence, while all the communities displayed some similarities to one another, the degree to which they had been touched by the modernization process divided them into two groups: communities which had been profoundly affected by this process and those which had been more lightly brushed by it. This division could be seen in the mentalities and values of the two groups of communities, in their attitudes about life, family, children, work, "outsiders," and the future. In some cases social institutions (such as the church and the male conversation groups) had changed and had come to take on different, though no less important, functions.

It is doubtful that outsiders like those at TVA fully perceived or understood what was happening to these communities. Indeed, many residents of the communities themselves did not. But before one can analyze the building of the Tellico Dam or measure the valley's complex responses to it, it is first necessary to understand the communities and people of the valley and the modernization process which was changing all of them.

It is just over thirty miles from the site where the Aluminum Company of America (ALCOA) built the Chilhowee Dam in 1955–57 to the point where the Little Tennessee River flows into the main channel of the Tennessee. Beginning at the Chilhowee Dam site and moving downstream,[3] the Little Tennessee meanders westward, banked by gently rolling hills which become slightly steeper the farther one moves from the river. However, after Bacon's Bend, where the river makes a 180° turn to head northwest, the valley floor fans out into a wide, fertile plain, a plain not visible from the river because of the steep bluffs along both sides of the channel. Beyond the plain, which is often over a mile wide, lie the inevitable rolling hills again, and beyond them rougher hills, some thickly wooded, barely accessible, and so remote that sightings of wild animals are occasionally reported. Through all this the river moves inexorably, fed by a number of tributaries such as the Tellico River, Bat Creek, Toqua Creek, Island Creek, Notchy Creek, and Ballplay Creek, and passing by Indian mounds and ancient village sites and around a number of small islands usually named for early settlers (McGhee, Calloway, Bussell, Toliver). The valley floor remains wide and fertile. Indeed, in a region not noted for rich, flat farmland, this is some of the best there is. On the south bluff, about twenty miles from the river's mouth, stands the palisade of Old Fort Loudoun, built by the British in 1756–57 during the French and Indian War and the scene of a horrible Indian massacre in 1760. Directly across the river at this point is the Tellico Blockhouse, constructed by the United States Army in 1794 to protect settlers from the Indians. Here the Little Tennessee River slows somewhat, winding toward the place where it flows into the main channel of the Tennessee River just south of TVA's Fort Loudoun Dam, completed in 1943. Rainfall in the area is among the heaviest in the nation, and it is not unusual for parts of the valley floor or the rolling hills beyond to be shrouded in fog or mist. Once the entire area was richly forested, but since commercial lumber companies thoroughly cut the region prior to the 1920s, hardwood trees are small and many of the pines are insubstantial. Until relatively recently, the river, its tributaries, and some dense forests acted as natural barriers dividing communities from one another. Yet in an Appalachia traditionally seen as beset with economic and social problems, life in the Little Tennessee Valley is comparatively good for many if not most.

In the years following the American Revolution, whites were generally slow to penetrate the Little Tennessee Valley. Land was plentiful in and around White's Fort, soon to be Knoxville, and the possibility of Cherokee attacks in the valley discouraged settlement there. Fort Loudoun stood unoccupied after the tragedy of 1760, in which the fort had surrendered to the Cherokees only to have the Indians attack and massacre the survivors three days after they had abandoned the garrison. Until 1807, confused negotiations with the Cherokees left settlers with dubious legal title to their land. Therefore, although a number of men received land grants in the valley for Revolutionary War service or had purchased land from either veterans or land companies, actual settlement was extremely slow.[4]

But in the early nineteenth century settlement increased considerably. Unlike the northern frontier settlement pattern, in which men went westward first to establish homesteads and then sent for their families, the southern frontier saw the influx of whole families—men, women, and children—creating communities which replicated those back East.[5]

Land near the river was claimed and occupied first, settlements including towns and large farms. Grants to the latter often encompassed over one thousand acres. McGhees, Jackson's, Kittrells, Akinses, Halls, and Blankenships were only a few of those fortunate enough to lay claim to the precious bottomland. Morganton, founded in 1813 by James Wyly, became "a thriving little river port with a growing population." Other towns were Madisonville (off the river, established in the 1820s), Decatur (located where the Tellico River runs into the Little Tennessee), and New Philadelphia (established 1828). Others would be founded later.[6]

These communities were quick to establish schools and churches. While the national government had set aside land to be used for educational purposes in the Northwest Ordinance (1787), families in the Little Tennessee River Valley often donated land and goods for their schools. Not unusual was the land grant made by J. S. Jackson for the lordly price of "$1.00 and the love and esteem we have for the cause of education." Churches were likewise supported, as one James M. Burton donated a portion of his property for the "love and affection which I have for the cause of Christianity and morality and for the further consideration of One Dollar." Teachers were hired and circuit riders contacted. Here men and women hoped to carve out their futures in the valley at the edge of the fast-receding frontier.[7]

Not surprisingly, the earliest settlers claimed the best land. Their large grants or purchases at once made them the new area's elite. Later arrivals were forced off the valley floor and onto the rolling hills or into the dense forests. Hence, even at the beginning, the valley had a class system based on the time of arrival, the quality of land possessed, and the amount of eastern "culture" imported. While 113 ladies of Madisonville were petitioning the state legislature in 1837 in the interests of temperance, communities were still being created which had little time or inclination to wage war against the demon rum. Those communities far from the valley floor still battled wolves, bears, and occasionally each other.[8]

Other practices soon separated the communities on the valley floor from those farther from the river. Sometime in the early nineteenth century, large landowners on the valley floor began to purchase slaves. The practice was not widespread, if we are to believe that slave graveyards discovered by TVA represented a majority of dead slaves. Only 82 graves were found. This leads to the hypothesis that slaves in the Little Tennessee River Valley were used largely as house servants and that the area did not witness many of the slave "gangs" prevalent elsewhere.[9]

By the 1850s it was clear that the communities of the Little Tennessee Valley had established certain patterns of class, economics, social life, and religion which rigidly differentiated them from one another. On the valley floor, large landholdings prevailed, with slaves employed to supplement family labor. Nearer to the river towns, those communities on the valley floor had more access to national news, outsiders, and innovations. Gradually they became less suspicious of newcomers, strangers, new ideas, and "culture." Communities up the tributaries or beyond them in the thick forests, however, seemed frozen in time, not radically changed since settlement. Landholdings were generally smaller, the lifestyle almost universally poorer, and news from beyond the valley was rarely received. While communities on the valley floor busied themselves chartering ferries and making roads to decrease their isolation, upland communities seemed unwilling to become part of the outside world. Strangers were greeted with suspicion and guns.

The rising national uproar over slavery in the 1850s affected the two types of communities in radically different ways. While not in the same class as the eastern and Gulf planters, landowners on the valley floor generally supported slavery, secession, and the doomed Confederacy. Yet all around them, the upland communities were hotbeds of Union and antislavery sentiment. Since the upland also was viciously hostile to blacks, one can surmise that much of their Union sentiment was in fact a deep-seated resentment of their "betters" on the valley floor. Although some rather nasty engagements took place in nearby Knoxville in late 1863, no record exists of anything more than some very unmilitary conflicts, mostly isolated shootings, in the Little Tennessee Valley.[10]

By the end of the Civil War, the Little Tennessee Valley was dotted with river towns and small rural communities. While no one now alive can recall the area's premodern communities in their purest form, fragments of recollections, orally transmitted history, folklore, customs, and careful work with census and other statistical data can be used to recreate a picture of life in the communities of the Little Tennessee Valley.

The communities that sprang up throughout the valley in the early 1800s were rarely named for founders or significant individuals. Instead they seem to have derived their names from peculiar natural features (Big Piney, Pumpkin Center, Mulberry, Poplar Springs, Corntassel, Lakeside), occasionally from older Indian names (Citico,

Talhassee, Toqua [which local residents came to pronounce To'Ko]), or from some activity that took place there (Ballplay). Except for the river towns, communities were made up of about 100–150 people, with a church or two, a general store, and occasionally a sawmill. Physical barriers and personal preference made contact between these communities rare, and boys from one community who called on girls from another were often greeted with verbal abuse and occasionally with stones.[11]

Several institutions served to maintain community cohesion and prescribe individual behavior. "Most social activities centered around the church," one person remembered. "Of course, there were no dances, but there were suppers, picnics, and other things. We used to go to church in a wagon, and in those days if you didn't get there early, you'd have to stand." If a community had more than one church, people often attended the one closest to their homes, visited the other church regularly, and saw no basic doctrinal differences between the two. Referring to baptisms, two people remembered Methodists as "sprinklers" and Baptists as committed to total immersion. "We used to go down to the sandbar yonder," remembered one, "and watch the baptisms. Everybody went." Hence the institution of the church, through its regular services, singings, revivals, "dinners on the ground," baptisms, and funerals defined the community and bound it together.[12]

Moreover, marriage inside the community was strongly encouraged, leading to a high degree of interrelatedness. By the early twentieth century, individuals with names such as Alean Millsaps Chambers McLemore were not so uncommon and even where interrelatedness did not appear in names, the connections were generally known and understood. While not unknown, marriages outside the community were not common. One man, whose home place had been in his family since 1861, revealed a persistent family rumor that his great-grandfather, upon marrying a Cherokee Indian woman, had been forced to leave the community but had fathered a son who later returned to rejoin the community and reclaim the family land.[13]

Marriage within the community solidified the community in a number of ways. It meant that land remained controlled by community members. Inheritance patterns reinforced this trend, for in the premodern years land tended to be divided equally among all children, male and female. A few examples are typical. When Jack Maynard died in 1918, his land was divided equally among six people: his widow, two sons, and three daughters (another six children had predeceased him). Similarly, when Sam Gray died, he willed his land equally to eleven children: three sons and eight daughters (Gray had two predeceased children). The will of William Kennedy bequeathed real property to his two surviving children (six had predeceased him) and to thirty-six grandchildren, those whose parents had died before William Kennedy himself. Obviously this kind of inheritance pattern could not continue indefinitely. But for most of the nineteenth century and in certain communities for part of the twentieth, this system worked fairly well, insuring widespread land ownership within the community and at the same time maintaining community unity and stability.[14]

Widespread land ownership did not, however, mean that these communities had no socioeconomic class structures. In fact, some families were able to use sagacious marriages and good business practices to maintain or increase their large holdings. These families owned much of the best bottomland on the valley floor, built fine brick houses, often owned slaves, and gave the communities some appearance of opulence. For example, J. C. Anderson came from Blount County to the Little Tennessee Valley in 1902, acquired three tracts of land (one of which, a 280-acre farm, was purchased by TVA in 1967 for $109,889.69), built a large brick home with walls a foot thick, and saw his three children marry well in the community. J. C. Anderson's son Bert even had electricity in his home, provided by his water-powered generator. In a classic understatement, J. C. Anderson's granddaughter, Charlotte Anderson Hughes, said, "I guess you could say our family was looked upon as affluent."[15]

Yet, while classes did exist, class lines were fluid, social mobility was prevalent, and rich and poor within a community seemed to get along well and treat each other with respect and courtesy. The wealthy do not appear to have expected or received deference at church, in politics, or in other aspects of community life. Those who flaunted their wealth were scorned. Speaking of one such example, a man sneered, "Why, they even built themselves a swimming pool! A swimming pool!" Then he spat into the dirt. In contrast, another wealthy man was spoken of highly: "Why, he'd sit down and drink coffee with you anytime." While in some communities, especially those near the valley floor, the disparities between rich and poor might be enormous, good manners, a rough-shouldered democracy of sorts, social mobility, personal and family ties, friendship, and restraint on ostentation counted for more than the very real economic barriers.[16]

In addition to church, marriage, landholding patterns, and interclass social relations, lighter social institutions and forms of entertainment also acted to reinforce community identity. In late spring and summer, for example, communities competed in "field ball" games, activities which consolidated their existences as communities. Field ball was a variation of baseball, in which all the men and boys in a community formed the team and all played at once, sometimes with over fifty people in the field. Women, girls, and small children all attended too, forming the "cheering section" and preparing food for a picnic.[17]

Just as eighteenth-century Americans had believed that only landholders had a true "stake in society," valley residents saw land ownership as terribly important. As one noted, "To my grandfather, wealth was land. Land was something that couldn't be taken from you." Any indebtedness such as a mortgage or bank loan was to be feared and, if possible, avoided. Indeed, even by the late 1960s when TVA began buying land for the Tellico Project, very little land in the valley was mortgaged, only 76 of approximately 750 tracts. Land also helped to determine a person's place in the community.[18]

Family life was extremely important. Although men often went to town by themselves and formed groups to work, hunt, or just make conversation, women tended not to go to town alone. "Women couldn't go much," one person recalled, "and you didn't see many women out except on Saturday or at church on Sunday." In fact, women had little time for trips, for women and girls had an almost endless round of chores to perform, including food preparation, washing, ironing, making soap and starch, making clothes ("We cut our own patterns from newspaper") and tending the children. At the same time, work, lack of outside recreation, and relatively limited transportation facilities threw families closely together in a manner that modern suburbanites would find uncomfortable.[19]

In addition to the value placed on land, freedom from debt, and family life, communities established unwritten codes of behavior which all understood. One person said that a "bad person" was one who committed murder, practiced adultery, or didn't mind his own business. But a host of other proscriptions shaped an individual's behavior in the community. Drinking was, and in many valley communities still is, publicly frowned upon. When the Coors Beer Company was considering the valley as a possible site for one of its breweries, the preacher at one of the valley's largest Baptist churches asked all in his congregation who disapproved of Coors to stand, and all save one did. Also disapproved were slothfulness, riding trains on Sundays, courting outside one's community, disrespect to parents, stealing, wasting money, parting with the land before passing it on to one's heirs, and a host of other major and minor transgressions. In a community so intimate, minding one's own business was one of the most important imperatives of all. "Minding your own business," one interviewee explained, "means not talking about other people, even if you knew their business."[20]

Practice, of course, might fall short of the ideal. Drinking was reproached, yet some drank, occasionally to excess. Stealing was forbidden, yet sometimes things were "missing." "Minding your own business" was the agreed-upon rule, yet a few men and women gossipped shamefully. In a society where most men owned guns, an occasional shooting occurred. And a community in which lives were so intimately intertwined could also be suffocating. A son's respect for his father was a given, but the process of declaring independence or "breaking away" could be difficult, wrenching, and occasionally violent. The fact that sons could be married and well into adulthood before inheriting their portion of land tended to perpetuate both deference and conflict. Hence, while these communities were both functional and satisfying for most people, they certainly were not the psychological or sociological Edens that many remember them to have been.[21]

In economic and demographic terms, the communities of the Little Tennessee Valley in the nineteenth and early twentieth centuries generally can be classified as premodern or transitional in nature. Yet by the early twentieth century forces were beginning to make themselves felt that, had economic and population trends gone on as before, would have put severe strains on these communities, particularly those in the rolling hills farther from the river.

The Little Tennessee Valley was almost exclusively engaged in agriculture. In the three counties which touch a portion of the lower Little Tennessee Valley—Blount, Loudon, and Monroe—only 391 people were employed off the farm in 1880 and only 963 by 1900. Even after numerous divisions, farms remained, by East Tennessee standards, fairly substantial, averaging 176 acres in Monroe County in 1880. Most of these farms were diversified, attempting to produce both for consumption and for the market. Farmers raised mostly corn, wheat, and vegetables and kept some pigs, poultry, and beef and dairy cattle (mainly family milk cows). Tobacco, an important crop later in the twentieth century, was hardly grown at all in 1880. Owing to prevalent agricultural practices, including lack of fertilization, crop yields were extremely low by modern standards: under 338 pounds per acre for tobacco in 1880 and under 16 bushels per acre of corn in 1910.[22]

Population was growing rapidly in the lower valley, due to a very high birth rate and fertility ratio. Even by 1930, when birth rates had begun to decline, the lower valley's crude birth rate was 30 percent higher than that of the United States as a whole and higher than that of the Southern Appalachian region. Judging from admittedly fragmentary cemetery records for the valley floor, infant mortality must also have been high. Yet the birth rate far exceeded it, and the population increased rapidly.[23]

Large families and comparatively low crop yields put increasing pressure on the land. These factors were probably less noticeable on the valley floor where soil was fertile and landholdings tended to be large. But in the uplands, this dual trend presented problems for communities which, had other factors not entered the picture, might well have reached a crisis point. Also, as population continued to grow, the traditional practice of dividing the patrimony equally among all heirs meant that the number of farms increased while average farm size declined. In Monroe County alone, the average farm size declined 13.3 percent between 1900 and 1910, from 120 acres to 104, and continued to fall in succeeding years. Moreover, with increasing population pressure, more and more farmland was being put under cultivation. In Blount County alone, the percentage of improved farmland rose from 19.7 percent in 1860 to 37.7 percent in 1910. Simultaneously, the number of persons per square mile in the area increased substantially, indicating that outmigration had not yet become significant.[24]

By 1900 some of the population appears to have slipped into tenancy. In that year 40.5 percent of Monroe County's farmers were tenants, up from 36.0 percent in 1880. Later evidence suggests that these tenants were not related to the owners for whom they worked. "Almost every farm of any size had tenants," recalls one person. "My father had four or five tenant families on his farms."[25]

There is little question, therefore, that by the beginning of the twentieth century increased population and comparatively low average crop yields had begun to put increasing pressure on land and resources at the lower end of the Little Tennessee Valley. To bridge the gap, an ever-increasing amount of farmland, some of it marginal in quality, was put into cultivation. Some farmers were losing their land and slipping into tenancy, mostly of the sharecropping type. Unless some new forces appeared to arrest these trends, clearly an economic and demographic crisis was on the horizon.

Yet that crisis never materialized. As in most societies confronted with economic and demographic problems, adjustments had to be made, and community values began to change as the forces of modernization intruded upon these troubled communities.

The Atlanta, Knoxville and Northern Railroad had caused no great stir until it was purchased by the Louisville and Nashville system in 1902. Connection to the L&N meant that the region now was tied to the "outside world." Moreover, it meant that more change was to come.[26]

In 1907 the Babcock Lumber Company of Pittsburgh, Pennsylvania, purchased enormous tracts from the Smoky Mountain Land, Lumber and Improvement Company. Lumbering had gone on in the valley prior to Babcock, but soon the giant combine swallowed up all its competitors. Babcock built its own private railroad to Maryville, tried to hire Italian immigrants to do the logging (they were frightened off by gunfire), and proceeded to employ local people to raze the rich timber of the lower valley. Tellico Plains, founded as an iron foundry town, quickly became a Babcock "company town," at once prosperous, crude, and rough. Logging camps with picturesque names like Rattle-snake Rock and Jeffrey's Hell were set up in the dense forests in the rolling hills far from the river, and timber was cut (some poplar was over four feet in diameter), hauled by oxen to the nearest railroad point, and transported into Maryville. By 1925, when a terrible fire roared through the forests, endangering the logging camps, Babcock had virtually denuded the valley.[27]

As Babcock was stripping the area of its good timber, the Pittsburg Reduction Company (later ALCOA) moved into the area. The potential of cheap water power and docile non-union labor had attracted the company to what it must have felt was one of the more remote areas in the East. In 1910 the company began investigating the area, and in 1913 it began to purchase land in Blount County for what was to be its enormous aluminum plant. Hence, not twenty years after wolf bounties had been ended in the Little Tennessee Valley, a huge industrial complex was being set up within thirty miles of where the wolves had roamed.[28]

In 1926 Stokely Brothers and Company, a canning firm which eventually became Stokely-Van Camp, began buying and leasing farm-land and setting up a canning factory. Not long after that, the Rural Electrification Administration (REA) brought electricity to the lower Little Tennessee Valley, making lights and radios common in an area where they had been all but unknown. And in a few years, World War II pulled men and boys from their communities to fight in Europe and Asia, places these people had heard of only in geography class — if they had gone that far in school. Indeed, probably because of its isolation, the area was selected as the site of a World War II German prisoner-of-war camp.[29]

The impact of modernization on the lower valley varied from community to community. Towns which, like Decatur, Philadelphia, and Morganton, were not fortunate enough to be on the railroad line became virtual ghost towns, while Upton Station (renamed Vonore in 1893), which was on the river and the railroad line, grew into a thriving little village with a depot, flour and sawmills, stockyards, and several shops.[30]

Larger towns outside the valley (Maryville, for example) and away from the river (Madisonville) were able to attract most of the large industries which came to the area. As early as 1903, the Madisonville Knitting Mill was advertising locally for child labor, to be recruited from the surrounding rural communities. Landless adult men and women found employment in these establishments, too. James "Buster" Carver remembers that many people commuted from the lower valley to work at ALCOA ("ALCOA paid better than Stokely"), at least until 1937, when a strike closed the aluminum plant. By 1920 the three-county area had 4,571 nonfarm workers; ten years later the number had risen astoundingly to 17,327.[31]

Even though regular contacts with people and institutions outside the valley were altering the attitudes, values, and traditions of some communities, other communities away from the river, in the rolling hills and hollows, were barely being touched by outside influences. Some communities were quite remote, making commuting outside the lower valley for work exceedingly difficult. Bears, cougars, and wildcats still roamed areas like Jeffrey's Hell and Coker Creek, and organized wild boar hunting (which attracted hunters from all over the nation) was still prevalent as late as 1936. Indeed, the modernization process widened the gulf between the upland communities and those on or near the valley floor.[32]

The Great Depression of the 1930s temporarily reversed the trend toward modernization. Many mills and factories in nearby towns (Loudon, Lenoir City, Maryville, Knoxville) were forced to close; the 1937 strike at ALCOA threw more people out of work; and commercial lumbering operations had virtually died out by the end of the 1920s. In Lenoir City alone, 889 families were receiving direct relief by 1934. Unemployed and desperate, many of the valley's nonfarm workers were forced back upon the land, and tenancy increased substantially to 1935. After 1935 these people began to desert the valley, and tenancy declined to below its 1925 figure. The depression, of course, would not last forever, and the dislocations, while sometimes wrenching and tragic, were nevertheless not permanent.[33]

More serious were the worsening problems of the area's agriculture. Farm size had continued to drop in the three-county area, and the proportion of acreage under cultivation had steadily increased. Moreover, owing to intensive cultivation, failure to rotate crops, and lack of fertilization, crop yields did not increase to meet rising needs. For example, corn yields in Monroe County, just under 16 bushels per acre in 1910, had risen only to 20.14 bushels per acre by 1940. Worse, TVA surveys of Loudon and Monroe Counties done in 1935 revealed that gullying and sheet erosion had damaged over seventy percent of the farmland in Loudon County and an equally significant amount in

Monroe. In fact, cultivation had reached such a point that twenty-five percent of the wheat consumed had to be shipped in. One can plainly see, therefore, that without the forces of modernization which had penetrated the lower valley and which had been virtually erased by the Great Depression, many communities of the lower valley already had reached the theoretical crisis point for premodern societies. Even the more prosperous communities on the valley floor were not untouched.[34]

In time, industry did return to the towns in and around the Little Tennessee Valley. Hence one can never be entirely certain whether the continued process of modernization or the people's responses to the premodern crisis were responsible for the economic, demographic, and ideological changes which began in the lower valley around the time of World War II and which profoundly altered the region. Perhaps the most important change was the significant decline in the crude birth rate and a concomitant decrease in the size of families. Though in 1930 the three-county area's crude birth rate (see Table 1) had been far above the national average, by 1960 it had dropped drastically, to a point very close to the national figure.

		Crude Birth Rates, 1930 and 1960			*Table 1*
	USA	S. Appalachia	Monroe	Loudon	Blount
1930	20.6	30.1	30.9	28.7	29.6
1960	22.7	23.0	23.8	21.4	21.3

Source: DeJong, *Appalachian Fertility Decline*, Appendix A, Table 1.

As Table 2 shows, the result was a significant decline in the area's average family size. The TVA land purchase records for the Tellico Project dramatically illustrate this trend toward fewer children and, hence, smaller families. For example, William Kennedy, who purchased land on the valley floor in 1918, had eight children. Only one of his children had that many, and the average number of children in his children's families was but 4.8. Similarly, Jack Maynard and his wife Julia had eleven children. Of those children who themselves had children, the average number of children was also 4.8. And those family sizes continued to decline steadily.[35]

Part of the reason for the decline in average family size in the lower valley was that young adults were leaving the valley to seek employment elsewhere. Between 1950 and 1960, outmigration from the three-county area totalled a staggering 15,523 people, *over half of*

Table 2   **Average Family Size, 1900–1960**

	Monroe	Loudon	Blount
1900	5.4	5.1	5.1
1930	4.8	4.75	4.59
1950	4.17	3.87	3.9
1960	3.8	3.5	3.6

Source: *U.S. Census: Family Characteristics*, 1900–1960.

them between twenty and thirty-four years old. Most of those out-migrants found jobs nearby, many in Knoxville, so it was not that families were being split apart by great distances. Contacts were regular. Nevertheless, an increased proportion of the area's population was composed of the very young and the aging. By 1960, Monroe County's "dependency rate" (the proportion of people zero to nineteen years old and sixty-five or over) was the fifteenth highest of Tennessee's ninety-five counties. Clearly the lower valley was losing young adults.[36]

Those who chose to remain in the lower valley often had to go farther and farther from home to secure work. By 1970, almost forty percent of the work forces of Monroe and Loudon Counties were working outside their counties of residence. "Some people would drive two hundred miles each day so they could work and still live here," said Ben Snider, owner and operator of a grocery store in Vonore and son of one of the first families in the valley. "There was something here that people wanted to hang onto." But "hanging on" was not always easy.[37]

To make ends meet or to ease the shock of periodic unemployment (which in 1960 was 8.2 percent for Monroe County males), a steadily increasing number of women from the lower valley communities found work outside the home. Between 1950 and 1960 Loudon County's female labor force grew 45.7 percent, while the male labor force *declined* 4.4 percent; a rising proportion of those women were married and had young children. By 1970, over 66 percent of the three-county area's female labor force were married women with husbands present; a good number had preschool children. This development profoundly altered family life and dynamics and served to keep families small. The increase in married women in the work force tended to raise the median family income in the lower valley, and those who grew used to that increased income would not want to abandon it.[38]

Even with smaller families, inheritance patterns were changing, too. No longer was it possible to continue to divide farmland equally among all heirs, and the traditional system was replaced by an interesting system of joint tenancy. All the children together would become heirs to the entire farm, but the farm itself would not be subdivided. Rather, one of the children, usually the son or daughter chosen to care for an aging mother, would continue to live on the "home place." For example, the Moser family had three children: Louise, Leona, and Thomas. When their father died, the three children, now adults, inherited the land. Title, however, went to Thomas, who agreed to stay on the home place, farm it, and care for his mother (she died in 1977).[39]

The people of the lower valley, then, were making changes in their practices, attitudes, and values as they were engulfed by the modern age. Family size, place and type of work, women's work, the role and dynamics of family life, separation of families through outmigration, increased family income, and inheritance patterns—all aspects of life had been altered, some of them profoundly. True, vestiges of the old ways still remained. For example, medical care was still very uneven in the lower valley. The nearest physician was in Madisonville, and the Vonore area was served by a retired doctor, Dr. Troy Bagwell. As one person remembers, "Dr. Bagwell used to drive his car into Vonore and would sit in the car while the patients lined up at the car window. He gave out medicine from his car, and if you needed a shot, you just stuck your arm or whatever through the window. The way his hand shook would make you nervous. But he was a good man and a good doctor."[40]

Modernization may also have contributed some less desirable features to community life in the lower valley. Rising rates of illegitimacy, suicide, and outmigration gauged the deterioration of community cohesiveness. For example, Charles Hall, owner of the Tellico Plains Telephone Company and the town's mayor, graduated in a high school class of forty-two, only five of whom remained in Tellico Plains. "There just was nothing here for them," Hall said sadly.

The new inheritance pattern, which was intended to keep farms together, made it difficult to dispose of property. "A lot of land was tied up in family trusts," commented Ben Snider, "and was almost impossible to sell." As in communities throughout the western world, modernization of the valley was a two-edged sword.[41]

As has been shown, the communities of the lower valley were neither passive nor unimaginative in the face of modernization. Many traditional practices and values were altered to meet the the people's changing needs. At the same time, to cushion the communities' shock when faced with such rapid changes, some traditional institutions began to take on new functions. Churches, for example, remained as symbols of community cohesion long after the actual unity had evaporated. Preachers spoke of the idyllic communities of the past, of present dangers, and of the dark, uncertain future. They called on their charges to hold fast to the old values even as church building committees replaced kerosene lamps with electricity and pot-bellied stoves with electric heat. No longer the center of the community, the church had become a buffer, a way to merge the old way of life with the new:

> You leave us to seek your employment[,] my boy,
> By the world you have yet to be tried,
> But in the temptations and trials you meet,
> May your heart to the Savior confide.
> CHO:
> Hold fast to the right, Hold fast to the right,
> Wherever your foot-steps may roam,
> O forsake not the way of salvation, my boy,
> That you learned from your mother at home.[42]

Churches were not the only social institutions which altered their functions in modern time. In the past, many had relied on hunting to put meat on the table. Most meat now, however, was purchased in grocery stores, and hunting became a symbol of male independence and his ability to provide. In the same vein, the male conversation groups became less a method of getting out of the house while women went to church than, like hunting, a symbol of male independence in an age when men were becoming less independent. Modernization did not eliminate these institutions, but it changed the reasons why people needed them.

As TVA penetrated the lower end of the Little Tennessee Valley to build the Tellico Project, it confronted communities which, in varying degrees, were already in touch with the opportunities and problems of modern society. To some outsiders, many of these communities appeared premodern, static, crude, remote, and backward. Admittedly, a few fit that description, but most, to a greater or lesser degree, were in the process of changing their lives and institutions so as both to profit from the new and to hold onto parts of the old. In short, many of the lower valley communities had become modern in a number of ways even before TVA offered even greater and more dramatic changes.

Interestingly, those communities least touched by modernization were the ones that embraced the new TVA project most warmly.

Exceptions to this rule, of course, were the larger towns (Madisonville, Loudon, Lenoir City), where a businessman-booster mentality generally held sway. But, by and large, the most modernized communities resisted the most. These communities, on or very near the valley floor, had always been the most prosperous, had early established regular contacts beyond the lower valley, and had therefore made the greatest strides toward modernization. These were precisely the communities which would be the most affected by the Tellico Project, either by altered highway and railroad connections or by total or partial land purchase and inundation.

So throughout the lower valley, on the eve of the most massive transformation since the original white influx, those communities stood poised between two worlds, unwilling to go backward (except in memory) to the dying past and uncertain about going forward into the terrifying shadows of the future. In male conversation groups and in churches, they ruminated, expressing their ambivalence, hoping that the changes they already had made had been the right ones. As the church songleader called out the next hymn, the members, like so many others, looked backward and forward at once:

> Precious mem'ries, unseen angels,
> Sent from somewhere to my soul;
> How they linger, even near me,
> And the sacred part unfold.
> CHO:
> Precious mem'ries, how they linger,
> How they ever flood my soul;
> In the stillness of the midnight,
> Precious, sacred scenes unfold.[43]

TVA never fully comprehended the complexity of the society into which it was about to intrude. Nor did the agency fully understand how the structure of East Tennessee communities had changed since the 1930s, when TVA first penetrated the Tennessee Valley. Ironically, TVA had been one of the agents of change in the valley. But the agency clung tenaciously to a view of the region as it once had been but was no longer. Had the people within TVA appreciated the complexity and diversity of the Tellico area, its people, its history, and its uneven road to modernization, they would have understood better what the Tellico Project was and was not capable of. Too, they would have understood better the local and regional opposition to the project which arose early in the struggle. But few people in TVA understood the real situation—at least not until it was too late.

4

STORM CLOUDS ON THE HORIZON:
THE RISE OF OPPOSITION

On the evening of September 22, 1964, Aubrey Wagner attended a public meeting at Greenback High School[1] to speak for the proposed Tellico Project. The meeting, sponsored by the Little Tennessee River Valley Development Association (LTRVDA), was ostensibly an "informational" gathering. But the real purpose seems to have been to soothe an anxious populace and silence the growing number of local critics and open doubters. Usually, Wagner was unequalled in situations of this sort.

However, the Greenback High School meeting almost turned into a disaster for TVA and a rout for Wagner. More than four hundred people attended, and one reporter later estimated that roughly ninety percent of them were opposed to the project. Wagner courageously stood up and spoke in favor of the dam, but he was interrupted several times by the audience, and no one else stood to support either him or the Tellico Project. In response to angry charges that TVA was shoving the dam down the local landowners' throats, Wagner reiterated his promise that "if the majority of the people don't want the Tellico impoundment, we won't build it." At this point, Judge Sue Hicks jumped to his feet and demanded a local referendum on the project. With but one way out of the corner into which he had painted himself, Wagner replied that the Tellico Project would affect far more people than just the local landowners and therefore no accurate referendum was either possible or justifiable. The crowd, already hostile, was visibly angry at Wagner's retreat, and one editorialist sourly noted later that "most persons believe that TVA will eventually 'decide what the people's choice is'." In short, while the Greenback High School meeting might have been "set up" by the opposition with the intention of abusing and embarrassing Wagner and TVA, the gathering also showed that the agency was not going to build the Tellico Dam without a fight.[2]

Since the TVA board's April 15, 1963 decision to seek congressional appropriations for the Tellico Project, a small core of opposition had formed among area landowners, businesses, and others such as the Fort Loudoun Association, who for various reasons were against the dam. That group had been small, never a majority of area residents or even of those landowners who would be directly affected. But they had been vocal and, as the Greenback High School gathering demonstrated, could pack public meetings and force TVA into making embarrassing *faux pas*. Moreover, local opponents had been bolstered by very small groups of state government officials, principally in the State Game and Fish Commission, as well as by a few Knoxville and Maryville businessmen. These groups did not always act in concert, even after the formation in 1964 of the opposition Association for the Preservation of the Little Tennessee River, approximately one month after Wagner's appearance at Greenback High School. But together they made it clear to a largely indifferent public that anti-Tellico dissenters were neither a collection of wild-eyed anti-TVA extremists nor an opposition that would easily buckle and roll over before the mighty presence of the Tennessee Valley Authority.

However, members of the opposition realized that they could expend mountains of paper writing letters to editors and government officials and pack all the local meetings ever held and still not stop TVA. What was needed, they concluded, was to carry the fight into the national arena and influence national public opinion on the issue. But who would care about a modest dam constructed in the middle of rural East Tennessee which, TVA was saying, would bring jobs and prosperity, halt outmigration, and be a boon to all? And who could combat the national reverence for TVA as the agency which had "saved" the Tennessee Valley and brought it into the modern age?

To focus national attention on what opponents believed was an ill-conceived project, anti-Tellico leaders invited Supreme Court Justice William O. Douglas to visit the area. That event was carefully choreographed by the opposition. The justice's carefully organized visit accomplished its purpose, and the Tellico Dam controversy briefly reached the national spotlight.

But TVA, initially caught off guard by the size and ferocity of the opposition, rallied quickly. The agency stumped the local area for support, used its tremendous influence to get important local personages and groups to come out in favor of the project, and attempted to discredit and silence key opposition leaders. And although local opposition to the dam never completely folded, without a permanent national base it could never stand for long against the agency. As one local editor told an opposition leader, "Around here you just can't fight TVA."[3]

It is exceedingly difficult to characterize the early opposition which arose in the Tellico area. Although the objectors never represented more than a minority of the local population, when the group gathered in force, as at Greenback High School, it appeared more homogeneous than in fact it was.

In general, the farther one got from the proposed dam site, the less local opposition there was to the Tellico Project. In Sweetwater (roughly eighteen miles from the site) and Madisonville (twenty-two miles), opposition was negligible, in Tellico Plains (over thirty miles) almost nonexistent. Yet even in the areas close to the proposed dam, the opponents, while considerably stronger, were still in the minority.[4]

Nor are other characteristics helpful in understanding why some opposed Tellico while others did not. For example, anti-dam leaders John M. Lackey and Judge Sue Hicks were somewhat elderly, but opposition leaders Thomas Burel Moser, Jean Ritchey, and Albert Davis were considerably younger. The ages of dam supporters varied similarly. The families of some opponents had held their land for generations, but so had the families of many supporters. And some opponents (like Ritchey and Philip Appleby) were comparative newcomers.

Years later the hard-core local opponents complained bitterly that their cause had been partially undermined by tenants ("who had plenty to gain an' nothin' to lose") and by landowners carrying steep mortgages ("they saw this as a way to get out from under"). It is true that the vast majority of tenant farmers did support the project, probably seeing TVA's promised industrial jobs as preferable to their present situations. Yet a few tenants opposed the dam, and, to make matters more complicated, some opponents were *both* owners and tenants. Moreover, tenants represented only a fraction of the population and in most cases were not influential forces in their communities.[5]

It is difficult to prove that landowners carrying mortgages were any more influential in undermining local opposition to Tellico than were tenants. For one thing, the vast majority of all affected lands were unencumbered by mortgages. In addition, while most people with mortgaged property did favor the project and readily sold their holdings, so did most landowners without mortgages. Finally, some property owners with mortgages held out to the bitter end, even to the point of having their lands condemned by TVA and taken by right of eminent domain. In sum, none of the traditional explanations— age, duration of land ownership, tenancy, mortgages, etc.—seems to show why some local people opposed the Tellico Dam while others — the majority — favored it.[6]

A somewhat more satisfactory explanation, if more difficult to pin down with precision, is the general ideology or "mindset" of those who opposed the project, in contrast to that of those who favored it. Attitudes about community and change per se appear to have been quite different among the two groups. For example, most people in business initially supported TVA, as did Ben and Carolyn Snider, who owned a grocery store in Vonore. The Snider family had been in the area for generations, the first Snider having been a Moravian circuit preacher who had ministered to his modest flock almost from the founding of the community. Ben and Carolyn, who was trained as a medical technician, hoped that the project would bring jobs and prosperity. They wanted their community to change, to become wealthier and more progressive, not just for their own sakes but for the benefit of everyone.[7] In fact, most supporters who gave reasons for their belief that Tellico would be "a good thing" cited jobs and increased prosperity. As supporters-turned-opponents the Sniders explained bitterly, "TVA did a good job of playing on people's dreams of jobs."[8]

On the other hand, those who opposed Tellico from the beginning preferred their community as it was to any vision of what TVA might make of it. While they did not oppose all change or change per se and were clearly aware of the problems their communities faced, many felt that much that they valued would be destroyed by "progress." Most people connected with the Fort Loudoun Association feared that their historic heritage would be obliterated by the TVA-induced changes. Many of the earliest opponents recalled with fondness the community's earlier days, usually those of their childhoods, the days of picnics, ballgames, baptizings, and family life. These nostalgic recollections, filtered through time, stood in stark contrast to the memories of many dam supporters who, like Tellico Plains businessman and Tellico booster Charles Hall, could remember only the problems — poverty, outmigration of young people, and the general feeling of having been stranded somewhere while the rest of the world passed by. In short, many of the dam's earliest local opponents shared a set of attitudes about community, change, and "progress" absent among the vast majority of local Tellico supporters. Hence, as opponents spoke with their neighbors and kin who supported the project, it was as if they were speaking on two different wave lengths.[9]

Of course, in the early 1960s vast numbers of local residents seemed to have had no opinion at all. Part of this rather curious reaction may have been due to the fact that, as the Sniders put it, "Nobody thought it'd ever happen." Even after April 1963, when the TVA board voted to seek appropriations for the Tellico Project, "nobody took it seriously." Another reason why many local residents seemed uninterested in the emerging controversy was that until 1967 it was not clear just how much land TVA intended to take. Most people in the area had little insight into TVA's true ambitions.[10]

So, for a number of reasons, opinion at Tellico was badly fractured over the dam issue. Though Wagner surely was engaging in wishful thinking when he said in September 1964 that "nearly all" the local reaction was favorable, the great majority was pro-Tellico or indifferent. The local opposition alone could not have stood for long against the mighty TVA, "the guv'ment," as some called it. The battle was too one-sided.

Early in the battle against the Tellico Dam, however, local objectors had several important allies whom TVA considered infinitely more dangerous than the locals themselves. These allies included the Tennessee Game and Fish Commission, some local industrialists in Knoxville and Maryville, the Tennessee Outdoor Writers Association, one staff member of the Knoxville Chamber of Commerce, and the editor of the Knoxville Journal. Working together, these people were able to mount a broad, coordinated opposition to the dam, based less on the emotional issues of seizure of land and fear of change than on a wide-ranging attack on TVA's claims for the project. Theirs was a formidable challenge.

Sometime before TVA's April 1963 announcement that it would seek congressional appropriations for Tellico, Bob Burch and Price Wilkins of the Tennessee Game and Fish Commission approached David Dale Dickey, director of area development for the Knoxville Chamber of Commerce and president of the Tennessee Outdoor Writers Association. Through contacts inside TVA, Burch had learned that the agency planned, contrary to its public stance, to go ahead with Tellico. Dickey, a quiet, slender, intense man with important connections among Knoxville's small social and political elite, had heard rumors as early as 1961 that TVA definitely planned to build the Tellico Dam, so he was hardly surprised by Burch's and Wilkins's news. The three men vowed to fight the project together and began looking for ammunition and allies.[11]

Burch, Wilkins, and Dickey were drawn together by their hope of keeping the lower end of the Little Tennessee River as a haven for trout fishermen. But they realized that their assault would have to be more broadly based than that. Believing that trout fishing attracted more tourists than would come to fish for bass in a lake, the three set about to prove that it would be better for the area economically to leave the Little Tennessee River as it was. In addition, Burch, Wilkins, and Dickey were willing to challenge TVA's claims that industry actually would come to the site after the dam was built and the reservoir created. Using some rather self-contradictory logic, the three opponents maintained that the uncertain future of the area since the early 1940s actually had *discouraged* industry from coming, but that even if the Tellico Dam were built, industry would not come.[12]

In their search for evidence to bolster their case, Burch and Wilkins invited Dickey to accompany them to Arkansas, primarily to study tourism along the White River in the hope of showing that a freeflowing river attracted as many visitors as did a lake. Burch had commented that lakes were so common in the Southeast that each new lake attracted proportionally fewer tourists. Later Dickey argued essentially the same thing in his study, "The Little Tennessee River as an Economic Resource," a thesis done for the University of Oklahoma's Industrial Development Institute. The thesis was published and distributed by the Tennessee Game and Fish Commission in September 1964.[13]

As they gathered evidence for what they believed would be a damaging case against Tellico, the three opponents also searched for allies. Although the number they recruited was small, it included some figures with both influence and power, people whose words they felt would be heeded in Tennessee and in Washington. Fred Stansberry, director of the Tennessee Game and Fish Commission and hence Burch's and Wilkins's boss, was almost a "natural." Commenting on Stansberry, Charles Chance, chief of TVA's Fish and Wildlife Branch, told TVA director of Forestry Development Kenneth Seigworth, "Stansberry has some feeling or obsession that the Corps [of Engineers] and TVA are just gobbling up every stream, and that somehow he is morally obligated to oppose the Tellico Project even though he admits that TVA will probably build it anyway."[14] Stansberry was an important recruit, primarily because, as Burch's and Wilkins's superior, he gave them a good deal of freedom to campaign against Tellico.

Of even greater potential significance, however, was the addition of Hugh McDade to the anti-dam cadre. McDade also was a member of the Game and Fish Commission but, even more important, he was an executive with ALCOA in Maryville and a member of the Maryville Chamber of Commerce. Undoubtedly a man of some influence, McDade could be counted on to lend the opposition forces an air of "respectability" among area businessmen. Moreover, even though McDade's positions on the Game and Fish Commission and with ALCOA—which, it might be presumed, would not fancy new industries competing for its labor pool—hardly made him a disinterested party, he was certainly better known in the area than Burch, Wilkins, or Dickey.[15]

Also recruited to the opposition, but of somewhat less importance to the cause, were Richard Myers, Jr.; Scott Mayfield; and several people at the Bowaters Paper Company. Myers, vice president of House Hasson Hardware Company in Knoxville, a large wholesale distributor, was an amateur archaeologist who had conducted digs in the Tellico area for twenty years. He opposed the dam on the grounds that the reservoir would destroy valuable historical and archaeological material. Since the professional archaeologists at the University of Tennessee were not interested at the time in the Tellico region, the opposition considered Myers a valuable addition to their numbers. To his credit, even though he vehemently attacked the project, Myers remained close friends with TVA lawyer and fellow amateur archaeologist Beverly Burbage. That kind of association would become unusual later.[16]

Initially the opposition hoped that Scott Mayfield and people at Bowaters would be great assets. Mayfield was the president of Mayfield Dairy Company, a large regional producer of dairy products which had pastureland in the Tellico area. The Bowaters Paper Company also had considerable holdings in the region, especially the Rose Island nursery, which would be inundated if the project succeeded. Yet neither Mayfield nor Bowaters became major forces in the opposition ranks. As their land holdings would be directly affected by the dam, perhaps they were perceived as too economically interested to be taken seriously. Also, it is possible that as area landowners they were more vulnerable to TVA pressure and therefore felt unable to speak as freely as they might have otherwise. Some opposition sources maintain that Mayfield and some Bowaters people secretly contributed money to the opposition coffers, although their open help to the anti-Tellico forces was modest.[17]

The local opponents of Tellico, then, formed a loosely-organized cadre. Their strategy was three-fold: 1) to join forces with the Tellico area opponents to convert more residents to their banner; 2) to furnish anti-Tellico information to influential Tennessee and federal officials who might be able to apply pressure on TVA to abandon the project; and 3) to raise public awareness in East Tennessee by packing public meetings, writing news releases and letters to editors, and influencing various groups who might come out against the Tellico Project.

Letter writing began in earnest in late 1963 and early 1964. Many of the protests were drafted by either Burch or Dickey. At this point neither had much hard evidence with which to attack the project. They did have some connections inside TVA who were surreptitiously passing along information, but in late 1963 TVA itself had not collected much in the way of hard data on Tellico. Moreover, almost no one within the agency knew how shaky the project's benefit-cost ratio actually was, so naturally that information was not channeled into the hands of the opposition. Hence the anti-Tellico letters that were printed by area newspapers were mostly appeals to save the freeflowing river for trout fishermen and attacks on TVA for undertaking a project that was neither necessary nor desirable.[18]

Dickey believed that it was important to get editorial support in the local newspapers. At the *Knoxville Journal*, the Tellico opponent was received cordially by editor Guy Smith. Smith had taken on a number of formidable opponents in the past, and there was nothing he liked better than a good fight. He had waged a running battle with the politically potent George Dempster, who had served a number of terms either as Knoxville's mayor or city manager in the 1930s and 1940s. Smith had never backed off, continuing to enrage Dempster with unfriendly, biting, and sarcastic editorials. Dempster loved to find Smith's car double-parked (a habit the fiery editor never broke) and order police to tow it away. But Dempster was only one of Smith's many targets; city councilmen, county commissioners, congressmen, senators, and even presidents had felt the withering blasts of his editorials. Yet Smith reserved his most scathing attacks for the advocates of "big government"—liberals, "free-spenders," "bleeding hearts" and their ilk. Mirroring the conservatism of the *Journal's* owner, Knoxville businesswoman and socialite Mrs. Ethel Lotspeich, Smith used the editorial page the way a back-alley pugilist used his fists.[19]

Smith's support of Dickey and the Tellico opposition, therefore, was natural. TVA represented everything Smith despised: big government, the domination of states and localities by Washington, and a swollen bureaucracy. Yet even as Smith promised Dickey his support, he warned the Chamber of Commerce staffer that it was a fight they would not win. "You can't beat the bastards," snorted Smith, even though he promised that he would try.[20]

Dickey was less successful, however, with Knoxville's other newspaper, the *News-Sentinel*. That paper, one of the Scripps-Howard chain, traditionally focused less attention on local and regional politics and tried to surpass the *Journal* in its national and international coverage. Politically more moderate than its competitor, the *News-*

Sentinel generally was friendly to TVA and took the agency's side in most disputes. In fact, some of the *News-Sentinel's* staffers, especially city editor Ed Smith (no relation to Guy Smith), were personally on good terms with members of the upper echelon at TVA, occasionally offering them advice and in one instance trying to intercede on the agency's behalf with the Knoxville Chamber of Commerce. Editorially the *News-Sentinel* supported the Tellico Project, and, while continuing to open its "Letters to the Editor" section to all opinions, the paper made it clear to its own employees that the paper would frown on anti-Tellico material. Staffer Carson Brewer told Dickey privately that "he wasn't as free to do all he might" on the growing Tellico story.[21]

While Dickey was busy gathering anti-Tellico evidence, writing letters, and attempting to enlist the support of area newspaper editors, his cohorts were trying to spread the word in Nashville. While there, they discovered another potentially important opponent of the project. Walter Criley of the Tennessee State Planning Commission had been disturbed when he had learned that TVA's plans for the Tellico area had been formulated without consultation with his commission. In Criley's view, relations between TVA and the commission were "strained." "TVA got along with the state government as long as we did what they wanted," Criley recalled, "and we weren't expected to disagree with TVA. . . . They really didn't want state input."[22]

According to Criley, many of the younger people at the State Planning Commission, "those who did not remember the 1930s," had serious reservations about the Tellico Project. They were openly dubious of TVA's claims that the project would bring a wave of industrialization to the region. Moreover, they were skeptical of the recreational benefits the agency was promising, believing that a lake at Tellico would draw people only from other TVA reservoirs. As one commission staff member asked rhetorically, "How many reservoirs do they need?"[23]

Criley and some other State Planning Commission people were particularly incensed by TVA's request that the State Department of Conservation build a state park at Tellico and that TVA use the benefits of that park in its own benefit-cost analysis. "State planners felt that we had all the flat water recreation that was needed," Criley explained, "and that what *was* needed was for TVA to conserve rivers."[24]

But even though a number of people in the State Planning Commission, as well as in the Department of Conservation, opposed the Tellico Project, that opposition was not to surface until 1971. For one thing, Tennessee Governor Frank Clement was an ardent TVA supporter, as was his heir apparent Buford Ellington. Hence, Criley and others felt that the climate was "unfriendly" to any Tellico opponents within state government. Moreover, State Planning Commission Director Harold Miller was temperamentally a TVA man. In the 1930s Miller had been a regional planner for TVA, and he continued to defend the agency staunchly. Miller often told his staff that the State Planning Commission did not have the technical expertise to challenge TVA, and that the commission should defer to the agency's greater technical prowess. Therefore, while there was a split within the State Planning Commission, Clement, Ellington, and Miller were able to keep Tellico opponents silent and impotent. Criley and others might secretly encourage Burch, Wilkins, Dickey, and the others, but they could not assist them publicly.[25]

Initially, then, the strategy of the Tellico opponents was less than successful. They had failed to weld the anti-Tellico forces together into a solid bloc with a coordinated plan of attack. Opponents in the immediate area of the proposed dam certainly were encouraged by the appearance of the Burch-Dickey-Wilkins cadre, and there were sporadic contacts between the two groups. But they remained two groups acting independently of each other. Moreover, while state and federal government officials had been given the anti-Tellico message, Governor Clement had smothered criticisms of TVA except in the Game and Fish Commission, which was an independent body. And the climate of opinion in Washington in the Kennedy-Johnson years was decidedly in TVA's favor. In the nation's capital, the agency's Marguerite Owen smoothed congressional feathers and made sure that the White House remained loyal to TVA. Chattanooga Congressman Bill Brock, a conservative Republican, was concerned about the possibility of TVA's acquiring land which ultimately would be sold at a profit to private interests, but he was even more concerned about the political clout TVA had in Washington and Tennessee. He wrote worried letters to TVA but apparently was soothed by Wagner's calm, heavily documented replies. Finally, letters to the editor did arouse some people, but the public attention span was short, and other national and regional problems competed for readers' attention. In sum, then, the anti-Tellico forces remained few, disunited, and weak.

This is not to say, however, that TVA did not take them seriously. At almost the first whiff of opposition, the agency wheeled out its heavy cannon. Carson Brewer's November 17, 1963, article in the *Knoxville News-Sentinel*, which reported the claims of Tellico opponents, caused a great deal of consternation within TVA. After all, Brewer was a local favorite, with his deft mixture of local history, homilies, chatty interviews, and essays on the natural beauty of the Smoky Mountain region. Hence, in TVA's view, Brewer's column could not go unanswered. Ken Seigworth prepared for the directors and the Office of Information a point-by-point rebuttal of the claims reported in Brewer's article. At the same time, TVA was identifying opposition leaders, drafting sample resolutions on Tellico for civic clubs, and trying to convince influential local people to climb on the Tellico bandwagon.[26]

In the past, TVA had been successful in working with established development associations in the Tennessee Valley. Indeed, ever since the Eisenhower years, it was supposed that these associations would initiate development projects and work cooperatively with TVA to bring those projects to fruition. Of course, most agency employees privately scoffed at working on equal footing with either state or regional development associations. Despite paying lip service to cooperative ventures, TVA was in fact loathe to take on any project which it could not dominate.

In the case of Tellico, however, no regional development association existed to request TVA's "cooperation," though there were industrial development associations in each of the three immediately affected counties. In order to weld influential local individuals and groups together to combat the growing opposition, TVA through its Office of Tributary Area Development (OTAD) created the Little Tennessee River Valley Development Association (LTRVDA). Not surprisingly, OTAD tried to keep a low profile in the LTRVDA, although its creation and domination of the association was hardly a well-kept secret. John M. Carson, Jr., was selected as the association's first president, and Marvis Cunningham was TVA's liaison with the new organization. Almost immediately after its formation, LTRVDA began distributing TVA-prepared literature on Tellico's projected benefits, writing letters to local editors, recruiting more members, and holding public meetings at which TVA personnel would extol the virtues of the Tellico Project.[27]

In the past, such tactics had routed any opposition to TVA projects. On the stump, TVA officials had been treated with awe and deference, had been touted as saviors of the Tennessee Valley and as bearers of electricity, prosperity, and economic modernization. Thus when LTRVDA arranged for Aubrey Wagner to speak at that open meeting at Greenback High School on September 22, 1964, the association had every reason to believe that the magnetic TVA director would strengthen local support and scatter the opposition.

But, as we have seen, the Greenback meeting had just the opposite effects. The opposition lay in wait for Wagner, cracked his normal composure, forced him into making some ill-considered statements, and then howled in anger when he tried to retreat. Years later, a psychotherapist who had attended the Greenback debacle recalled that Wagner's personal pride had been badly bruised and that his defensively stubborn "reaction to the people" both at Greenback and later was the result of his rough handling by the opposition at that public meeting. True or not, after Greenback, Wagner chose "safer" audiences for his pro-Tellico messages.[28]

The Greenback High School gathering had damaged the LTRVDA and been a real tonic for the opposition. In an effort to keep the initiative, twenty-two local opponents held a secret meeting at the Bowaters Rose Island nursery on the evening of October 13, 1964. Judge Sue Hicks of the Fort Loudoun Association, who had called Wagner's bluff at Greenback, "told the group that a larger number of people had signed petitions in favor of the [Tellico] project and that something had to be done quick to oppose the project." Reproductions of a map showing TVA's proposed taking line (which, through administrative confusion, had been secured from TVA by R. D. Ackard of Greenback) were distributed, confirming the opponents' worst fears that TVA was bent on a major purchase of land above the projected reservoir.[29]

Those who attended the Rose Island meeting responded to Hicks's appeal by immediately forming the Association for the Preservation of the Little Tennessee River (APLTR) and vowing to raise $1,000 to buy newspaper advertisements to oppose the project. Hicks reportedly pledged the first $100, and he and John Lackey formed a committee of two to collect the rest. Another probable topic of discussion was how to gain control of the pro-Tellico LTRVDA, since a directors' meeting at which new officers would be elected was scheduled for October 15, two days away.[30]

Apparently a good deal of politicking went on in the two days between the Rose Island meeting and the LTRVDA directors' meeting on October 15, 1964. Surprisingly, Judge Hicks was nominated and elected president of the body, and anti-Tellico men Philip Appleby and R. E. "Bob" Dorward also won offices. Cunningham later reported to TVA that it "was very apparent the election of Judge Hicks and Bob Dorward was a cut and dried proposition." Hicks immediately took the chair.[31]

The opponents had worked out their strategy well. First, John Lackey read a "prepared resolution in opposition to the project, and made a motion for it to be adopted." Appleby seconded. The following discussion was warm, with the most perceptive comment made by pro-Tellico Joe Downey, who said "he thought the association would be split if it took a stand." When Marvis Cunningham, TVA's observer, tried to speak, Hicks refused to grant him the floor. A motion to table Lackey's motion split the eight directors in half, with Hicks, Appleby, Lackey, and Clayton Curtis blocking that attempt. John Cardwell of Lenoir City objected to Hicks, as chairman, voting on the motion to table, but Hicks refused to recognize him. With the two sides already deadlocked, no vote was taken on Lackey's anti-Tellico motion. Again Cunningham tried to speak "but was interrupted by Judge Hicks and forced to stop." The new president then launched into an anti-Tellico diatribe, after which the meeting fizzled and died.[32]

The October 15 meeting effectively signaled the death of LTRVDA. With Hicks and his opposition cadre in control of the association which TVA had itself created, no real cooperation with the agency was possible. And since TVA was determined both to proceed with the dam and to maintain the fiction that it was doing so in response to local appeals, it could not afford to give further legitimacy to LTRVDA. Hence the agency simply cut the organization adrift and, on November 11, 1964, assisted in the formation of the more malleable Tri-Counties Development Association. The purpose of this new association was to "obtain support for the project." Cooperation with local groups, then, would continue, but only on TVA's terms.[33]

In TVA's eyes, LTRVDA had failed because it had allowed Hicks and his cadre of opponents to capture control. No such mistake would be made with the Tri-Counties Development Association (TCDA). All of its initial officers were strong Tellico supporters. Two were chairmen of their respective county associations "for TVA and Tellico"; one, Mrs. James G. Carson, had provided Marvis Cunningham with the details of the Rose Island meeting; and two were actually TVA employees. Fearing the agency would be embarrassed by the presence of TVA employees serving as association officers, Wagner wrote to Van Mol, "I wonder if it wouldn't be advisable for all offices to be non-TVA employees." OTAD's Kilbourne agreed, noting that his office "has suggested" that TVA keep a low profile in the association because of the potential "adverse" effects on the agency's "public image." The two TVA employees were encouraged to resign from their posts, and both did. Once that minor problem was out of the way, TVA began making suggestions as to how the new association could best lobby for the Tellico Project and providing materials to help it do so.[34]

Members of the TCDA yearned for change. They saw their own communities as economically moribund, and they agonized as young people abandoned the region to seek better opportunities elsewhere. Believing that only good could come from change and progress, they shared a desire to bring their region into the mainstream of modern

American life. To them, TVA offered a shortcut to growth and prosperity. And although most of them stood to benefit personally, either economically or politically, from the Tellico Project and the changes they hoped would come in its wake, their vision reached beyond their individual interests to the growth and prosperity of the region itself. In this, they seemed to see themselves as "founding fathers" working to regenerate a troubled homeland.[35]

This mentality was best expressed to the authors by Charles Hall, the Tellico Plains mayor and president of the Tellico Plains Telephone Company. Speaking with a soft Appalachian accent, Hall likes to affect the image of an easygoing, likeable country boy. Hunting rifles and shotguns are much in evidence; in his office a huge picture of a covered wagon crossing a mountain fork bears the title "Wagonmaster Hall." Hall is so congenial and courteous that he was liked even by his bitterest anti-Tellico opponents. Beneath his exterior, however, is a quick and shrewd businessman, an unabashed "booster," and a man totally committed to TVA's vision of regional prosperity.

The son of a local merchant, Hall scoffed at the idea of bringing regional prosperity through agriculture. "Today modern farming is suited to the Great Plains," he explained. "We fought so hard for Tellico because our young people were leaving. . . . Just the dumb ones like me stayed around." And in response to a question concerning the problems inherent in industrialization and economic growth, Hall shot back, "We'd like to face those problems—we *want* growth." Unshakable in his belief in growth and utterly convinced that TVA offered the best chance at achieving those objectives, Hall was the epitome of the Tri-Counties Development Association with which TVA, after late 1964, would "cooperate."[36]

There is little question that the TCDA, in its support of the Tellico Dam, reflected the opinions of a majority of local residents. In mid-October the *Monroe Citizen* released the results of a straw poll which showed that 63.3 percent of its readers who responded favored the dam. In addition, TVA was keeping a close count of letters, petitions, and group resolutions it was receiving both for and against the project. As of October 19, 1964, the agency had received numerous pro-dam letters and twelve group resolutions favoring the project. On the opposition side, TVA had received only ninety-five letters, one petition with thirteen signatures, and one group resolution (from the Vonore Lions Club, whose past president was John Lackey and present head was R.E. "Bob" Dorward). And while the furious campaigning and politicking of both sides make these figures considerably less than scientific, there is no reason to believe that in late 1964 a majority of area residents did not favor the Tellico Project.[37]

Yet in spite of the comparative weakness of the opposition, TVA was plainly worried. Indeed, since the Greenback embarrassment, the agency had become paranoid about those who opposed Tellico. Opposition, it believed, was part of a conspiracy, well financed and well organized by private business interests, ideological reactionaries determined to bring TVA to its knees. After all, hadn't such groups opposed TVA all along, from the fight with private power interests in the 1930s to the conservative attacks during the Eisenhower years? Hence, TVA viewed the relatively small opposition groups as a serious menace and behaved accordingly.

Many within the agency were convinced that the opposition was being secretly financed by ALCOA and the Bowaters Paper Company. Their motive, TVA thinking went, was to have the labor pool all to themselves and therefore not have to compete with new industries for workers. In an October 1964 report to the general manager, Assistant Public Information Director "Pete" Stewart admitted that the "connection between Bowaters and ALCOA on one hand and our Tellico project is hard to put your finger on." Even so, Stewart laid out a good deal of circumstantial evidence. As for Bowaters, Stewart reported that R.E. "Bob" Dorward, a Bowaters employee, was an officer of the Association for the Preservation of the Little Tennessee River and was president of the Vonore Lions Club when that organization passed its anti-Tellico resolution, with Dorward himself casting the deciding vote. Moreover, the October 13 meeting, at which the APLTR was formed, was held at the Bowaters nursery on Rose Island. Stewart also reported that Bob Edgar, head of the Bowaters Woods Department, was "outspoken against the project" and that "Bowaters people were spreading the report that TVA would buy Tellico land for $60 an acre and sell it for $600."

The position of ALCOA, Stewart continued, was "clearer, but still well covered." To begin with, Hugh McDade, head of public relations at ALCOA, was a good friend of Tellico opponent Bob Burch. Reporting on a telephone conversation between himself and McDade, Stewart wrote that McDade and ALCOA opposed a major land purchase policy and wondered what the effects of such a policy would be on Blount County, site of the ALCOA plant, "since ALCOA would not want to be in a position of subsidizing a TVA project." "This was," Stewart explained, "a swords-point conversation . . . and I have puzzled as to the purpose of the call. He [McDade] has done us a favor in telegraphing the kind of punch he is likely to throw. . . . I can only guess that the key is his aggressive, provocative tone by which he may have sought to elicit statements as to TVA authority or intentions which could be used against us."

Continuing his indictment of ALCOA, Stewart passed along "Mike" Foster's report that "he was in the Blount County Chamber of Commerce offices and heard one end of a conversation in which an unnamed ALCOA official insisted that the Chamber meet at an early date and adopt a resolution against Tellico. This was accompanied by a threat to withdraw ALCOA's membership in the Chamber unless action were taken." Finally, on another front, Stewart noted, "Porter [Taylor] feels that, from his own experience with the Knoxville Chamber, ALCOA has a hand in its [the Knoxville's Chamber's] work and may be behind Dave Dickey's activities."

The Stewart memorandum and a later attachment to it by Stewart's boss Paul Evans, in which Evans reported that Bowaters's true motive was saving the Rose Island nursery, make it clear that TVA was convinced that the agency was about to be the victim of a "well covered" conspiracy to stop the Tellico Project. To counter that conspiracy, great speed was necessary to increase support for the project, damage or discredit the opposition wherever possible, and bring the Tellico Project to Congress before the opposition gained additional strength.[38]

TVA moved quickly to block opposition from farm groups. Although not politically strong at the local level, farm groups carried a good deal of national lobbying weight in Congress as well as in Lyndon Johnson's White House. Farmers and their lobbyists had long been sensitive to the use of eminent domain to scoop up farmland. Since the Tellico Project proposed to take roughly thirty-nine thousand acres of prime farmland out of production, TVA anticipated that the project would encounter national opposition from farm groups.

The Monroe County Farm Bureau had already passed an anti-Tellico resolution, thanks (editor Joe Bagwell reported to TVA) to the efforts of Hicks and Lackey. But other local bureaus had not yet acted. Moving quickly, the Tri-Counties Development Association's Pat Maroney contacted the presidents of both the Blount County and Loudon County Farm Bureaus and extracted pledges that no such resolutions would get on their respective agendas. Moreover, the head of the Blount County Farm Bureau promised to block any anti-Tellico resolution introduced by Monroe Countians at the upcoming state convention. In spite of those efforts, a strongly worded anti-Tellico resolution was adopted by the Tennessee Farm Bureau Federation late in 1964. Yet that resolution seems to have had relatively little effect nationally, since it was not one of the farmers' top priorities.[39]

Agricultural interests within TVA also had to be silenced. In late December 1964, TVA's Agricultural Resource Development Branch, headed by Gerald Williams, circulated a report which infuriated agency brass. Authored by Roger Woodworth, the report calculated the effects of taking thirty-nine thousand acres out of production. Approximately four million dollars in farm income per year would be lost. And that loss would ripple out, affecting farm supply businesses, processers, retailers, and others.[40]

Fearing Woodworth's findings would only stir up more trouble among farmers and farm groups, TVA's Office of Information demanded that Woodworth's figures be "checked." Williams actually went into the field to sample twenty-seven percent of the local farmers. The results of his survey were most disheartening to TVA leaders: Woodworth's calculations had been almost perfect. Worse, roughly four-fifths of the farmers Williams surveyed voiced strong opposition to the project. Defending himself, Williams noted, "Please understand that our people did not run a poll to determine who was in favor of the project and who was not. In every case, farmers contacted during the collection of the field data strongly voiced their opinions. We did not ask for them, we kept count."[41]

In response, TVA virtually ignored its own Agriculture staff—as indeed it had done for years, since the end of World War II. Industrial prosperity, Paul Evans countered, would far outweigh the agricultural loss and, he even argued, would "stimulate" the agricultural economy, presumably through an increase in nonfarm population. "The most prudent use of the land resource," another TVA person loftily stated, "is to devote it to issues which will improve the economic opportunities of the people and alleviate the hardship that presently exists." To TVA that meant "factories in the fields," industrialization, nonfarm modernization. It could be no other way.[42]

As TVA was trying to neutralize opposition from farm groups, the agency was pressing hard to get endorsements of the Tellico Project from area business leaders. Much of the latter effort centered on the Greater Knoxville Chamber of Commerce. Approval from the Knoxville chamber was important for a number of reasons. TVA could use an endorsement by that large and influential body to offset what many believed were ALCOA's efforts to elicit negative resolutions from the Maryville and Blount County chambers. Moreover, a positive resolution by the Knoxville chamber might serve to silence David Dickey, who, as the chamber's director of area development, was a salaried employee of the Knoxville group. Indeed, to TVA the Knoxville chamber's endorsement was crucial.

Many Knoxville businesspeople, however, were dubious about the Tellico Project. For one thing, some feared that new industrial sites on the proposed Tellico Reservoir would attract industries for which Knoxville also was competing. The 1950s had brought difficult economic times to the city. Between 1948 and 1960, roughly 1,700 jobs had been lost in the city's manufacturing sector alone. Unemployment was high in Knoxville throughout the decade, reaching a disturbing 9.7 percent in 1958. The rise in real per capita income had been meager, and suburbanization was luring retail businesses from the downtown area. In truth, Knoxville desperately needed new industries, new jobs, and new money. Any competitor on the horizon was a potential threat.[43]

Troubled by the city's inability to attract new industries, in 1962 the Metropolitan Planning Commission and the Knoxville Chamber of Commerce had hired the Washington, D.C., consulting firm of Hammer and Associates to analyze the city's problems and offer solutions. The Hammer Report made it clear by implication that Knoxville and Tellico would be competitors for new industries, that the new industries of the 1960s and 1970s would not need barge shipping and could locate wherever there were sites, that these new industries would not be so closely tied to natural resources, and that therefore "TVA's role in local economic development is no longer a critically strategic one." The Hammer Report concluded by urging Knoxvillians to develop industrial parks as sites for new industries, to begin to train a pool of skilled labor, and to build crucial interstate highway connections.[44]

For these reasons, a number of Knoxville chamber members opposed a positive resolution on the Tellico Project. William Yandell, a Southern Bell executive who in 1964 was president of the Knoxville chamber, recalled that there was "considerable doubt whether that [Tellico] was a proper site for industrial locations." Nevertheless, TVA actively courted the chamber and tried to get it to take a positive stand. There were several lunches and "informal" meetings of the chamber and the TVA directors, especially Frank Smith and A. R. Jones. As Yandell remembered it, "We had a lot of pressure on us to take an official stand. . . . Certain members of the [TVA] board wanted us to take a stand . . . there was strong feeling about it." The chamber did appoint a special committee to study the Tellico Project but, as Yandell recalled, "we never could get from the [TVA] directors any real assurance that the Tellico Dam would be of any real help."[45]

Discrediting Dickey was another problem altogether. In summer 1964 Dickey had attended and received a certificate from the University of Oklahoma's Industrial Development Institute. His thesis, "The Little Tennessee River as an Economic Resource," was an attack on TVA's Tellico Project and an appeal for preserving the river. A twenty-six page version of Dickey's thesis had been published by the Tennessee Game and Fish Commission and, according to TVA's J. Porter

Taylor, "is being distributed widely and is the basis for much of the criticism that has been directed toward the Tellico Project in letters to the press, articles by outdoor columnists, and letters to TVA."[46]

Taylor wanted to attack Dickey's thesis head-on. He was angered by the fact that when "Mr. Wagner met with the special committee of the Knoxville Chamber of Commerce to present the case for Tellico, more than half of the members of the committee present had copies of the booklet." Taylor appealed to TVA's general manager, explaining that "we believe that publication of an analysis of the summary [of a TVA reply to Dickey] . . . is needed." Van Mol, however, disagreed, telling Taylor and the directors, "I don't think publication of an analysis of Dickey's thesis, as suggested by Porter Taylor, is necessary or desirable; it would only serve to focus attention on the thesis and its invalid claims." A. R. Jones agreed in a margin comment on Van Mol's memo: "We should stick to a positive approach in our publications." Hence TVA preferred to deal with Dickey's thesis by ignoring it, hoping that it would not receive much publicity and would soon be forgotten.[47]

TVA could not ignore Dickey himself, however. In early 1965 the Knoxville Chamber employee had compiled a "fact sheet" for the Association for the Preservation of the Little Tennessee River, largely derived from his thesis and other anti-Tellico research. Copies of this fact sheet were well circulated, and one was included in a letter Dickey wrote to President Lyndon Johnson to inform the president of the "extreme threats it [Tellico] poses for the free enterprise system."[48]

Whether TVA openly pressured the Knoxville chamber to restrain Dickey is still a matter of dispute. TVA has consistently denied that it did so, and Yandell in 1983 could recall no such pressure. Yet Dickey strongly claims that Wagner called officers of the Knoxville chamber to his office for the purpose of getting Dickey fired. "Wagner tried to do me personal harm," Dickey maintains. Whether Dickey's claim is correct or not, it is known that Wagner "was not at all pleased" and that he did complain to the Knoxville chamber, asking "whether David Dickey was speaking for the chamber." Of course, Wagner knew very well that he was not, and the TVA chairman's complaint must be interpreted as a show of TVA's displeasure with Dickey's independence and a hope that the chamber would rein Dickey in.[49]

Director A. R. Jones was still more eager to put pressure on the chamber concerning Dickey. Infuriated by Dickey's letter to Lyndon Johnson, Jones fired off a memo to the other two directors, saying, "I think David Dale Dickey's . . . letter to the president on Tellico calls for a strong 'official' protest to the Knoxville Chamber of Commerce." Dissuaded by Paul Evans from making a protest, Jones then dictated an angry letter to Ray Jenkins, a pro-Tellico local lawyer with strong Washington connections as well as influence in the Knoxville Chamber of Commerce. Jones's "personal and confidential" letter to Jenkins complained that "said official [Dickey] is, in my opinion at least, attempting to obstruct the industrial development of an area on the Little Tennessee River which obviously is in the Knoxville trade territory. . . . It is incomprehensible to me that a responsible member of the Chamber of Commerce [Jenkins?] would allow an employee of theirs to use falsehoods and misleading statements for apparent personal reasons to the detriment of his employers and neighbors." The implication could not have been lost on Jenkins. But the aging lawyer, while agreeing with Jones that Dickey's letter was "asinine," refused to be drawn into the effort to silence Dickey. Ultimately Dickey and the Knoxville Chamber of Commerce did part company, although TVA's role in that break will never be fully known. Dickey was passed over for a promotion within the chamber which he wanted; one person told the authors that he had become "too controversial." Nevertheless, it is clear that TVA considered Dickey enough of a threat that it made some efforts to get the chamber to muzzle him.[50]

Although TVA saw the opposition to Tellico as a well-financed and coordinated conspiracy which involved ALCOA, Bowaters, the Game and Fish Commission, farm groups, outdoor writers, and obstructionists like David Dickey, the agency reasoned that the opposition would simply collapse once Congress approved funds for the project. So TVA breathed an almost audible sigh of relief when, on January 25, 1965, President Lyndon Johnson submitted his 1966 budget to the House of Representatives. The president had included in his budget nearly $6 million to begin the Tellico Project. True, opponents had rushed to Washington to plead their case before the Public Works Subcommittee of the House Appropriations Committee. But the agency had rounded up an impressive collection of witnesses in its own behalf. Moreover, even though TVA always insisted that the agency was nonpolitical, it had an effective Washington lobby and had rarely been disappointed in a Congress controlled by Democrats. Lyndon Johnson's 1964 landslide victory over Senator Barry Goldwater had left the president's political party with more-than-comfortable congressional majorities. In the flush times of the "Great Society," TVA had every reason to believe that it would get what it wanted. If national media attention could be kept away from Tellico, TVA believed the project would encounter little trouble.[51]

Therefore, the entrance of Supreme Court Justice William O. Douglas into the fray was the source of considerable alarm to TVA.

While Douglas probably had been too damaged by past controversies to be of much help, his visit to the Tellico area in early April 1965 made "good copy" for the national press and briefly focused national attention on the Tellico Project. Moreover, Douglas's appearance in East Tennessee could weld various opposition factions together as well as dramatize their appeal to stop the dam.

The Douglas visit was carefully staged by a small collection of Tellico opponents—Dickey later said that "no more than a dozen were involved." Harvey Broome, president of the local Wilderness Society, formally invited the justice. Bob Burch of the State Game and Fish Commission (dubbed by one pro-TVA editor "the wordy protector of fish") was able to convince the Eastern Band of Cherokees from Cherokee, North Carolina, to send a delegation. Although never in the forefront of the opposition (they hoped that the project would be killed without their having to get involved), some Cherokees were unhappy over the fact that the proposed reservoir would inundate ancient towns and burial sites. Dickey provided copies of his "fact sheet," and Mack Prichard of the State Parks Commission supplied more information for the press. Mrs. Alice Milton and Judge Sue Hicks of the Fort Loudoun Association arranged a reception at the fort. John Lackey brought in the Tellico area opponents. In sum, Douglas's visit was planned as the ultimate "media event," and its sponsors hoped that national press coverage would result in such an outcry that Congress would shrink from funding the project.[52]

In a ceremony designed to maximize publicity, Douglas met with the Cherokee Indian delegation at the supposed site of the ancient Cherokee village of Chota. The Supreme Court justice accepted a petition from the Indian representatives, replying loftily, "Tell your chief I will be very happy to carry his message to Washington and present it personally to the president and express the hope that this beautiful valley will not be destroyed by the hand of man." The Cherokees then presented Douglas with a Sioux war bonnet (which, Cherokees reasoned, most whites believed all Indians wore) and the justice posed for photographers. There are few, if any, non-Indian politicians who can appear in a Sioux war bonnet without looking foolish, and Douglas was no exception. Then Douglas and his wife went out on the river to do a little trout fishing. Rumors circulated that the Game and Fish Commission had stocked the river so that the justice's efforts would be rewarded.[53]

Wagner was very upset by Douglas's visit. Media attention was just what TVA did not want at this crucial juncture, and the agency was getting plenty of it. Moreover, once back in Washington, Douglas announced that he planned to write an article on the Little Tennessee

River for *National Geographic*. Wagner drafted a remarkably restrained letter to the Supreme Court justice, circulated it in TVA, and then never sent it. Instead, the TVA chairman flew to Washington for a breakfast meeting with Douglas to plead TVA's case. Whatever transpired at that breakfast, it did not change Douglas's mind or lessen his ardor. Later, in print, Douglas condemned TVA as the "worst offender" against the environment, and in a curt letter to TVA Director Frank Smith, stated, "I hope your plans to despoil [the river] are never fulfilled."[54]

With Douglas so obdurate in his opposition to Tellico even after the Wagner breakfast, TVA turned its attention to *National Geographic*. In October Paul Evans drafted a letter to the prestigious magazine for Wagner to sign. In a memo to the TVA chairman, Evans noted, "A letter, as attached, may encourage *Geographic* editors to avoid going off the deep end." That same day Wagner sent the letter to *National Geographic* President Gilbert Grosvenor. Two months later George Crossette of the magazine informed Wagner that *National Geographic* had "postponed" plans to publish Douglas's article, a postponement which eventually became a rejection. The Supreme Court justice's attacks were quickly forgotten by the general public, as were other assaults by the early opponents of the Tellico Dam. Thus, by 1965 it appeared as if the TVA had won.[55]

In retrospect, the opposition to the Tellico Project that arose in the mid-1960s never stood much of a chance. Never a majority of the local population, the objectors were poorly organized and financed. They had insufficient access to the media to argue their case. Opponents in Tennessee's state government had been effectively silenced by their higherups; local farmers could not get the attention of the national farm lobbies; men like David Dickey had simply been worn down, in his case driven to near-exhaustion by the cause and by threats to his livelihood. Tired and battered, he withdrew from the arena.

True, opponents had badly embarrassed Wagner at Greenback High School. They had captured control of the LTRVDA and had left it in wreckage. They in turn had roused some farmers' groups, the historic preservationists at Fort Loudoun, the cautious Cherokees, and Justice Douglas. But TVA was always there—to respond to the farmers' ire, to negotiate with the Fort Loudoun Association, to soothe the Cherokees, to discredit Douglas, to troop an impressive array of witnesses before Congress. Whenever the opposition moved, TVA appeared to meet and envelop it.

Part of the early opposition's problem was that it simply did not know enough about TVA's Tellico Project. Information on the actual "taking line" was difficult to come by (the 1964 map obtained by Ackard was only a clue). That was largely because TVA itself did not know how much land it would eventually want to take. More important, TVA kept the crucial benefit-cost figures a tightly guarded secret. As it turned out, those calculations would have been most damaging. Most early opponents never knew how vulnerable TVA's benefit-cost analyses really were. Ultimately, these figures would become a major issue.

5

THE SEARCH FOR JUSTIFICATION:
THE BENEFIT-COST PROBLEM

When things were going his way, Sen. Allen Ellender of Louisiana could be both gracious and witty. A shrewd mixture of southern gentleman and "good ole boy," the diminutive former lieutenant of Huey Long was famous in Washington for his gumbo dinner parties and his penchant for world travel at government expense (After one "investigative" junket to Italy, he mourned that the "canals in Venice are filled with water"). A mainstay in the Senate since 1937, Ellender had hobnobbed, chatted, and sipped whiskey with two generations of national leaders.[1]

When angered or aroused, however, Ellender could be a dangerous enemy—wily, caustic, and abusive. Like most southern politicians, Ellender maintained a fierce love-hate relationship with the federal government. A rigid segregationist (in 1962 he had been a diplomatic embarrassment to the Kennedy administration with his offhand remark that Africans were incapable of governing themselves), he staunchly opposed federal programs designed to end poverty and to bring full political rights to blacks. Moreover, as a fiscal conservative, Ellender liked nothing better than hacking a federal bureaucrat to pieces over what he considered an unnecessary federal appropriation. Yet as chairman of the Senate Agriculture Committee, he had done his best to take care of the nation's farmers, brutally attacking anyone who stood in his way.[2]

Thus, when Ellender faced Wagner in the Senate subcommittee hearings on Tellico in mid-1965, the senator was confronting a man who represented everything he despised. This massive federal agency was proposing to take land from farmers and turn it over to industrialists. Wagner was a Yankee who hoped to rebuild the South in a new image, a man who was using federal revenues from other states, including Louisiana, to create a "bonanza" (Ellender's term) in the Tennessee Valley. Wagner was in for a rough time.

What made matters even worse for the TVA chairman was the fact that Ellender had suspected how porous Tellico's benefit-cost ratio actually was. The Louisiana senator may have enjoyed playing the country boy, but he was no fool. Drawing Wagner into a trap, Ellender asked him whether TVA had followed the Corps of Engineers' methods in determining Tellico's benefit-cost ratio, which Wagner was claiming was 1.4:1. No, the TVA chairman replied, TVA had used a different formula. However, Wagner asserted, trying to recapture the initiative, if the Corps of Engineers' methods had been applied to Tellico, the benefit-cost ratio would be 1.9:1.[3]

Not true, Ellender rejoined, closing the trap. In fact, the senator announced, he had asked Gen. W. P. Leber of the Corps of Engineers to make an independent study of the Tellico Project (Wagner had been aware of the study) and that if the "same yardstick that is used by the Corps of Engineers were applied to [Tellico], I doubt that the benefit to cost rates would be more than 0.8 to 1."[4]

Ellender had uncovered what the early opponents of Tellico had missed: its highly speculative benefit-cost ratio. For over five years, a small army of people at TVA had been working furiously to find a better economic justification for building the Tellico Dam. But, in spite of Wagner's claim, the agency was no closer to a defensible benefit-cost ratio than it had been in 1960. Ellender had sensed that potentially fatal flaw. If he and others were heeded, Tellico would never get off the ground.

At TVA, as early as 1963, Chief Engineer George Palo plainly was worried. Wagner had made it clear that he wanted to take the Tellico Project to the board of directors in mid-1963. Yet, despite months of work, the agency was having great difficulty coming up with a feasible justification for the project. Since the late Eisenhower years, publicly-financed projects had had to demonstrate to the Bureau of the Budget and to Congress that a project's anticipated benefits outweighed its costs. Both benefits and costs had to be expressed in dollars, and the two juxtaposed to form a benefit-cost ratio. In order to be deemed acceptable, a project had to show a benefit-cost ratio of at least 1:1, and probably higher.[5]

What worried Palo was not that TVA would fail to demonstrate an acceptable benefit-cost ratio for Tellico, but rather how that favorable ratio would be generated. Previous TVA construction projects had been justified on the basis of the agency's "traditional" benefits—power, navigation, and flood control. Other benefits, such as the enhancement of land values adjacent to a reservoir, increased recreation income to the region, and jobs created by industries moving to the reservoir area, had been viewed simply as bonuses, "extra" benefits over and above the traditional ones. Yet, as Wagner had explained in his 1959 "more dams" speech, "traditional" benefits would not be sufficient to justify future projects. Further, he strongly implied that the "extra" benefits should be calculated and included in justifications of all future undertakings.[6]

Palo was uncomfortable with this kind of thinking. A man enormously respected in both TVA and Knoxville, he had come east from his native Wisconsin to earn a degree at MIT in 1928. One of the few TVA engineers who had had broad experience in the private sector, he had joined the agency in 1934 as head civil design engineer and moved rapidly up the ladder to the position of chief engineer by 1959. Courtly and cultured, Palo and his wife Anne frequently entertained and actively supported the theater and the public library. Honest and blunt, Palo was regarded by most who knew him as a gentleman, an "engineer's engineer."[7]

In February and again in August 1963, Palo shared his doubts about Tellico with General Manager Van Mol. The chief engineer was especially concerned about the degree to which Tellico's justification had become dependent on the purchase of extensive land above the projected shoreline. "The purchase of land with the expressed intent to resell it at a profit," Palo wrote somberly, "has made opposition and inevitably someday will be tested in the courts." Later he warned Van Mol, "If only the benefits to navigation, flood control, power, and shoreline development are included in the allocation, the benefit-cost ratio . . . would be 1.0:1. Furthermore, if only the benefits to power, navigation, and flood control are considered, . . . the benefit-cost ratio for this circumstance would be 0.9:1."[8]

Palo's warnings make it clear that the Tellico Project could not be justified to either the Bureau of the Budget or Congress on the basis of traditional benefits and probably was still not acceptable when some non-traditional benefits were included. Furthermore, although the FUDAR Committee had been trying since 1959 to devise ways to quantify non-traditional benefits, little progress had been made by mid-1963. In short, the Tellico Project probably could not be justified economically.

But within TVA many, and especially Wagner, were already convinced that the project would not only be an economic boon to the lower Little Tennessee Valley but would also be the model for Wagner's "more dams" initiative. Those working on Tellico's benefit-cost ratio were under enormous pressure to come up with a way to demonstrate what Wagner and others already believed to be true. Hence, in search of a good benefit-cost ratio, the Tellico Project was changed and expanded. The tail of justification began to wag the dog. Worse, there is considerable evidence that, under extreme pressure, some within TVA either purposely or accidentally altered some of the figures to create a favorable benefit-cost ratio. George Palo had good cause for alarm.

Several problems confront anyone who would trace TVA's tortuous efforts to come up with an acceptable benefit-cost ratio for the Tellico Project. To begin with, Wagner liked to have several offices work independently on a problem, each group reporting either directly to him or to the general manager. Often various offices worked blindly, in ignorance of what others were doing on the same problem. This led not only to a good deal of overlap, a "reinventing the wheel" syndrome, but also to difficulty in determining how particular calculations were arrived at or who made them. Wagner picked and chose from various reports those things which pleased him or confirmed his own predilections, often ignoring underlying assumptions or calculations. Hence it is nearly impossible to establish any firm chronology for the evolution of Tellico's benefit-cost ratio.

More worrisome is the fact that benefit-cost ratios are neither "real" nor tangible. Rather, they are based on econometric models—theoretical projections of future economic behavior which are highly speculative in character and founded on assumptions which are open to question. These models are used by economists to predict the economic outcomes of certain decisions, such as lowering the prime interest rate, increasing taxes, or building a dam. To be fair, however, a prediction of what will happen if those decisions are not made must also be calculated. The difference between the economic outcomes with and without the decisions being made would be the net economic benefits of the decisions themselves. But, as noted above, no model is "real" or can be applied with certitude. In addition, most models are designed to predict economic behavior in the short run, one year at most. But TVA used its models to project economic trends over decades, a dubious undertaking at best. Indeed, many economists themselves avoid econometric models altogether as too speculative and too suspect to be of much value. TVA was virtually forced to use them,[9] but many at TVA remained uncomfortable about doing so. Worse, the more the agency employed these models, the more many came to trust that the models predicted what definitely would happen. TVA was building a dream world within its web of econometric models and then living inside it.

Finally, much of TVA's work on Tellico's benefit-cost ratio was veiled in secrecy. Not only must the opposition not know how economically vulnerable the project actually was, but the Bureau of the Budget and Congress similarly must be kept in the dark. Those who, like economist Kerry Schell, objected or refused to cooperate virtually were told to go along or leave TVA. To be sure, many who worked on the benefit-cost problems kept their doubts and misgivings to themselves and rarely if ever spoke about them publicly. For example, as we have seen, George Palo had serious reservations. Yet he was a loyalist who would never have thought of taking his doubts to the general public. Those who *would* talk openly about Tellico's benefit-cost problems were almost entirely outside the agency and not privy to some of the methods TVA ultimately used to come up with an acceptable ratio.

Ever since 1959, when the FUDAR Committee had begun grappling with the benefit-cost dilemma, those within TVA who were close to the problem had realized that, in order to justify future projects, the agency would have to return to its original policy of purchasing substantial land above the high water markers and then claiming as benefits all development that occurred on those acres. Indeed, this realization very likely underlay the board's 1960 decision to reverse itself on the small purchase policy. Some, like Palo, feared that taking land for ultimate resale would inevitably be challenged in the courts. Others, like J. Porter Taylor of Navigation, though favorable to overpurchase, thought the new requirement of a favorable benefit-cost ratio was "a lot of hogwash" and that projects should be built simply because they were good projects, a view which was popular among some engineers but impractical given the new climate of opinion in Washington. Still others, like Assistant General Manager E. P. "Phil" Ericson, tried mightily to create favorable benefit-cost ratios without using non-traditional benefits—in vain. Hence, even though the board had approved a large purchase policy, many within TVA were hesitant to engage in large land purchases to bolster benefit-cost ratios.[10]

One reason for hesitancy was that the non-traditional benefits of land enhancement, recreation, and industrialization seemed to defy precise quantitative measurement. The 1961 FUDAR report virtually admitted the committee's failure to do this, even though it had worked for over four years, had help from outside consultants, and was privy to similar efforts by the Corps of Engineers. But to an engineer like Wagner, a problem without a solution was unthinkable. Through Van Mol and Palo, he spurred the agency to even more furious efforts, announcing through Van Mol that he was determined to bring Tellico before the board in mid-1963.[11]

Another reason for TVA's difficulty in quantitatively justifying large land purchases to capture non-traditional benefits was its staff's general lack of familiarity with long-range econometric models. Indeed, in some cases these models did not even exist and would have to be created. Moreover, econometrics was the offspring of a younger generation of economists, men and women more comfortable in the comparatively new computer age. To some within TVA, econometric models were strange tools, not always easy to understand and often very difficult to use. Yet if an acceptable benefit-cost ratio for Tellico was to be found, these models, TVA recognized, would have to be used.

In constructing or adapting econometric models to find a healthy benefit-cost ratio for the Tellico Project, TVA began by making five general assumptions which were to form the basis for all its subsequent work. Unfortunately, all of these assumptions were false, sometimes dangerously so. These general assumptions were:

1. *The Tellico area would remain economically static without the project.* In fact, outmigration had slowed down dramatically, nonfarm employment had increased, money brought into the area by recreation (mostly by trout fishing) was rising, and there were some indications that new industries might locate in the area.

2. *All economic progress that occurred in the area after completion of the project should be attributed to the project.* As shown above, this assumption was patently false. Yet, in order to calculate the *real* benefits of the project, TVA would have had to estimate what would happen to the area if the project were *not* built. The TVA staff was even less willing to engage in this counterfactual exercise than in the other projections.

3. *If an economic benefit could take place at Tellico, then it would take place.* This assumption artfully turned possibilities into certainties, ignoring what was going on elsewhere in the Tennessee Valley. Tellico would have to compete for recreation and industry with many other areas, but TVA generally paid little attention to this fact.

4. *The Tellico Project would not detract from any economic benefits already being enjoyed in the area, but would only add to them.* This simply was not true. Agricultural benefits would be heavily damaged, as would trout fishing. Towns like Vonore would be hard hit by the cutting off of vital transportation links. In short, there were numerous negative economic impacts and disbenefits which TVA did not measure.

5. *Once set, project costs would not rise faster than the normal annual inflation rate of the early 1960s (three to four percent).* As the Tellico Project grew in size and scope, costs inevitably mounted. Much of that growth was to capture more benefits, so a dangerous spiraling effect was created. Increased costs meant that benefit claims had to be correspondingly inflated, until by the mid-1970s the whole process was becoming divorced from reality.

TVA's prodigious effort to construct a good benefit-cost ratio for the Tellico Project rested on these dangerously false assumptions. Even if an acceptable benefit-cost ratio was found, it would be nearly as parlous as the assumptions on which it rested.

In early 1963 it was by no means sure that an acceptable figure could be constructed at all. In February, Assistant General Manager Phil Ericson confessed to Van Mol that, in his opinion, "Tellico was not markedly superior" to Tims Ford, Sugar Creek, or other projects TVA had on the drawing boards unless the non-traditional benefits could be calculated. If large purchases of shoreline property around the projected reservoir were made, benefits could be boosted from an anemic 1.07:1 to 1.53:1. How Ericson arrived at the latter ratio is not entirely clear. What is clear is that his figures were frightfully shaky.[12]

If Van Mol and Wagner needed convincing that large land purchases were crucial to the project, it is likely that Ericson's calculations convinced them. On May 7 Van Mol announced to Palo that Wagner was determined to begin the project in 1965 and that "substantial purchases" of land would have to be included to make the project palatable to Washington. Yet, Van Mol confided, work must go on to get a "sounder" benefit-cost ratio, and that work must be done as quickly as possible.[13]

Ever since the early days of the FUDAR Committee, it was clear that the two areas which would most dramatically improve a benefit-cost ratio were *land enhancement* and *recreation.* Land enhancement involved claiming as a benefit of the project all increases in the value of land surrounding the proposed reservoir. Recreation benefits had to do with money brought into the area either by tourists or by investors who would build facilities and employ people to serve tourists. But how could those projected benefits, while undoubtedly substantial, be given dollar values sufficiently precise to satisfy the Bureau of the Budget and Congress? Worse, it was obvious from the beginning that land enhancement and recreation benefits actually *overlapped* in places. How could they be adequately separated?

On the surface, land enhancement seemed to many to be the

easiest non-traditional benefit to quantify. After all, one needed only determine the amount of land that would be taken above the normal elevation of the reservoir, establish how long TVA would hold that land, and then raise the land value annually by an agreed-upon inflationary multiplier. Yet, as with so many issues involved in the benefit-cost ratio quest, what appeared easy turned out to be an economist's nightmare, with results satisfying almost no one.

In April 1963, Howes had admitted that "at this stage of research the method [for calculating land enhancement] is not reliable enough to be used in decisionmaking." Calling the results of his months of work "promising but inconclusive," he asked for another year to collect data and do more research. The problem, according to Howes, was that the area land sales he had used for comparisons had been those of "developed land and recent land sales (1955–1962)" which had sold at relatively high prices and had not increased measurably in value since. In another year, Howes assured Van Mol, better data would be able to produce higher land enhancement figures.[14]

But neither Wagner nor Van Mol could wait a year for Howes's recalculations. Instead, memos went out asking Navigation, Project Planning, and other divisions and branches to get involved in the search for a way to nail down land enhancement benefits. By late June, Palo reported to Reed Elliot that a benefit-cost ratio with "reasonably close" dollar amounts was almost completed, with over twelve percent of the total benefits coming from projected land enhancement. Palo calculated that 16,500 acres above the reservoir would have to be purchased at an estimated cost of $4.58 million. If TVA retained the land for ten years, the land at the end of that time, he claimed, would be worth $11.4 million, an increase of roughly 10 percent per year. Although it is not entirely clear how Palo reached the cost and multiplier figures he did, to many within TVA his early work was a step in the right direction and a vast improvement over that of the more fastidious Howes.[15]

A few days later Reed Elliot's Division of Water Control Planning improved upon Palo's land enhancement benefit when it added $5.2 million for land held by the agency beyond ten years and then included an additional $5 million for the "enhancement of adjoining land" not held by TVA but enhanced by the presence of the project. That virtually doubled Palo's more modest figure, from $11.4 million to $21.6 million. Some within TVA even spoke of buying more acreage at Tellico to increase the land enhancement benefit still further, the earliest evidence that the project's benefit-cost analysis was beginning to dictate the size of the project itself. Ultimately additional parcels were purchased, including additional acreage to expand the "model city" of Timberlake, and the "taking line" became a constantly shifting and often confusing gauge of TVA's appetite.[16]

Palo had wondered earlier about the constitutionality of a public agency's taking land for resale, and Wagner himself as late as July 7 had shown the same concern, asking, "Can we legally buy extra land to capture enhanced value?" Still, land enhancement had become an important ingredient of Tellico's evolving benefit-cost ratio and simply could not be eliminated. That became clear when, on August 28, 1963, Palo submitted to Van Mol a report which Palo boasted was both "detailed and up-to-date." On land enhancement, the chief engineer tried to moderate and calm a process that was threatening to get out of control. He refused to include as a benefit the roughly $5 million which "would ultimately accrue to interests other than TVA." He did raise projected land costs by 19 percent from his June figure of $4.58 million to $5.45 million. At the same time, he scaled down the benefits by roughly 9 percent, from $11.4 million to $10.5 million.[17]

Yet while Palo and others might have felt more comfortable with the new calculations, clearly Van Mol, his staff, and the board did not. At a September 5 meeting, Van Mol and his staff tore the land enhancement figures apart. In his notes on the meeting, Van Mol wrote that the general consensus was that the land enhancement figures were so dubious and baseless that Director A. R. Jones was "fighting" them. And, Van Mol implied, he was right to do so.[18]

While the chief engineer's estimates of land costs were too low, on the benefit side of the ratio Palo was characteristically cautious and restrained. He had brought costs more into line with present realities and had reduced the inflation multiplier from 10 percent to under 7 percent to reflect more accurately the area's true rise in land values. Yet he still was able to bring in a 1.4:1 benefit-cost ratio for the project. If anything, Van Mol, his staff, and the board should have been pleased.

But what transpired makes it clear that the chief objection to Palo's August 28 report was that it had not been bold enough. Early in October 1963, Van Mol asked Wagner if he could bring a new set of benefit-cost figures before the board at its October 7 meeting. In assessing land enhancement, Van Mol had used Palo's $10.5 million as a base. To that he added the $5.2 million which Palo had taken out, representing enhancement of land not owned by TVA. Then, instead of adding these to the benefit side of the ledger as Palo and others had done, Van Mol subtracted them from the project's costs, reasoning that sales of these lands would be money which could be applied against costs. It is likely that Van Mol and his staff were trying to calm a board which had seen the cost of the project rise, in Palo's August estimate, to a disturbing $60 million.[19]

Then, inexplicably, the projected land sales figures were applied *a second time* to reduce net costs further, to under $31 million. And then, having used anticipated profits from the sale of land twice to reduce costs, $12 million in "general economic benefits" (these could only have been land enhancement, since everything else imaginable had a separate category) was added to the *benefit* side of the ratio. Land enhancement had been counted *three times,* twice to reduce costs and once to pump up benefits. It was either an accident or an act of fiscal legerdemain which Howes, Palo, and others never could have carried out. Under intense pressure to find a better benefit-cost ratio for Tellico, some people had acted carelessly, foolishly, or duplicitously.[20]

Yet land enhancement was only one component in Tellico's benefits package and, as it turned out, a comparatively modest one. The traditional benefits would, it was anticipated, bring the project's benefit-cost ratio to approximately 0.9:1 (Palo's August 28 figure). Land enhancement would tilt the ratio slightly toward the benefits side. That left recreation benefits as the additional magic component which would scatter the critics and help the Tellico Project sail smoothly past the Bureau of the Budget and Congress.

In July 1963 Wagner, hoping to strengthen Tellico's flabby benefit-cost analysis, asked, "How does it shape up if we leave out recreation but include [land] enhancement?" To those who had worked to shore up Tellico's justification, the answer was very simple: recreation benefits were desperately needed in order to bring in an acceptable benefit-cost ratio for the Tellico Project. The project could not, in fact, be launched without them.[21]

But how could those benefits be assigned dollar values? Howes and his FUDAR Committee had wrestled unsatisfactorily with this problem for months. The Corps of Engineers used a simple formula whereby they estimated how many visitors (tourists, fishermen, hikers, etc.) would come to a project, guessed at how much each visitor would spend in the area, and then multiplied that figure by the estimated number of visitors. The whole process was an exercise in wishful thinking, but the Corps had been successful in getting projects approved by using this highly questionable calculation, and some within TVA believed that this kind of estimate would work for Tellico, too.

But how would TVA go about estimating the number of visitors to Tellico or how much they would spend there? Twenty years later, TVA's David Pack described the process:

In making recreation estimates for the Tellico project, TVA used the accumulated knowledge of past experience in the recreation utilization of the reservoirs of the Valley. Because of such factors as water surface area, length of shoreline, and geographic location, it was felt that the Tellico project would most closely assume the characteristics of similar TVA reservoirs in relation to recreation use and development. Projections of recreation use were based on this concept.

The estimate in the Tellico environmental impact statement of 1,750,000 annual visits within six years after project completion originated in 1963. It was based on 1962 visits/shoreline mile to all TVA reservoirs multiplied by Tellico shoreline mileage. This estimate included fishing and hunting use.[22]

Having "determined" how many visitors would come to Tellico each year, it was almost child's play to estimate how much each would spend in the area. Simply put, TVA guessed. In his June 24, 1963, memo to Elliot, Palo had estimated that visitors would spend an average of fifty cents per visitor per day in the area. How had Palo come up with that figure? Other federal water projects had calculated anywhere from seventy cents to two dollars per visitor per day. Palo simply lowered the figure arbitrarily to fifty cents, "because land enhancement is a credit (at least partly) to recreation." In other words, fully aware that there was a serious overlap between land enhancement and recreation benefits, Palo simply pulled the fifty-cents-per-visitor figure out of the air. His motive—to cut down double-counting of the overlap between land enhancement and recreation—was admirable, but his method was somewhat less than scientific.[23]

Once the idea of fifty cents per visitor per day emerged, it seems to have taken on a life of its own. For his part, Howes thought the figure "as good as any" and went on to apply the figure to *all* TVA reservoirs, emerging with a benefit-cost ratio of 0.95:1 for the whole TVA system, *based on recreation alone!* For a man who had spent his career at TVA arguing for more recreational facilities, these were heady calculations indeed.[24]

But a number of people at TVA were not as sanguine as Howes. Fletcher Riggs of the Office of Tributary Area Development believed that any number of visitors the agency came up with would be "quite inflated," since a large number would be local individuals and families who usually did not spend very much money on their outings. Others worried that the numerous nearby manmade lakes (twenty-two within a fifty-mile radius, sixteen of them TVA reservoirs) would draw off visitors from the proposed Tellico Reservoir, especially since many of the visitors presently coming to the region were trout fishermen and river boaters, and what they came for would be destroyed by the Tellico Dam.[25]

The trout fishermen constituted a particularly sensitive issue. These fishermen, who had raised some of the first cries of opposition, considered the area "unique" and spent a good deal of money there. Indeed, a joint study by TVA and the Tennessee Game and Fish Commission estimated that 24,000 trout fishermen visited the lower Little Tennessee River annually and spent over $80,000 in the area each year. According to Bradley Weeks of Chattanooga, they came from all over the eastern states, attracted by the superb tailwaters, the Little Tennessee's limestone base (which reduced acidity and therefore prompted insects and fly hatches, on which the trout fed) and the fact that it was a stream "with a regulated minimum flow of cold water, so there were no periods when it dried up." "Good tailwater," summarized Sam Venable, outdoors editor for the Knoxville News-Sentinel, "is like diamonds—if diamonds were as common as pebbles, they wouldn't be valuable." Hence any recreation benefit claimed by TVA for Tellico would have to demonstrate who would come to the area and what they would do once the trout fishermen abandoned the region.[26]

Moreover, if TVA were to assert that other fishermen would take the place of trout anglers, the agency would have to prove that the Tellico Reservoir would produce good fishing and that people would come there for it. But this was not so easy to do. Inside TVA, forest and fishery people were "professionally astounded" that the agency would destroy the trout fishing to create another lake, and one TVA biologist quietly predicted to Sam Venable that Tellico could end up even worse than Melton Hill, a TVA lake known for poor fishing. Venable explained why: "Manmade reservoirs have a definite productive cycle. . . . Flooding a river bottom that contains heavily fertilized land produces a rich aquatic community, much like a floral community on the land. The reservoir's food chain is kicked into high gear, and the spawning is tremendous." But gradually the fishing success decreases.

Venable classes the fishing on new reservoirs as "predictable," with a fairly long period of productive fishing after impoundment. Norris and Kentucky Reservoirs had very productive fishing for a long time after impoundment, but after a couple of years "Tellico fishing has gone straight to hell in a handbasket," even though the first summer after impoundment it was "awesome." Venable feels that Tellico proved a "flash in a pan" in terms of sports fishing. Could TVA have known this when it pressed for recreation benefits? Venable feels the geography of Tellico is all wrong for reservoir fishing. The best reservoir lakes have shallow broad waters with lots of obstructions which attract the numerous warm-water species of the region.

But at Tellico, the broad shallow part near Chilhowee Shoals is where the coldest water is, and the warm water downstream is where the reservoir is narrowest. The water conditions and geography, said Venable, are at odds with one another.[27]

But TVA desperately needed recreation benefits to shore up Tellico's benefit-cost ratio. One former TVA economist recalls that he and others were virtually ordered "to justify the flimsy recreation benefits." In order to do this, the agency took its calculated recreation benefits (based on projected visitors and their per-day expenditures) and added to that figure additional projected visits by fishermen and hunters. Since many of those already included in the recreation benefits *were* fishermen and hunters, clearly TVA was to some extent double-counting visitors. Then the agency, based on experience at other TVA reservoirs, projected a 200 percent increase (from 29,349 to 170,000) in the annual number of visitors after the Tellico Reservoir was impounded, *without deducting the trout fishermen* who presumably would no longer visit the area. Recreation benefits were inflated, then, until they comprised roughly 40 percent of the total project benefits. Moreover, it was a critical 40 percent, for it took the benefit-cost ratio for Tellico considerably above the minimum 1:1. At the very least, the process was an exercise in wishful thinking. At worst, it was a mathematical fabrication.[28]

But land enhancement and recreation benefits were so loosely calculated that TVA, according to one former employee, "didn't know if it [Tellico] could fly." Even more benefits were needed to bring the project's benefit-cost ratio to around 1.5:1, a figure judged acceptable to the Bureau of the Budget and the Congress.

Power and flood control benefits were comparatively small and not elastic enough—and the Office of Power already had complained that the claims for power benefits were too high. Navigation benefits were more flexible and thus more susceptible to inflation. Here TVA calculated the amount of transportation savings to industries which might have located at Tellico if the reservoir had been in existence, devised a per-acre savings, and then multiplied that figure by the projected 5,000 acres on the Tellico shoreline. By assuming that all 5,000 acres would be occupied by water transportation–oriented industries within ten years of the dam's completion, the agency was able to claim $400,000 annually in navigation benefits. Even then Wagner complained that the benefits were underestimated ("looks too small to me"). But on navigation benefits, no one felt there was much profit in going further.[29]

TVA had always claimed that the primary reason for building the Tellico Dam was the economic rejuvenation of the region, an area which, as has already been shown, the agency continued to insist would not prosper *without* the Tellico Project. Indeed, according to one TVA economist, once the dam was completed, Tellico would possess all the factors needed for rapid and sustained industrialization: good sites, an ideal navigation channel, large quantities of pollution-free and silt-free water, good railroad and highway linkages, low-cost electricity, and a good labor supply.[30]

Why not, then, claim the fruits of this industrialization as part of the benefits package? This is precisely what TVA Navigation head J. Porter Taylor proposed to do. Assessing the entire history of industrial development along the Tennessee River, Taylor argued that 0.8 jobs per acre had been created. He then applied 0.8 jobs to the 5,000 acres of potential industrial land created by the Tellico project, to get 4,000 jobs. Since for every 100 new manufacturing jobs, he reckoned, 65 nonmanufacturing jobs were created, Tellico's potential was 6,600 anticipated jobs. Using 1958 mean annual pay for manufacturing jobs ($3,838) and nonmanufacturing jobs ($2,764) applied to the potential 4,000 manufacturing jobs and 2,600 nonmanufacturing jobs, Taylor's figures generated annual wages and salaries of $22,500,000. The assumption was that all 5,000 acres would be occupied and that all 6,600 positions would be filled by currently resident agricultural workers, who had earned $645 per year as hired workers in 1958. This mean wage of the 6,600 future employables would have been $4,300,000 had they remained in agriculture, leaving an annual net economic benefit to the area's inhabitants of "about $18,000,000 excluding added increases in profits, rents, and other non-wage income."[31]

Not all within TVA shared Taylor's enthusiasm or agreed with his attempts to create a "jobs benefit" to add to Tellico's benefits package. Howes, who believed that the project would produce a general economic benefit to the region, pointed out that the new jobs projected by Taylor were "not likely to be filled by people in agriculture." Instead, Howes maintained, those people would be *displaced* by immigrants lured by the promise of industrial employment. In fact, the *region* might be economically "saved," Howes implied, but some of the people who most needed help might be hurt in the process.[32]

Taylor was undeterred by such critics. The only people in TVA who questioned the benefit-cost process or his industrial jobs benefits, he later said, were some "high-falutin' economists. They told me, 'That's crazy—you can't do it.' Well, in effect, we *did it!*" In response to his critics, Taylor reverted to a well-worn TVA theme: Tellico was to be a "convincing demonstration" of the "importance of developing shoreline lands." That, of course, might well be true. Indeed, most TVA people were convinced that it *was* true.[33]

In sum, TVA was able to create an acceptable benefit-cost ratio for the Tellico Project of 1.4:1.0. To do so, the agency had to keep project costs unrealistically low, while benefits were grossly inflated. The question was not whether benefits would occur—most people inside and outside TVA were convinced that they would. The problem was that, since the agency could not find econometric models and formulas to use in quantifying benefits, many of its figures were flabby, open to considerable argument, or simply wrong. Almost surely the pressure to bring forward a healthy benefit-cost ratio caused the agency to claim too much for the project. That in turn made some TVA employees cynical, a new feeling at an agency which heretofore had had much to be proud of. The project revived fissures within the ranks and cast aspersions upon TVA's once uncontested record. In one sense TVA bartered its unity and reputation for a project which was becoming increasingly untentable.[34]

As TVA labored over Tellico's benefit-cost ratio, key groups within the agency were virtually ignored. The cries of the forest and fisheries people, who pointed out that recreation benefits would be lost if the dam were completed, went unheard. But a much more significant loss than the recreational one would be that of agriculture, as several thousand acres of prime farmland would be inundated and several thousand more would be acquired by TVA and turned over to industries. The protests of TVA's Agriculture Division and the responses of those in power suggest how determined TVA's leaders were to build the Tellico Dam, even in the face of staunch internal opposition.

When David Selznick wrote his now-classic *TVA at the Grass Roots* (1942), the Agriculture Division was, next to Power, the most influential entity within the agency, potent enough to affect TVA policies in planning, land purchase and use, and population removal procedures. Working with the University of Tennessee's Agricultural Extension Service, the Agriculture Division had earned an international reputation and had wrought a "Green Revolution" in the Tennessee Valley. Harmful row cropping, soil erosion, hog and corn subsistence farms, and "slash and burn" cultivation had been all but eradicated and, through demonstration farms, increased fertilizer usage and other programs, valley farms had achieved a heretofore unknown prosperity. The "down corn, up horn" movement, which emphasized dairying, beef cattle, and stock breeding, was beginning to take hold, promising a "new age" for farmers of the Tennessee Valley.

But as TVA's attention shifted more and more toward attracting industry and tourism, the Agriculture Division lost power. The return to the large purchase policy was one sign of its waning prowess. Another was Tellico. As Paul Evans, director of the TVA Office of Public Information, wrote, "TVA [at Tellico] is going to substitute industrial/recreational activity for agricultural." Thus, while TVA was setting fire to its own grass roots, those in the Agriculture Division seemed powerless to do anything about it.[35]

But Agriculture did not take its predestined defeat lying down. Roger Woodworth of TVA's Agricultural Resource Development Branch wrote a stinging report on Tellico, designed to put the division's best foot forward. His conclusions were that the Tellico area could compete nationally in agricultural production, that some of the prime farmland in the East would be either flooded or taken by the dam project and that to take 39,000 acres out of agricultural production would entail a net farm production loss of $4 million annually. Another TVA agricultural economist, verifying Woodworth's figures, actually raised them by $100,000.[36]

TVA anticipated that its large land purchase for the Tellico Project would create land enhancement, recreation, and general economic benefits which it counted upon to carry its case for congressional approval. But the withdrawal of productive farmland would create a "disbenefit" which should have been entered on the "cost" side of the ledger—the annual revenue derived from the farmland retired by TVA from agricultural use. Agricultural economists in the Office of Agricultural and Chemical Development (OACD) were acutely aware of the significance of this disbenefit, and one of the most outspoken was H. A. Henderson.

Henderson's office at Muscle Shoals, Ala., in 1983 was decorated with a framed cartoon. In it, a buzzard is perched on an abandoned tractor, and a caption reads, "Who let all the farmland get away?" He spoke purposefully and carefully: "All the factors changed during Tellico. Before, no farmland problem; at the end, there was one. Before, a surplus and plenty of farmland; at the end, poverty. Before, production for the U.S. market; at the end, production for the world market." The "new issues," he argued, forced a recognition that industrial development and general welfare were not synonomous and that setting limits on agriculture constituted setting a limit on growth. In the 1960s, said Henderson, information was being developed to show that agricultural production could be managed in such a way as to provide a greater economic multiplier than industry in many regions. There was a need to evolve regional agricultural markets and put more land into cultivation, not take it out. The conversion of one percent of all farmland to other uses, according to Henderson, increased the public's food costs seven percent. In financial terms, the conversion cost per acre per year of farmland was $1,500.[37] And yet at Tellico TVA was going to flood thousands of acres of prime farmland for industrial development. If, say, 28,000 acres of prime land were inundated, the initial conversion cost would be $42,000,000, which should properly have come into the "cost" side of TVA's benefit-cost analysis.

Though information on agricultural disbenefits was being disseminated in the 1960s, said Henderson, "TVA would not accept the new information." During nearly two decades of debate on Tellico, TVA avoided using any type of agricultural "disbenefits" in its assessment of the Tellico Project. To do so would have made a tenuous project completely untenable.

As TVA prepared for budget hearings, agency leaders flatly denied the argument of its own agriculturalists, claiming that advanced farming techniques would require *retirement* rather than expansion of land in the future. "The most prudent use of the land resources," a TVA spokesman argued, "is to devote it to issues which will improve the economic opportunities of the people and alleviate the hardship that presently exists."

But head agriculturalist L. B. Nelson disagreed: "We should avoid implying that 'the hardship that presently exists' includes all farms in the project area . . . over 60 percent of these farms are commercial operations. For the valley as a whole, such farms represented only 46 percent of all farms in 1959."[38]

Ironically, while TVA was emphasizing at Tellico that industrial development *was* economic development, the national trend had already shifted, and the rapid loss of farmland was coming to be seen as a major ecological problem. As a later TVA document would put it, "Until about 1960, current doctrine indicated that industrial development, total development, and general welfare were almost synonomous. More recent studies have indicated that the size of agriculture may provide a limit to total growth and welfare."[39] Such shifts, although known to TVA agriculturalists, were virtually ignored by TVA Tellico proponents. When these viewpoints did surface, it was only as the project appeared totally halted because of the Endangered Species Act, and TVA was then willing to consider "alternatives" to Tellico. Describing TVA as a "victim of changing times," Henderson argued that there were too many people in TVA who, instead of giving management all the facts, gave it only those it wanted: "TVA was victimized by sycophantic supporters who would not tell the board anything but what the board wanted to hear."[40]

The studies on Tellico agriculture which did appear show how high the *real* cost of Tellico was when agricultural disbenefits were counted. In 1964, before TVA started to purchase land, farm sales on 24,370 acres amounted to $1,877,000. Of this amount, livestock sales accounted for 80 percent of the total. This income was adjusted to reflect tri-county trends on a ten-year basis using 1974 prices. The projected sales were $3.7 million per annum (24,370 open acres). Using Rapid Adjustment Farm demonstration experiences in Middle and East Tennessee, total farm sales could be boosted to a projected $6.8 million, with as high as $15 million per year projected if farm enterprises were restructured for regional market networks. Mixing, say, 6,500 acres of vegetables, 800 acres of perennials (apples, strawberries), 150 acres of tobacco, and "such livestock and poultry enterprises as dairy, broilers, and feeder pigs," it was estimated that, with intensified management and marketing techniques, the Tellico land purchased by TVA could have been made to yield an enormous per-annum benefit from the land alone and to have provided in the intensive phases of rapid adjustment and high value enterprises, thousands of man-years of labor. If these disbenefit income figures had been added over the calculated life of the project, the costs would have been prohibitive.[41] So potentially explosive was this issue that, though it never grabbed headlines or created a stir in Congress, Henderson ironically suggested, when it looked as if Tellico wouldn't be completed, that "so much national policy has been spent on it that we ought to turn Tellico into a place to study national resource development based on *true* information."[42]

Late in 1964 TVA took the Tellico Project to the Bureau of the Budget with the hope that the project could be included in President Johnson's budget for fiscal 1966. Along with Tellico, the agency also proposed the Tims Ford Project and the continuation of a $150 million steam power unit at Bull Run, Tennessee. Both Tellico and Tims Ford had been under consideration by TVA for years, were at roughly the same planning stage, and had similar benefit-cost ratios of 1.4:1. Even though Johnson's anti-poverty program and the Vietnam War were cutting deeply into other federal programs and the government was facing its sixth straight deficit, apparently TVA felt it would have little trouble securing appropriations for all these undertakings.[43]

But Johnson had instructed the Bureau of the Budget to be "prudent" and to keep expenditures down. Taking its charge literally, the bureau told TVA that it could not fund both Tellico and Tims Ford, and that the agency would have to choose between them. For TVA the choice was easy: Tellico was to be the showcase of the "new mission," whereas Tims Ford had a lower priority. Not surprisingly, TVA chose to send up Tellico.[44]

In January 1965, Lyndon Johnson submitted his "prudent" budget to Congress. Included in it was $5,775,000 to begin the Tellico Project. That item went to the House Public Works Subcommittee, chaired by Rep. Joe L. Evins of Tennessee's Fourth Congressional District. Evins, generally regarded as pro-TVA, was not expected to give the agency much trouble.

But TVA's Washington lobby had misread the shrewd congressman. True, Evins had been a TVA loyalist, but only when that loyalty had helped his district. The Tims Ford Project was in Evins's district, and the congressman was irked that TVA had withdrawn his pet project in favor of Tellico. Therefore he made no effort to stop his colleagues when they attacked Wagner for taking a great deal of land above the reservoir for industrialization. One congressman forced the TVA chairman to admit that industrialization at the earlier Melton Hill Dam near Oak Ridge had been disappointing, while another pried out of Wagner the fact that TVA already owned 3,607 acres of prime industrial land elsewhere on the Tennessee River system and that another 23,740 acres of prime frontage was in private hands and clearly developable. Wagner's lame reply that TVA would have to acquire additional acres at Tellico anyway (for navigation, flood control, and power and to "avoid unceremonial severance of tracts") was palpably untrue.[45]

It was then that Evins made his move. In one of the more sagacious pieces of congressional wizardry in modern memory, the congressman was able to convince his subcommittee to shift the appropriation for Tellico to Tims Ford. The House Appropriations Committee supported its subcommittee in June 1965, and the whole House subsequently went along.[46]

Later Evins defended his action by saying that the Tellico Project had aroused much local opposition, while the Tims Ford Project had not. He claimed that "the committee was flooded with letters, telegrams, resolutions, petitions, and editorials for the Tellico Project and against the project." Hoping that the conflict might resolve itself, the subcommittee "postponed [Tellico] without prejudice." Of course, that was another way of saying that Evins had gotten his own pet project approved and funded first. Whatever the reason, TVA probably had made a tactical error in trying to push Tellico over Tims Ford before Evins's subcommittee. Given its commitment, however, the agency could do nothing else.[47]

TVA was stunned. Although Tims Ford was also an agency project, TVA believed it was vital to get Tellico underway before costs rose and destroyed the delicate benefit-cost ratio and before opposition became stronger. In an effort to reverse the House's decision in the Senate, TVA encouraged private citizens and groups to endorse the Tellico Project and put pressure on Congress. Forty-two local businessmen and dignitaries, including Knoxville Mayor Leonard Rogers, were flown to Washington to plead Tellico's case before the Senate, while the agency began pressuring local groups for endorsements. Some members of the Knoxville Chamber of Commerce reported to one of the local newspapers that unseemly pressure had been applied, and Wagner made a personal appeal to the Knoxville City Council for its endorsement.[48]

TVA's only chance was to reverse the House decision in the Senate. The appropriations bill, however, went to the Senate Subcommittee on Public Works, where Evins's Senate counterpart was Allen Ellender. As we have seen, Ellender was unfriendly to TVA for a host of reasons. Moreover, Ellender knew a good deal more than most of his colleagues about benefit-cost ratios, having been chairman of the subcommittee which had drafted Senate Document 97 in 1962.

The senator began his attempt to undercut the Tellico Project by stating that, since TVA had claimed power benefits for the project, it should be financed by revenue bonds. Wagner replied that, since there actually were no power facilities at Tellico, the law did not allow the agency to use bonds to fund the project. But Ellender was obdurate, insisting that the TVA chairman was holding to the letter of the law in order to deny its spirit. On financing Tellico through TVA bonds, the senator concluded, "I say that is what we ought to do, and that is what I am going to suggest to the [Senate Appropriations] Committee before we spend any more Federal funds in the Tennessee Valley. . . . You cover just a small area of the United States [and] you have a bonanza there."[49]

Having placed Wagner on the defensive at the outset, Ellender then lured the TVA chairman into the benefit-cost trap. As noted earlier, Wagner tried to parry Ellender's thrusts by claiming that, had TVA used the Corps of Engineers' methods, Tellico's "true" benefit-cost ratio would be 1.9:1. With tongue in cheek, the senator told Wagner that he should have used those figures to make Tellico "more attractive" to the subcommittee. But Ellender stated acidly that the project was grossly overrated even at 1.4:1, that the claims of "general economic benefits" ($15 million) were much too soft, and that in sum Tellico was a bad project.[50]

Finally Ellender attacked the fact that TVA was planning to purchase land ultimately to be sold to industries at a profit. "Why don't you let private enterprise do some of that, instead of you getting into business?" the senator asked caustically. "It is very seldom that land is purchased from federal funds, and resold for profit, and the profit goes to the government." Of course, Ellender knew very well that the large purchase and resale was crucial to Tellico's shaky benefit-cost ratio, a point which he hammered home again and again.[51]

To try to reverse the momentum Ellender had created in the subcommittee, TVA pulled out all the stops. Sen. Albert Gore and Sen. Ross Bass of Tennessee spoke, strongly supporting Tellico, and Rep. John Duncan of Tennessee's Second District, which included both Knoxville and the Tellico area, lent his endorsement. Then Duncan yielded to Knoxville attorney Ray Jenkins (whom TVA had flown to Washington to testify), a man famous for his oratorical stemwinders.

Jenkins had gained national notoriety as the special counsel for Senator Joe McCarthy's Permanent Subcommittee on Government Operations in the 1950s and as an ardent defender of the senator in the Army-McCarthy hearings. The Knoxville lawyer, whom Richard H. Rovere once called "a veteran defender of moonshiners and husband-shooters in Tennessee," began his encomium by stating that he could not possibly "warm up to the subject" within the time limit of twelve minutes imposed by Senator Gore. The latter, who knew his man well, responded that Jenkins could not warm up to that "or any subject" within twelve minutes. And, indeed, Jenkins' speech was florid enough to make even Wagner blush. To make a point about how jobs were needed, he spoke of the "fine mountain people" of his community who had raised money for a shirt factory; even widows contributed. Jenkins had been "violently opposed" to the Tellico Project at first, he said, because he owned a farm that would be inundated, land which was "the source of more pleasure and satisfaction than anything I have ever owned." But after talking to the "distinguished directors" of TVA, Tellico became "a matter close to my heart." Jenkins then argued that Tellico would contribute $300 million to the economy in the next twenty-five years, a figure he said TVA supplied. As long as Jenkins stuck to the script supplied to him by TVA, he was more or less reasonable. But at times he became almost unintelligible; in one instance he linked Tellico's recreation to crime prevention.[52]

The subcommittee, over Ellender's protest, reversed the House's decision and switched the funds from Tims Ford back to Tellico. Thus the full Senate, which ultimately approved the subcommittee's action, virtually ignored the Louisiana senator's revelations about Tellico's suspicious benefit-cost ratio and bowed to the wishes of Tennessee's senators. Ellender again objected, but the Senate held firm.[53]

In the House-Senate conference committee on the bill, Evins again held sway. Once again the funds were switched from Tellico to Tims Ford and, in what was generally considered a slap in the face to Wagner's pet project, the conference committee restored a Senate-deleted item of $100,000 for a study of "the possible adverse effect of the Tellico Dam on pollution in the Little Tennessee River." Tired of wrangling, both houses finally deferred to Evins's desires. Then, in what he must have considered a magnanimous gesture, Evins told a group of disappointed Tellico advocates that he would support their project in 1966.[54]

Opponents of the Tellico Project should have been delighted, and, indeed, some were. J. Fred Moses, Jr., a Knoxville attorney, said the decision had reaffirmed his faith in the democratic process. But most opponents were not so sanguine. They realized that it was not they, but Evins, who had stopped the dam, and that the stoppage was only temporary. Indeed, to many people, even to many of the project's opponents, Tellico seemed an inevitability. In his 1966 testimony before Ellender's committee, a tired, sad, defeated John Lackey would voice this note:

> LACKEY: Mr. Chairman, down in Tennessee, if you are against TVA in politics, in Tennessee that is a bad word. . . . But our trouble has been TVA people are just smarter than we are, I guess.
> ELLENDER: They have the money.[55]

In the months between Tellico's shelving by Congress and the renewal of the appropriations fight the next year, TVA appears to have made little or no headway in improving the project's benefit-cost ratio. C. W. Nash of TVA's Division of Reservoir Properties had been working diligently to improve and nail down land enhancement benefits, but with little success. In fact, the conscientious Nash reported to Van Mol that, in his opinion, the land enhancement benefits TVA was claiming *already were too high*. He contended that a significant increase in land values would occur only within about one-tenth of a mile from the shoreline, not the two miles the agency was claiming. Also, he felt that no available models could predict the amount of land enhancement beyond ten years, yet TVA was claiming twenty-five years. Obviously irritated, Assistant General Manager Phil Ericson retorted that Nash's model "has not *actually* been applied to the Tellico case," and that even if it were it would tend to underestimate land enhancement benefits resulting from "unusually great recreation potential (such as Tellico)." Indeed, for an agency determined to push the Tellico Project through, Nash's conclusions were difficult to take, for his work threatened to undercut the very economic justification of the project. Ultimately, then, Nash was ignored, and TVA returned to Congress in 1966 with its Tellico position essentially unimproved.[56]

Before TVA could even get Tellico back to the House subcommittee, the Bureau of the Budget, to slow down inflation, slashed over $2 million from the agency's project request. And once it did get to the House, Tellico was roundly attacked both in the Public Works Subcommittee and on the House floor. The apparent failure of Melton Hill was brought up again, as were the questions of whether East Tennessee's economic growth depended on still another TVA dam, and of the extent to which Tellico would pollute area waters. Replying to Arizona Rep. John Rhodes's testy question of what harm would be done if the project were delayed "for a year and a half or so" or until these questions were resolved, Wagner stiffly replied, "Of particular concern is the fact that the development of the Little Tennessee area would be set back just that much. The people who called on Mr. Evins the other day are interested in pushing this thing through. There is a limit to how many times you can throw a wet blanket on a group of enthusiastic people and expect them to still maintain their enthusiasm." The chairman's answer to Rhodes disclosed a new line of thought at the agency, a line probably emanating from Wagner himself. The reasoning went something like this: TVA would build the Tellico Dam *because the people of the area wanted TVA to build it.* In fact, many within TVA gradually came to believe that it was the people themselves who had *initiated* the idea of the project, and that the agency was merely acceding to the public will. Though that view was palpably erroneous, it was a view that would continue to grow within TVA.[57]

Joe Evins, however, was as good as his word, and he was able to ram the Tellico appropriation through his subcommittee and onto the House floor. In addition to the $3.2 million set aside for Tellico, $9 million was approved for Tims Ford, prompting Ellender to remark later, "I see my friend, Mr. Joe Evins, has been very busy here." Indeed, Evins further enhanced his reputation as a pork-barreler by receiving TVA's promise that it would also study the possibility of erecting dams on the Emory and Obed Rivers, both in Evins's district, in return for his support of the Tellico appropriation.[58]

Once on the House floor, however, the Tellico appropriation ran into resistance which Evins could not tamp out. John Dingell of Michigan and Richard Ottinger of New York both attacked the project, Ottinger going so far as to call TVA's proposal "an unwise one." For his part, Dingell drafted a bill which, had it passed, would have required TVA to consult with the Tennessee Game and Fish Commission and the U.S. Fish and Wildlife Service before proposing any new dams. Dingell and Ottinger both felt the wrath of TVA Director Frank

Smith, who called Ottinger's objections an "indiscriminate attack" and Dingell's bill bureaucratic meddling.[59]

Undoubtedly Dingell and Ottinger were the early harbingers of the conservationist-environmentalist movement of the late 1960s and 1970s. Most congressmen and senators were not yet concerned about the environment. Nor, apparently, were they particularly interested in Tellico's dubious benefit-cost ratio. Ellender did his best to revive that issue in the Senate ("I think I brought that out last year") and to form an unlikely marriage of convenience with the National Wildlife Federation, the Wilderness Society, and other conservationists, but to no avail. The Tellico appropriation passed handily in both House and Senate, and Lyndon Johnson signed the federal budget on October 17, 1966. Tellico construction was scheduled to begin early in 1967. At last, it appeared, the Tellico Project was to become a reality.[60]

In the end, when Congress made its final decision, the unreliable benefit-cost ratio did not play a very large role in the deliberations. And the Louisiana curmudgeon who, for reasons of his own, tried to awaken his Senate colleagues to the problem did not live to see how correct he had been—he died in August 1972. As costs mounted almost uncontrollably and benefits shrank under scrutiny, Tellico's benefit-cost ratio looked worse and worse. As one former TVA employee said, "I would never even start a project that couldn't begin with better than a 2 to 1 ratio, just to cover a margin of error."

How many within TVA had realized just how bloated the agency's benefit-cost claims for Tellico actually were? Surprisingly few—probably only those who had worked directly on the problem. Howes had suspected that something was amiss and said so, very early. Certainly Palo was uncomfortable. C. W. Nash and some others had told their superiors what they did not want to hear and had been ignored. And, according to one former TVA economist, Mike Foster must have known. Whether Wagner, his board, General Manager L. J. Van Mol, or others in high places fully realized how weak the agency's claims were may never be known. Their belief, which they shared with most within TVA, was that the Tellico Project would bring enormous prosperity to the region. This belief was so strong that the leaders may well have been blind to the warning signs emanating from below. By 1966, their ability to view the Tellico Project objectively, if it had ever existed, had long since vanished. What remained was their faith in the agency and in themselves and their talents. Surely, they believed, later generations would see their decisions as prescient, wise, and bold. The smoke of opposition would be blown away by Tellico's success.

Above: The Little Tennessee River Valley, looking north from the site of Fort Loudoun (foreground), downstream to the Highway 411 bridge; *below:* closeup aerial view of Fort Loudoun, built by British troops in 1757.

The Little Tennessee River Valley, looking north at Calloway Island, at approximately Mile 24. Toqua Indian mounds are on the mainland, at left.

Above left: Gordon Clapp, chairman of the TVA Board of Directors, 1947–54; *above right:* Arthur Jandrey, assistant to the TVA general manager, best known for his report on TVA land acquisition; *below left:* General Herbert Vogel, chairman of the TVA Board of Directors, 1954–62; *below right:* Aubrey "Red" Wagner, TVA general manager, 1954–61; TVA board, 1961; chairman of the board, 1962–78.

Above left: Gabriel O. Wessenauer, TVA Chief of Power and one of the most powerful figures in the agency; *above right*: Louis Van Mol, TVA general manager, 1961–70; *below*: an aerial photograph of the Tellico Project site, with TVA's plan of the project overlaid.

Above left: J. Porter Taylor, director of TVA Division of Reservoir Properties; *above right:* George Palo, TVA chief engineer; *below left:* Reed Elliot, director of TVA Water Control Planning; *below right:* Robert Howes, director of TVA Reservoir Properties and chairman of the Future Dams and Reservoirs Committee.

Above left: Paul Evans, director of TVA Office of Information; *above right:* Richard "Dick" Kilbourne, director of TVA Office of Tributary Area Development; *below left:* E. P. "Phil" Ericson, assistant to the TVA general manager, responsible for overseeing the budget for the Tellico Project; *below right:* Don McBride, TVA Board of Directors, 1966–75.

Above: Minnard "Mike" Foster, director, TVA Division of Navigation Development and Regional Studies; *below:* Justice William O. Douglas being saluted by three Cherokee youths, 1964.

Above: Directors Frank Smith, "Red" Wagner, and Don McBride confer with University of Tennessee archaeologist Alfred "Ted" Guthe on Indian remains taken from the Tellico area; *below*: aerial view of one of the archaeological digs at the Tellico site.

Worth Greene (with hat) showing excavation of a Cherokee burial site at Tellico to two Cherokee visitors.

Above: A. J. "Flash" Gray, chief of TVA Regional Planning staff during early years of Tellico planning; *below left*: James Gober, chief of TVA Regional Planning staff during the Timberlake planning period; *below right*: F. B. "Red" Williams, head of the Boeing Corporation team that helped to plan the Timberlake experiment.

Above: TVA model of the planned city of Timberlake; *below:* Lynn Seeber,
TVA general manager, 1970–78.

Wagner cheered by TVA employees on his last day at work.

Above: TVA model of the Tellico Dam, constructed at the agency's model shop; *below:* TVA Board of Directors, 1984, from left, Richard M. Freeman, Charles H. Dean, Jr., and S. David Freeman.

6

ATTACKS AND COUNTERATTACKS:
THE TELLICO DEBATE GOES NATIONAL

On April 12, 1971, TVA Director of Information Paul Evans sent a memorandum to General Manager Lynn Seeber concerning an important shift in tactics by the opposition to the Tellico Project. In addition to its many other duties, the Information Office had been given the task of keeping track of the project's opposition and keeping Seeber informed about it.

Paul Evans was one of the most intelligent, talented people ever to work for the talent-rich TVA. A native of South Dakota and descendant of a Methodist minister and Indian fighter in the Dakota Territory, Evans graduated from Dakota Wesleyan University in 1937 and began his long career in journalism as a reporter and sports editor for the *Mitchell (S. Dak.) Daily Republican.* After holding several editorial positions, spending some time as a public information officer for the U.S. Department of Agriculture, and taking a year's leave to accept a Nieman Fellowship at Harvard University where he studied economics and political science, Evans came to TVA in 1951. The next year he was appointed director of information, a post he would hold for twenty-three years.[1]

By 1975 Evans was probably the most respected nonengineer at TVA. Tall, thin, with a rapidly receding hairline, he looked as much like a college professor as the tough, honest journalist he was. Evans rarely handwrote or dictated anything, preferring to punch out his thoughts and ideas on an old typewriter, the noise of which carried far beyond the door of his modest office in Knoxville's New Sprankle Building.

According to an unnamed "correspondent," Evans reported in his April 12 memo to Seeber, the opposition to Tellico was "mounting and regrouping along environmental lines." A group which included John Locke, a local landowner and vocal opponent of the project, was scheduled to meet on the following Saturday at Vonore High School with some "New York lawyers." Moreover, a "war chest" of $100,000, of which $15,000 already had been raised, was being sought to battle the project.[2]

There is little doubt that the general manager took Evans's report seriously. The next day, April 13, Seeber related the information director's findings to the TVA directors and to Dr. O. M. Derryberry, TVA's director of health and safety. To Derryberry he added, "This makes an Environmental Statement for Tellico more urgent than before."[3]

Evans's memo noted an important shift in the opposition to the Tellico Dam. Although Justice Douglas's brief 1965 visit had been a well choreographed "media event" designed to focus national attention on TVA and the Tellico Project, by the start of the dam's construction in March 1967 national interest had all but evaporated. Tellico opponents could not hope to compete for national attention with the war in Vietnam, urban riots, President Johnson's post-election troubles, and the resurgence of a remarkably resilient Republican party. By then, the Tellico foes were as few, localized, and internally divided as they had been at the outset. Nor could they seem to agree on where the Tellico Project was most vulnerable — TVA's taking line policy and land use plans, the project's suspect benefit-cost ratio, or the agency's destruction of a beautiful river which was revered by sportsmen and nature lovers alike. In short, in 1967 it appeared that TVA, as it had done so often in the past, had weathered the opposition and won.

But by 1971, barely four years later, the situation had changed dramatically, as Evans explained in his memo to Seeber. Most important, the nation had become concerned with the environment, and it was only natural that environmentalists (Evans once referred to them as "neo-environmentalists") would scrutinize projects like Tellico closely. Environmentalist attacks on the project not only revived the spirits of anti-Tellico landowners, sportsmen, and farmers, but also added numerous upper-middle-class men and women to the opposition. These people knew how to organize, how to pressure politicians, how to deal with the media, how to marshal public opinion, and how to use the judicial structure. This new group almost immediately assumed the leadership of the opposition, deftly unifying landowners, environmentalists, archaeologists, and sportsmen, drawing in the Eastern Band of the Cherokees, flooding government officials and the press with articulate letters and well-subscribed petitions. Finally they drove TVA to court and stopped the project. As Paul Evans remembered, "They were a very skillful and well-organized opposition."[4]

TVA had never before encountered such opponents. These were not parochial landowners, grumpy sportsmen, or conservative retrogrades who saw TVA as the first wedge of the socialist onslaught. These were liberal academicians who usually fought their most important battles in sleepy faculty senates, business people who usually thought it "bad form to stand out from the suburban crowd," college students, and many others. In response to their attacks, TVA shifted its approach, claiming for itself the title of "balanced environmentalists." Simultaneously the agency used its vast network in the business community to try to overwhelm opponents' efforts. And, as will be seen, the "model city" of Timberlake was created in an attempt to dampen criticism by talking about the jobs, housing, and industries that the Tellico Project would produce. But this time the opposition did not buckle.[5]

The beginning of construction clearly had demoralized the small and badly divided opposition. Landowners, even some who had bitterly opposed the project, began selling out, as TVA appraisers and land purchase officers moved slowly through the valley. Only a few, like farmer Wade Swafford, openly resisted. According to Swafford, after one land purchase officer had visited him several times, "I took a tobacco stick [an approximately six-foot long stick, pointed at both ends, used to hang tobacco for curing] and chased him into the road. He was never rude to me or nothin', but he knew I meant business." Swafford was the exception. Many sold willingly, amazed at the prices TVA offered for farms that had been owned by families for generations. Those few with heavy mortgages, usually contracted with local banks to buy equipment or make improvements, saw a quick and easy way out of their financial problems. And, according to last-ditch opponents, the elderly also sold quickly, either confused by the whole process or afraid to stand against a power like TVA. "With some of their neighbors selling, and confused and afraid as they was," recalled one longtime Tellico foe, "those old people never had a chance." In short, while some landowners held out, among the farmers whose land was to be taken significant opposition virtually collapsed. Indeed, subsequent public opinion polls of local landowners consistently showed that a high percentage favored the project.[6]

The temporary waning of opposition, however, allowed TVA to turn its attention to other Tellico-related issues. One question was whether landowners whose property would front on the Tellico Reservoir would be permitted to retain nonexclusive access rights to the reservoir. Most TVA people opposed sharing reservoir access rights with anyone. However, as Elliot commented, "There may be exceptional cases that should be considered in the light of local special needs." In a memo to Assistant General Manager John Rozek, Elliot conceded that leaving the access right of some landowners would save money, perhaps cut down on condemnation cases, and create "better public relations."[7]

What Elliot meant is a matter of some conjecture. Clearly he and the Water Control Planning staff opposed any policy that would attach strings to any land purchase. Perhaps he believed that any controversy at all would revive landowner opposition. Or perhaps there were other political reasons for leaving some access rights in place. In any case, landowners already opposed to the project believed that TVA showed favoritism to some landowners. It was the kind of nagging accusation that would survive for years. Nevertheless, the TVA board chose on May 25, 1967 to adopt the policy that "owners of several tracts should be permitted to retain nonexclusive access rights to the water." The exception was shoreline specifically designated for any program such as industrial development or even, in some cases, recreation.[8]

Once the Tellico Project was begun, some within TVA quickly realized that more land above the reservoir was needed and that the taking line would have to be extended. A private consultant to the agency, Robert Gladstone, recommended the purchase of an additional 350 to 400 acres. Other planners within TVA added 800 to 900 acres to Gladstone's recommendation for recreational purposes, including a golf course. Still others wanted 58 acres more at Vonore for industrial development. Even as land appraisals and purchases were continuing, TVA's appetite was growing.[9]

The elastic taking line presented a number of problems. Landowners were in an uproar. Some who had been told their land would not be needed for the project now were told that it was. "My daddy and I went up to the land office to try to get some maps," recalled an irritated Alfred Davis, "and every time we'd go up, the taking line would be different." This might have been an overstatement, but changing the taking line did make landowners insecure and to some extent revived local opposition.[10]

Apparently no one within TVA had ever done a careful study of how many families would have to be removed to make way for the Tellico Project. Even after the purchase of property had begun, the agency still had no idea of the size of the affected population. And the ever-changing taking line complicated the task even more. In July 1967, Gilbert "Pete" Stewart, Evans's second-in-command at TVA's Information Office, sent out an emergency memo for a "hurry-up" figure on the number of affected families. In contrast to earlier reservoir projects, where meticulous studies had been done of the population to be removed (at Norris, TVA's first project, the agency knew the figure down to the person), the best information Stewart could get was a very rough estimate—which turned out to be wildly incorrect. Project Planning staffers reported to Stewart through Branch Chief Don Mattern that approximately 600 families were within the line, whereas the actual figure was closer to 350. While changing procedures surely accounted for some of this inexactitude, probably uncertainty even within TVA over the location of the taking line and the acreage to be acquired was equally responsible.[11]

TVA was well aware that an expanded taking line would increase project costs, perhaps significantly. With the benefit-cost ratio such a delicate problem, the agency was loathe to let costs creep upward. By June 1967 it was public knowledge that project funds were tight and that construction was being slowed down for lack of money. Very likely the agency, in an effort to get a good benefit-cost figure, had seriously underestimated the costs of land acquisition and construction. Even so, rising costs and the resulting project slowdown caused some within TVA to reassess the more ambitious land-taking policy. The Land Branch in Chattanooga proposed trimming 1,766 acres from the total project to save an estimated $600,000. Yet, on the other side, Mike Foster, recently made chief of Navigation Resources Branch, was simultaneously pushing to extend the taking line even farther. In short, the ever-shifting taking line created uncertainty and confusion both inside and outside TVA.[12]

TVA seems to have been unable to control either the scope or cost of Tellico. The project was already being forced to consider a slowdown because of lack of money. True, national inflation during the late Johnson years was partly responsible for the problem. But TVA was also unable to "freeze" the project, to stop readjusting its scope, and all this was tremendously costly in the long run. Indeed, it was as if Tellico was to be Wagner's and TVA's *perfect* project, the one that would prove TVA's worth as the agency approached its 50th anniversary, the one which would correct all the errors of past projects, the one that would demonstrate that TVA could solve every regional problem. As OTAD's Kilbourne mused, "Everything that everybody wanted, we could do."

In September 1968 Robert Wetherholt of the Planning Section, Division of Navigation Development and Regional Studies, openly expressed grave concern over Tellico's mounting costs, which has spiraled from approximately $43 million in 1966 to a worrisome $67 million by mid-1968. Foster, Wetherholt's boss, also was concerned, as was Chief Engineer George Palo. "It should be possible," Palo remarked caustically in January 1969, "to define [Tellico's] scope with accuracy to endure to the completion of the project." That same month Assistant General Manager Phil Ericson expressed the hope to Foster that a regular intra-agency review of the project "will *not* throw the door open on Tellico . . . scope and cost."[13]

At this same time, costs of the war in Vietnam were seriously affecting national domestic programs, sapping their effectiveness and increasing the inflation rate. In 1968 Congress appropriated only $7.9 million to Tellico for fiscal year 1969. The next year that appropriation was cut back to $4.6 million. With an average monthly payroll of $160,000, and with land acquisition and construction costs rapidly mounting, it was almost inevitable that construction would slow down and the completion date would be extended. In November 1968 TVA glumly announced the inevitable: construction would have to be either slowed or halted, possibly for up to eighteen months. By January 1969, the agency had reset the completion date from 1971 (a June 1967 estimate) to 1974. With only twenty-five percent of the land acquired by February 1969, land purchases temporarily stopped.[14]

The slowdown at Tellico outraged local supporters of the project, some of whom incorrectly assumed that TVA was losing its ardor for the whole project. In February 1969, Loudon County Judge Harvey Sproul called together the membership of the Tellico Area Planning Council to discuss the unwelcome turn of events. Sproul reported that he and Wagner had met the previous December to discuss the status of the project. Sproul had expressed great concern over the funding crunch. Wagner had agreed, adding that "if Tellico is to get additional funds, TAPC and all other interested groups will need to push for this by working through their congressmen." Taking Wagner's broad hint, the council passed a resolution asking TVA for an estimate of the amount of money necessary to keep the project going for another fiscal year. In turn, TAPC promised it would lobby for additional appropriations. Before the council's next meeting in late March, TVA told TAPC that $13 million was needed to keep the Tellico Project on schedule. Like the malleable organization it was, the council accepted the figure and set out to lobby a list of senators and representatives for additional appropriations. In the early Nixon years, however, such efforts had little effect, and the original appropriation stood at an amount less than half of what TVA said it needed. Both the agency and local Tellico supporters were disconsolate.[15]

Worse, the slowdown in construction gave the weakened opposition forces new hope. As early as mid-1967, local opponents had shifted their tactics, zeroing in on the expanding taking line and the large amount of land being taken which would not be inundated. Local landowner Elmer Giles nicely summed up part of the opposition argument in a handwritten letter to Congressman John Duncan:

Our main complaint is not what the water is taking, but land back of the water line. Several of us have land that doesn't cover, but will apparently be bought by the T.V.A. Some of the places aren't touched by the water whatsoever. . . .

We have heard this is because of the Industrial Development and so forth. This couldn't possibly be right because it would be 10–15 miles long with thousands of acres of land.[16]

TVA's response to the taking line outcry was to keep plans and changes as confidential as possible. The actual taking line was still being changed, and only confusion would result from any premature publicizing of precisely what land TVA wanted. Too, the agency did not want to repeat its 1964 error of allowing R. D. Akard to "borrow" a project map and distribute it to Tellico opponents. In 1968 when Akard, who apparently had shifted ground and now supported the Tellico Project, again showed up at TVA, he was given the cold shoulder—and no information.[17]

The issue of purchasing or condemning nonreservoir land for resale to encourage industrial development, residences, and recreation facilities was the point on which TVA felt it was the most vulnerable. Almost from the time the agency agreed on a large purchase policy at Tellico, some people within TVA were worried about that policy's legality. Although supporting evidence is not now available to scholars, it is likely that the agency's Office of General Counsel anticipated such an action. An inquiry from Congressman Bill Brock produced the following memo from the general manager's staff:

The use now planned for this property [the Asa McCall farm, about which Brock was inquiring] is for industrial development. However, the staff would prefer not to specify this as the sole reason for acquiring the property. In the event it becomes necessary to file a condemnation case, the Division of Law would prefer to keep the reasons for the taking on a broad base.

One University of Tennessee economist, himself hardly neutral on the Tellico issue, believed the issue was not an economic but a constitutional one; some angry former landowners later insisted that eventually they would file suit to test this very point.[18]

One wonders whether the opposition realized at the time how vulnerable TVA was on the issue of purchasing or condemning private land for resale. The question was never tested or resolved. Ironically, in April 1969 Supreme Court Justice William Douglas reappeared in the Little Tennessee Valley, grabbed the headlines with a well-publicized fishing trip down the "Little T" (which the Nashville Tennessean claimed again had been specially stocked by the Tennessee Game and Fish Commission), and launched verbal salvos against the Tellico Dam.[19]

The Douglas visit and subsequent events served to shift the opposition's attention away from the issue of land acquisition and toward the Tellico Project's effects on the environment. Had the condemnation and purchase issue been pursued to its limits, the opponents of the Tellico Dam might have been able to stop or at least vastly alter the project. On the other hand, Douglas drew to Tellico the attention of the growing national environmental movement, a loose coalition which would be enormously influential in the 1970s. While the avocations of these middle-class suburbanites and professionals may have been fishing, boating, canoeing, hiking, nature study, or simply preserving the environment, it was their vocational skills in business, advertising, writing, science, engineering, and a host of other fields that made them so potentially threatening to TVA. Opposition biologists could stand toe-to-toe with the agency's fish and wildlife people; TVA's engineering designs could be called into question by experts; the project's benefit-cost ratio could be challenged by professional economists and businesspeople; the opposition's use of the media could match TVA's sagacious manipulation of news. In short, the second Douglas visit threw the Tellico Project into conflict with the national environmental movement which, at the time, seemed to hold out to many opponents the best possibility of defeating the Tellico Dam.

By 1969, Douglas himself had publicly referred to TVA as the "worst offender" in the destruction of the environment and had had accepted by *True Magazine* the article on the Little Tennessee River and the Tellico Dam that he had originally planned for *National Geographic*. In one sense, then, Douglas's return to the region was a publicity tour for his upcoming article; wherever he went, the justice was interviewed and photographed offering off-the-cuff anti-TVA accusations to anyone who would listen.[20]

Douglas's article appeared in the May issue of *True*. Titled unsubtly "This Valley Waits To Die," it was an emotional assault on the Tellico Project and spoke in almost reverential tones of the Little Tennessee Valley itself. Outraged, pro-Tellico Charles Hall, now president of the Tri-Counties Development Association, shot off an angry protest to the editor of *True*, which began, "There is something about Spring that brings forth both the sap and the saps. . . . In the latter category one finds such philosophers as Supreme Court Justice Douglas whose most recent tirade again faults construction of TVA's Tellico Dam." In one of the angriest statements ever issued by this fairly even-tempered man, Hall persisted, "one would hope that the well-financed and well-organized lobbies of the antis will be overcome by the outpouring of mail and telegrams from Mr. Average Citizen." Even as it opened its retaliatory attack on Douglas, TVA, through Director Don McBride, made sure to send a warm note of appreciation to Hall. "Congratulations!" wrote McBride. "TVA appreciates your concern."[21]

The Douglas article caused a great national stir. In spite of Hall's hope, the overwhelming majority of responses were extremely anti-Tellico and pro-Douglas. Senators such as Philip Hart (Michigan), Herman Talmadge (Georgia), and Jacob Javits (New York) were so overwhelmed with letters that they inquired of TVA as to what the devil was going on. Several letters to senators, congressmen, and President Nixon either enclosed copies of the *True Magazine* piece or referred to it. Organizations such as the Appalachian Anglers and the Southern Field and Creel Club went on record as opposing the project, and long-term opponents such as Trout Unlimited, the Tennessee Conservation League, and the National Audubon Society revived their anti-dam stands with stinging resolutions. By mid-1970, when TVA made a survey of groups opposed to the Tellico Project, of the forty-one organizations the agency indentified, fifteen were outdoorsmen's and sportsmen's groups and six were conservationist or environmentalist organizations. Clearly the Douglas article had had a great effect.[22]

TVA saw the attacks of Douglas and the conservationists-environmentalists as highly threatening. The national pressure which these comparatively new opponents could bring to bear against the embattled agency was one source of worry. But, in addition, the 1969 passage of the National Environmental Policy Act (NEPA) required that all projects such as Tellico submit an Environmental Impact Statement to prove that they were not damaging to what many considered the highly fragile environment. Projects which failed to file such statements or those whose statements were considered inadequate were subject to legal action by the government or private citizens and could be closed down. Hence, while Douglas's article in itself probably would have been insufficient to halt the Tellico Project, in conjunction with the rise of the national environmentalist movement and the passage of NEPA, it was potentially an important document.

TVA was caught in an increasingly awkward position. Even as the agency was expanding the taking line in order to add certain features to the project, costs, which had been badly underestimated, were mounting dangerously. Construction had already been slowed to a crawl due to rising costs and meager congressional appropriations. Local landowners were in open revolt. Local dam boosters like Charles

Hall, Harvey Sproul, the Tri-Counties Development Association, and the Tellico Area Planning Council were becoming discouraged, probably causing TVA to fear that these agency-initiated organizations were in danger of disintegrating. Now Douglas's article had brought the project to the attention of the national environmentalist movement, which, armed with NEPA, might make a great deal of trouble. Some were badgering TVA to file an Environmental Impact Statement for Tellico. And, if all of this wasn't bad enough, even within TVA some were questioning the wisdom of stubbornly continuing. According to one TVA employee, Van Mol called a series of meetings at which he effectively announced that if anyone disagreed with TVA board policy, he or she should consider finding other work. Some did. And although Tellico per se was never mentioned by Van Mol, the implication was quite clear. It appeared that the Tellico Project, on the boards since 1936, was once again in trouble.[23]

But Wagner had not become chairman of the TVA board by wilting under fire. In the face of increased opposition to the Tellico Project both inside and outside TVA, Wagner moved decisively to unify the agency and its supporters behind the project while at the same time covering up Tellico's weaknesses and presenting a firm front against the environmentalists.

Inside TVA, Wagner sought to stifle criticism of Tellico. Undoubtedly Van Mol's meetings with TVA employees bore the imprimatur of Wagner. The chairman also further centralized the administration of the diverse Tellico Project. For this, he needed a man he could trust, a man of unswerving loyalty to the project and to TVA and one who could see the project in the broadest and grandest possible terms. In early 1969 Wagner chose Minard I. "Mike" Foster to coordinate the Tellico Project. Foster was an economist, with a Ph.D. from the University of Florida. After joining TVA in 1954, largely due to his talent, his ability to work with the private sector, his constant advocacy of bringing industry to the Tennessee Valley, and his close relationship with Wagner, he had risen quickly to division head by 1968. At Foster's death in 1979, Wagner referred to him as "a grand, warm human being," and Foster was that. But he was also a tough boss whose tenacity, when it came to Tellico, matched Wagner's own. In choosing Foster to coordinate the project, Wagner further demonstrated his determination to push the project through to completion, no matter how strong the internal grumbling and opposition might grow.[24]

Wagner moved decisively in other ways as well. To shore up the steadily diminishing benefit-cost ratio, he ordered a study of the feasibility of putting power generating facilities at Tellico in addition to the canal to carry water to the generators at Fort Loudoun Dam. Undoubtedly he reasoned that additional power benefits could be included in the ratio, while the Office of Power would assume a portion of the construction costs. The idea was a desperate one, ultimately impractical.

To mollify local politicians who supported the Tellico Project but who also questioned TVA's constancy, Wagner cleverly included them in meetings and conferences while keeping the real power in TVA hands. Too, the chairman hoped to neutralize the Eastern Band of the Cherokee Indians,[25] whom Dickey, Burch, Douglas, and others had tried to manipulate into opposing the dam, by cooperating with the more conservative among them to develop the reservation's resources. If aroused, Wagner and others at TVA reasoned, the Cherokees could become a real problem for TVA. As Evans strongly implied to Mike Foster, cooperation with the Eastern Cherokees would reap tremendous public relations benefits for TVA.[26]

Wagner's greatest trial, however, was the challenge of the environmentalists, a challenge which pained him personally since he considered himself an environmentalist, too. Soon after Douglas's attack on TVA in True Magazine, letters from environmentalist-conservationist groups began to pour into Washington insisting that the Tellico Project be stopped, since, not having submitted an Environmental Impact Statement, TVA was in violation of NEPA. John Franson of the National Audubon Society had brought that point to the attention of Congressmen Richard Ottinger and John Dingell, both instrumental in the passage of NEPA and already unfriendly to Tellico. Ottinger passed the word along to Russell Train of the Council on Environmental Quality (CEQ) in July 1970. Train in turn contacted TVA, asking whether TVA intended to file an Environmental Impact Statement. Dingell also wrote directly to Wagner, informing the TVA chairman that he considered the statement mandatory and asking whether TVA had filed one with the CEQ.

In late July and early August, Wagner wrote personally to Dingell, Ottinger, and Train. While TVA supported NEPA and had "developed procedures for compliance," he claimed, TVA had no intentions of filing an Environmental Impact Statement for Tellico. He argued that since the Tellico Project had been initiated prior to the 1969 passage of NEPA, the agency was not required to file one. Undoubtedly many lawyers would have agreed with Wagner's interpretation. But in light of the national concern for the environment, the preservation of endangered species, and pure food and water, Wagner's reply seemed that of an arrogant businessman saying "to hell with the environment."

Although in truth Wagner was sensitive to many environmentalist-conservationist concerns, it is little wonder that environmentalists like John Franson saw Wagner and TVA as the enemy. "If the project is well-conceived," wrote a disgusted Franson to Ottinger, "as Mr. Wagner says it is, then the TVA should have no objection to filing . . . [an environmental impact] report."[27]

Even though his response had been controlled and logical, privately Wagner seethed. In his determination to push ahead with Tellico, he saw the environmentalists as unreasonable obstructionists, wealthy suburbanites who wanted their sports and recreation facilities protected even if it meant forfeiting economic growth and job opportunities for others. In typical Wagner fashion, the chairman stubbornly repeated his refusal to submit an Environmental Impact Statement for Tellico, even after it became clear that the refusal was damaging TVA's prestige both inside and outside the agency. Even when he gave in on this point a year later, Wagner did so grudgingly, doggedly maintaining that TVA had reached a suitable balance between environmental protection and economic growth. Needless to say, few conservationists and environmentalists took him seriously.

Unlike previous groups that had opposed TVA projects, the environmentalist opposition did not buckle and cave in. Rather, it gained in strength, fed by the power of the national environmentalist movement and the growing suspicion, held even by many liberals, that government had become bloated, impenetrable, and impervious to the public will. Without doubt part of this frustration with government was triggered by other national dilemmas: the bewilderment of liberal whites over what was to them a rather unpleasant turn in the civil rights crusade (Black Power, Black Panthers, Malcolm X, urban riots, explusion of liberal whites from "the Movement"); the agony of Vietnam, in which the government appeared arrogant and insulated against all reason and protests; dying cities; growing poverty; the glut of college graduates; the inflating economy. In short, TVA and the Tellico Project made convenient and highly vulnerable scapegoats which many seized upon in their frustration and anger.

This is not to say, as some within TVA did, that the environmentalists who attacked Tellico were insincere. Admittedly, for a few, "environmentalism" may have been a mask for other motives. But Tellico certainly provided many people with a "safety valve" for their rage over other issues. Unable to change the course of events in one area, many people shifted their ire to what they considered to be a "winnable" alternative: bringing down the Tellico Project.

To Wagner and to many within TVA, the most infuriating thing about the environmentalists was their unwillingness to compromise, to sit down and "reason together," to play the age-old game of give and take. As with many social reformers of the late 1960s and early 1970s, theirs was an all-or-nothing battle, a conflict in which compromise was seen as surrender and "reasoning together" a form of prostitution. Hence, each time TVA boasted about the number of jobs that would be created or the number of young people who would not have to leave the area to find work, the environmentalists dug their heels in deeper. That they were using the same tactics as TVA (media manipulation, appeals to Congress, tarring their opponents unfairly as plutocrats and destroyers of the environment) mattered not at all to them. After all, *their* cause was pure.

Within TVA itself, there were those who surreptitiously supplied the Tellico opposition with information not intended for public release. Ever since the Tellico Project's revival in the mid-1960s, some within the agency had thought the whole affair poorly conceived, ill-timed, and unnecessary. Some were committed environmentalists and so shared common ground with some of TVA's most implacable foes. Clearly some of these TVA employees passed information to the opposition, which rendered that opposition all the more dangerous to the TVA leaders.

Thus, between 1969 and 1971 the opposition gained strength. Not used to losing battles, these people, collecting themselves under the umbrella of the environmentalist movement, were ready to take on TVA in a struggle from which only one side could emerge triumphant.

Douglas's article poked the environmentalist hornet's nest. Almost from the moment of its appearance, anti-Tellico letters began to flood into Washington.[28] Environmentalists refused to let Douglas's accusations be forgotten. A widely circulated newspaper column by Ralph de Toledano essentially repeated the *True Magazine* charges. Moreover, well-known naturalist and conservationist George Laycock, in his book *The Diligent Destroyers* (1970), charged that TVA was "moving into the real estate business," an obvious reference to TVA's buying and taking land at Tellico for eventual resale. That charge hit home with many conservatives who joined with environmentalists to form a strange but highly effective opposition.[29]

Once unleashed, the attacks on TVA and Tellico came from all directions. In April 1971 State Senator Fred Berry introduced a series of resolutions aimed at stopping the project. Berry, a Knoxville funeral home director, was hardly the most progressive Tennessee legislator (a resolution had once been introduced in the state legislature to name him "official state fossil"), but as a conservationist and a political conservative he was irritated by the Tellico Project. And Berry recognized a genuine grassroots political movement when he saw one. Of the three resolutions he introduced, perhaps the most interesting was the one which sought to amend the 1969 Tennessee Scenic River Act to include the Little Tennessee River, a move which would have stopped TVA's construction. John Hart, a local television executive and president of the Greater Knoxville Chamber of Commerce, was livid and fired off numerous letters attacking Berry and his resolutions. Although successful in the state senate, Berry's efforts never got further than the House of Representatives.[30]

Momentum clearly was with the opposition. To keep it going, UT botany professor Ed Clebsch and Oak Ridge economist Duane Chapman organized the float trip on May 9 to highlight the river's natural beauty. Billed as a "Save the Little T Float Protest," the event received the active support of the Tennessee Citizens for Wilderness Planning and broad media coverage. The event certainly kept the pressure on TVA, maintained regional and national attention on the project, and helped to weld the fragmented opposition together.[31]

Within TVA many were stung, confused, and angry. Why should the YMCA Indian Guides oppose them? The local chapter of the Daughters of the American Revolution? The Knox County Young Republicans? The Knoxville Men's Garden Club?[32] TVA employees were being harrassed by their neighbors, and in June 1971 Director Frank Smith reported to General Manager Lynn Seeber that some teenagers had thrown eggs at his door. Friendships were strained as the Tellico issue intruded itself into local clubs, churches, neighborhoods, and Sunday School groups. TVA had always attracted criticism, but this kind was disquieting and painful. For those at TVA who privately had doubts about Tellico but who publicly maintained the agency's "solid front," this unwelcome turn of events was truly agonizing.[32]

To reverse the momentum and quiet the rising opposition, an Environmental Impact Statement was crucial, and both Paul Evans and Seeber knew it. After all, they reasoned, why was a financial war chest being put together if not to sue TVA for violating NEPA? And what were the agency's chances in a national atmosphere dominated by "ecology freaks"? Even though TVA consistently had maintained that, since the Tellico Project was begun in 1967, it was not bound by the 1969 federal statute, Seeber and others recognized that TVA would have to reverse itself, would have to come up with an Environmental Impact Statement, and would have to do it fast.[33]

Actually, all the time that Seeber was maintaining publicly that TVA would not file an Environmental Impact Statement, a small group at the agency was working frantically to put one together. By May 21 a draft statement was ready for internal review, and on June 18 the statement was mailed to various federal and state agencies for their reactions. Even then, TVA stubbornly continued to maintain that no such statement was required and that the agency had created it voluntarily. But clearly the reason for filing an Environmental Impact Statement on Tellico was the anticipation of a lawsuit. With that fear uppermost in their minds, TVA staff worked overtime to make the Seeber-imposed June 18 deadline.[34]

While waiting for reactions to the Environmental Impact Statement, TVA sought to deal with the rising crescendo of protest. News releases on the "model city" of Timberlake proliferated, as did agency claims that Tellico would create 25,000 new jobs but would not damage fishing. The TVA figures on the benefit-cost ratio mysteriously climbed to 3:1 by June 1971, though an internal memo the previous January had set the ratio at 1.3:1. To offset opposition letters, the Information Office sought "to generate letters to Knoxville editors. . . . We should talk with each and emphasize [the] importance of counter letters." Pro-dam people received material from the Information Office to assist them in their efforts. Indeed, the public relations arm of TVA was moving into high gear.[35]

But each of the agency's news releases brought forth a storm of protest greater than the one before. Each of its public relations ploys was widely and openly mocked. Each of its thrusts and parries was met with a better maneuver from the opposition. While not completely routed, TVA's public defense of Tellico was clearly deteriorating.

In August 1971, the Environmental Defense Fund filed suit in federal court to stop the project. And while the U.S. District Court declined to issue a temporary restraining order, it was evident that that decision would be appealed and that a long round of legal battles was in store, battles which threatened to bring to the surface some unwelcome facts about the project. Moreover, initial reactions to the agency's Environmental Impact Statement had been adverse, as both state and federal agencies found much in it to criticize. More organizations and influential individuals were aligning themselves with the opposition, including the Southeastern Indian Antiquities Association and state parks naturalist Mack Prichard, who was to become a leading foe. More serious in the short run (since it could affect impending lawsuits) was the August announcement that a University of Tennessee economics professor, Keith Phillips, had assigned as a class project a scrutiny of Tellico's benefit-cost ratio and, it was rumored, was about to blow those calculations to dust. If the benefit-cost ratio was destroyed, some reasoned, the Tellico Project would simply collapse in a puddle of false claims and bloated rhetoric.[36]

Keith Phillips was a professor who didn't mind controversy. Indeed, some would say he courted it. A native of the Rocky Mountains, he enjoyed puncturing easterners' pretensions by referring to their beloved Smokies as "foothills." Mountain climber, outdoorsman, and a regular fixture at the university gym, Phillips was a fierce competitor who loved to win battles, whether physical, verbal, or scholarly. Moreover, he was an avowedly conservative economist, a libertarian, and an admirer of Milton Friedman in a profession which traditionally had tipped its academic hat to John Maynard Keynes and John Kenneth Galbraith. "I've always been outspoken," said Phillips in classic understatement.[37]

Phillips' suspicions about the Tellico Project were first aroused by his examination of the benefit-cost ratio of the earlier TVA project at Melton Hill. "I had done background reading on Melton Hill," Phillips remembered, "and it was pretty clear that the benefits TVA claimed there simply had not materialized." Meanwhile, several factors were drawing Phillips to Tellico. His libertarian economic views had given him a "healthy suspicion" of government-instituted planning projects. Too, Phillips was familiar with the Little Tennessee River, having fished the lower end for trout. Although he himself had never belonged to any organized group which fought the Tellico Dam, his personal friends included David Dickey and David Etnier, a fellow outdoorsman and avowed enemy of the Tellico Project. "We had discussed the possible implications of a rare fish or something being found down there," Phillips recalled.[38]

Hence, when Phillips was looking for a project for a class of four graduate and seven undergraduate students in summer 1971, an analysis of Tellico's benefit-cost ratio almost naturally came to mind. The class met weekly throughout the summer. Each student was assigned one part of TVA's benefit-cost study for Tellico and was required to do documentary research at TVA. Other projects, both TVA and non-TVA, were selected by Phillips for comparative purposes. The graduate students supervised the work of the undergraduates, with Phillips himself acting as a coordinator who pulled together the disparate research and provided suggestions rather than pushing students to reach certain conclusions.

According to Phillips, TVA was anything but cooperative. Although some individuals within the agency privately expressed their reservations about the Tellico Project and "did help students find things," in general TVA "made it difficult for us to find data [and] made it difficult for us to withdraw material from the TVA Library, often claiming it was 'checked out'." With little doubt as to what Phillips and his students were doing (one of the students was a close friend and neighbor of Wagner's), the agency clearly was not happy about this unwelcome turn of events.[39]

Although the students largely wrote the paper, the class's analysis of Tellico's benefit-cost ratio was informally dubbed the "Phillips Report." To those who had followed the tortuous path of TVA's quest for an acceptable benefit-cost ratio, the result could hardly have been surprising; the Phillips Report was a point-by-point attack on one of the Tellico Project's most vulnerable points. To begin with, the report accused TVA of using a discount rate for benefits that was "substantially and unrealistically below market rates of interest." In other words, the students claimed that any public project that could not yield as much return as if that money had been invested in other ways was a submarginal project and "should be abandoned."[40]

This argument easily could have been countered and overridden, but the report went on to assert that TVA's claimed benefits for Tellico were exaggerated in nearly every instance. On shoreline development, the Phillips group questioned whether TVA could count on selling Tellico shoreline at the agency's estimated prices, given the amount of available shoreline property that already existed. Moreover, the report strongly implied that this benefit was an exercise in double-counting, since other land-related redevelopment benefits were claimed by TVA elsewhere. On recreation benefits, the report countered TVA's assertions by stating that the general area was "saturated with similar recreational facilities," that the visitors-per-day figure projected by TVA was too high, and that the visitors expected would be ones drained off from other reservoirs. On TVA's pet notion of industrial development, Phillips and his students argued that the agency's claims were badly bloated. On and on the Phillips Report went, filling 106 pages with potentially damaging claims concerning TVA's negligence, overoptimism, naivete, and purposeful deception.[41]

According to TVA, the agency was never sent a copy of the completed Phillips Report, though it did obtain one. Phillips, however, mailed copies of the report to approximately forty individuals, among them Senator Howard Baker, Congressman John Duncan, and Justice Douglas. Undoubtedly the intent was to stop the Tellico Project. The *Knoxville Journal*, a longtime critic of TVA and foe of Tellico, broke the news of the report in a way which made those unfamiliar with the issue believe that the points made in the Phillips Report were unquestioned facts. Also, Phillips himself became much sought after as a speaker to opposition groups, saying later, "I thought as a citizen I owed society more information."[42]

The agency could not permit the Phillips Report to go unanswered.

TVA staffer Marguarite McGloghlin was assigned to investigate Phillips himself, looking especially for evidence of his earlier opposition to Tellico and his links to anti-Tellico groups. The investigation turned up nothing. Responding to public comments in the press that the Phillips Report was unbiased, an angry Mike Foster wrote to Flash Gray, "Let's draft the idea of an editorial asking, what unbias? And see if we can interest one of the area newspapers in running it." A friendly editor in Madisonville agreed to use the piece.[43]

More to the point was TVA's efforts to counter the claims made in the Phillips Report. Indeed, an entire volume of its final Environmental Impact Statement was devoted to that purpose. The agency maintained that the Phillips Report was full of "errors and distortions," that the figures were misleading, that the report was based on economic theories long rejected, and that if similar reports had been issued at the times of the purchases of the Louisiana Territory, Alaska, or the Panama Canal Zone, those enormous benefits never would have been realized.[44]

Clearly the pressure of opposition was beginning to tell on TVA. Some within the agency became prickly and short-tempered. Director Frank Smith wrote to Environmental Defense Fund (EDF) board member Lee C. White, warning him that continued opposition to change and progress should be stopped, lest the EDF "becomes too firmly established as a branch of the John Birch Society."[45] Commenting on an attack by Duane Chapman, a long-time opponent who had recently been appointed to a position at the New York State College of Agriculture at Cornell, Paul Evans wrote to Seeber, "I would derive a great deal of pleasure by responding to this letter with the suggestion that if the independent research opportunity which prompted his move to New York is as faulty as his research at Tellico, he is being grossly overpaid." Wisely, that comment was never mailed. But in an uncharacteristically unguarded moment, Wagner had publicly referred to Tellico opponents as having "rocks in their heads." Seeber had attempted to silence Mack Prichard by complaining to Tennessee Commissioner of Conservation William Jenkins that Prichard was abusing his position as state naturalist by speaking against Tellico "in his official capacity." As the trial date for the EDF lawsuit neared, the agency possibly used its influence to discourage some opponents from testifying. One such person may have been Sam Venable, the outdoors editor for the *Knoxville News-Sentinel*. Reporting on the possible witnesses, J. Porter Taylor informed Foster that "We understand his [Venable's] employer will not allow him to testify against the project." Where Taylor got his information is not clear, for Venable angrily denies that his editor prevented him from testifying. What is evident, however, is that TVA had talked with someone at the newspaper about the "Venable problem." As Venable puts it: "I was told I *would not* column on Tellico." Queried as to whether or not that meant the *News-Sentinel* was unwilling to criticize the project, Venable stated: "Their policy was to *promote* the project." While the above developments are not in themselves damning, together they give a picture of an agency responding to intense pressure by "circling the wagons" and lashing out at attackers.[46]

Obscured by the noise of battle was the fact that the project was continuing relatively smoothly, if slowly. Approximately two-thirds of the land had been acquired by the end of 1971, with a minimum number of condemnations. The concrete portion of the dam was completed, and the earthen part and the canal were well under way. A telephone poll conducted by a state legislator had revealed that support for the project in the Tellico area remained high, even among affected landowners (of course, that was where the local opposition centered, too). Influential area businessmen, bankers, and political figures were still warmly in favor. With some notable exceptions, the *Knoxville Journal* being the most prominent, the media seemed to be on TVA's side. Presumably referring to the EDF, the *Maryville-Alcoa Times* complained, in not atypical fashion: "Who gave New York groups the right to fight Tennessee growth?"[47]

Moreover, as a federal agency TVA had at its disposal resources not available to its opponents. For example, one of the early condemnation proceedings was against Harold DeLozier, a Maryville dentist who owned a cattle farm within the Tellico taking line. Judge Robert Taylor of the U.S. District Court had ruled earlier that disputes between TVA and individual landowners would be settled by a three-person commission, from whose decisions there could be no appeal. Hence when DeLozier refused to abide by the commission's ruling, TVA began condemnation proceedings, using the DeLozier dispute as a test case for the Tellico Reservoir. DeLozier argued that he and his wife had "realized very substantial earnings from their property" and that the price offered by TVA was insufficient. The agency, however, requested and was able to secure from the Internal Revenue Service copies of DeLozier's federal income tax returns for 1966 through 1970, which cast doubt on DeLozier's claims. In sum, while national environmental opposition was strong and apparently growing stronger, TVA had access to enormous resources inside and outside the agency to strengthen its bulwarks.[48]

But the environmentalists would not go away. In early December 1971, Tennessee Governor Winfield Dunn publicly came out against the Tellico Project, marking the first time a Tennessee governor had done more than whisper mild criticism of TVA. Formerly a Memphis dentist, Dunn was a Republican who had never held a public office prior to his gubernatorial race in 1970. But against a badly fragmented Democratic party whose standard bearer was a man rumored to have engaged in shady business dealings in a fast-food chicken franchise, Dunn had been able to slip to victory. Although some saw Dunn's attack on Tellico as politically motivated, publicly at least his opposition was on conservation and environmental grounds. Wagner sent a seven-page letter to the governor outlining the many benefits of the Tellico Project, and TVA encouraged a letter-writing campaign to the governor. But Dunn was unmoved.[49]

Governor Dunn's opposition was an important setback for TVA. Coming days before the EDF lawsuit was to be heard, the timing of the announcement could not have been worse. Moreover, Dunn's public stance made it clear to state employees (especially in conservation, game and fish, wildlife, state planning) that they too were free to attack the project as state employees and that the governor would protect them. Dunn's opposition, TVA correctly sensed, was a disaster.

Worse yet was a public statement by the U.S. Department of the Interior less than a week before the lawsuit was scheduled to begin. In a news release, Interior asserted that TVA's Environmental Impact Statement for Tellico was incomplete. The draft statement had been circulating since mid-June, and the final statement, which TVA claimed would incorporate criticisms and suggestions, was not due to be released until February 1972. Yet many who read the department's statement in their local newspapers must have jumped to the conclusion that the *final* report was being criticized. Again, the timing could not have been worse. And TVA's response, which said that "TVA does not regard Interior as final authority on either the content of the environmental statement or the merits of the Tellico Project," appeared arrogant, stubborn, and exceedingly self-serving.[50]

By the time the lawsuit commenced in Birmingham in mid-December, Trout Unlimited and local landowner Thomas Burel Moser, whose property had been condemned, had joined the EDF in its petition for an injunction to stop the Tellico Project on the grounds that no Environmental Impact Statement had been filed. Much of the early part of the proceedings was taken up with TVA's petition for a change of venue from Birmingham to Knoxville. Since TVA had offices in both states, either site would have been proper. But TVA wanted the case to be heard by Judge Robert Taylor in Knoxville, not Judge Clarence Allgood in Birmingham. Taylor was widely known as a friend of TVA, and his prior decisions in agency-related cases had done nothing to change that impression. Change of venue was granted, and both sides readied for a battle scheduled to begin in early January 1972.[51]

In an interview with political scientists Stephen Rechichar and Michael Fitzgerald, former TVA General Counsel Beauchamp Brogan confessed that "TVA was so sure that it [Environmental Impact Statement] was not required because of Tellico's partial completion that it didn't take the threat of a suit too seriously." Perhaps Brogan truly had been ingenuous, and perhaps the agency was so confident of its position that it saw no need to exert itself unduly. Perhaps, also, Judge Taylor's pro-TVA track record had lured the agency's lawyers into a false sense of security. Perhaps by 1971 TVA simply did not take the opposition very seriously, agreeing with State Senator Ray Baird that the opposition was mostly "eggheads from elsewhere ... [who] have never known poverty and ... just like to join 'causes' against time-proven and beneficial 'establishments' such as TVA."[52]

The hearing lasted only two days and was over before TVA realized what had happened. The brunt of the opposition's argument was that TVA had not formally filed an Environmental Impact Statement (the agency was to do so in February) and that the draft statement which TVA had been circulating since June 1971 was woefully inadequate. David Etnier, a University of Tennessee zoologist, testified that two species of fish, the dusttailed darter and the smoky madtom, were in danger of becoming extinct if the dam was completed and the river destroyed by the reservoir. Ed Clebsch, who had led the opposition float trip in May 1971, argued that the reservoir would destroy some rare species of vegetation. Destruction of the trout stream, TVA's dubious benefit-cost ratio and claims that the project would bring jobs, and the flooding of ancient Cherokee sites were all challenged. The best TVA could do was to repeat its standard assertions that the project was not subject to NEPA and that to stop the project now would hurt the watershed and cause massive erosion.[53]

Taylor's decision, then, "astonished" TVA. He issued a limited restraining order which prohibited TVA from continuing work on the earthen sections of the dam. Work on new roads could go on. In addition, Taylor ruled that the Tellico Project was subject to NEPA's provisions. TVA appealed Taylor's decision, but in March 1972 the Sixth Circuit Court of Appeals, with Taylor presiding, upheld the injunction to stop construction on the dam.[54]

Opponents were delighted by the rather unexpected decision.

In a press release, Governor Dunn noted that he was "pleased" with the temporary injunction and saw a good chance that the Tellico Project would be stopped permanently. Dunn also announced that he would push the Little Tennessee River for "scenic status." Other opponents were equally optimistic. News had leaked out that the magazine *Sports Afield*, a prestigious journal for outdoorsmen, was ready to feature the Tellico Project in an upcoming issue, accusing TVA of pushing the Tellico Project for its own purposes and ignoring the interests of conservationists, environmentalists, and sportsmen.[55]

At first TVA was stunned by the decisions. Yet in few places within the agency was there even the slightest suggestion that the project should be scrapped. As Director Don McBride put it, "We have lost round one, but we will win the fight and the Tellico Project will become a reality." McBride's opinion reflected that of the TVA "establishment"—that the federal court decisions were but minor setbacks and that, as usual, TVA would triumph in the end. After all, the injunction was only a temporary one, and TVA could continue with all phases of the project *except* the dam. Roads could be built, the grading and canal work could continue, and property could still be bought or condemned. Indeed, in the twenty-one months following the injunction by the federal court, TVA continued to purchase and condemn land at Tellico, acquiring roughly 7,723 acres and paying out approximately $2.4 million.[56]

While TVA made it clear that it had no intention of abandoning the Tellico Project, opponents saw the federal court decisions as an opportunity to shut the controversial project down for good. Opposition speakers fanned out across the "rubber chicken" circuit, appealing to social, political, and civic clubs to add their voices to the growing cry to stop the project permanently. Keith Phillips briefly returned to the limelight, and various members of the Association for the Preservation of the Little Tennessee River were much in evidence. Especially active was the association's president, Kenneth Elrod, who was actively campaigning for the spot on the TVA Board of Directors to be vacated by the retiring Frank Smith. But the most sought-after opposition speaker was David Etnier, whose testimony in the lawsuit had given a new dimension to the anti-Tellico argument. Etnier repeated his charges that filling the Tellico reservoir would destroy a number of already endangered species. In an address to the university's chapter of the Young American for Freedom, Etnier expanded his accusation by stating that if TVA's Environmental Impact Statement were complete and totally factual, then there would be no Tellico Dam.[57]

If the project was to be stopped permanently, the opposition realized, three things were going to have to happen. First, Phillips's and Etnier's assertions had to be followed up, to keep regional and national awareness high. Second, the amassed evidence and reports would be worthless unless they resulted in increased political pressure both at the state level, to bring state government into opposition, and at the national level, to cut off congressional appropriations. Finally, the Cherokee Indians would have to become involved so as to create a truly emotional issue of national scope. Correctly sensing that TVA eventually would be able to overcome the temporary injunction, the opposition set its sights on these three objectives.

Republican Governor Winfield Dunn had already gone on record as opposing the project, had shown ill-disguised pleasure at Taylor's temporary injunction, and was moving to block the Tellico Dam permanently by placing the Little Tennessee on the state's list of scenic rivers. In turn, he had been named "Conservationist of the Year" by the Cherokee chapter of Trout Unlimited. Moreover, Dunn's stand had brought other Republicans into opposition, most notably State Senator Fred Berry, whose resolutions against Tellico had preceded the governor's own statements; State Representative Victor Ashe (scion of a wealthy Knoxville family, Yale graduate, and a young man who many believed had a bright political future); and freshman U.S. Senator William Brock. Although the low-keyed Brock had claimed to be neutral when it came to Tellico, in March 1972 he urged the Environmental Protection Agency (EPA) to make a thorough analysis of Tellico's Environmental Impact Statement and left no doubt that, in his opinion, TVA had not adequately answered the questions put to the agency by the dam's opponents.[58]

In spite of the fact that its governor and both of its U.S. senators were Republicans, Tennessee nevertheless was still considered a Democratic state. Although voters did not register by party, the majority of the state's voters, except in East Tennessee, were Democrats, as were both houses of the state legislature. Therefore, the opposition reasoned, some anti-Tellico support from Democratic politicians would be useful. For that reason, Tellico opponents were delighted when Congressman Ray Blanton, campaigning for his party's nomination to oppose U.S. Senator Howard Baker, began making anti-dam statements. During a speech at the University of Tennessee Law School, Blanton remarked that Tellico may well be "a bureaucratic blunder." Although the congressman had cagily added the almost impenetrable caveat that neither extremist view on the project "is the answer for economic progress," clearly the damage to TVA and Tellico had been done. And though Blanton was anything but steadfast in his opposition to the project, at the time he was seen as another example of Tellico's deteriorating political position at both state and national levels.[59]

To keep pressure on the politicians, the opposition organized a nationwide petition and letter-writing campaign. In this they were aided by the March 1972 appearance of *Sports Afield*'s anti-dam article, "This River Is Waiting To Die."[60] Titled almost exactly like Douglas's *True Magazine* piece, the article triggered letters from sportsmen to President Nixon.

Petitions signed by over seven thousand people virtually inundated Nixon's Deputy Special Assistant Roland Elliott. Almost every state was represented, with especially heavy returns from Pennsylvania, Ohio, Illinois, Florida, New York, Michigan, and Tennessee. An analysis of the zip codes of the Knoxville, Tennessee, signees shows that a high proportion were from the upper-middle class suburbs of West Knoxville and from neighborhoods near the University of Tennessee campus.[61]

The early 1970s was the "golden age" for political pollsters. With the increasingly widespread access to computers, polling became something of a mania, with public opinion samples being taken on nearly every conceivable issue and non-issue. So it was almost natural that the opposition, TVA, and others would want to use them to bolster their respective positions on the Tellico controversy. At least a half-dozen polls were taken, some shockingly unscientific and some comparatively judicious. The results varied rather badly. In early 1972 Congressman LaMar Baker took a straw poll in his district which indicated that 37 percent of the respondents opposed the Tellico Project, 36 percent favored it, and an astounding 26 percent had no opinion. To counteract that survey, TVA released the results of its own poll, which showed that 70 percent favored the dam, 14 percent were opposed and 16 percent had no opinion. Were that not enough, the Boeing Corporation hired Charles W. Roll, Jr., a former academician who had founded his own polling firm, to do a careful and "scientific" survey. Roll reported "overwhelming" support for the project in the Tellico area. While Roll's report was probably the most accurate, the mass confusion surrounding the conflicting poll results allowed anyone to quote one of the polls to support his or her opinion.[62]

The opposition still needed an emotional issue in order to keep national attention focused on the Tellico Project. Petitions were only temporarily effective; opinion polls were confusing and contradictory; economic studies and debates over the benefit-cost ratio often were exceedingly complicated and dry; local property owners were badly divided; politicians were notoriously inconstant. Opponents realized full well that TVA was preparing a massive effort to get the temporary injunction lifted. Conservation, environmental, and sportsmen's groups remained firm, but nationally many other campaigns threatened to diffuse their efforts. A new front was needed in the battle against Tellico, an emotionally charged issue which would continue to focus national attention on the project and simultaneously be one that politicians would be embarrassed to oppose. That issue was presented by the Cherokee Indians.

The Tellico Reservoir would inundate several long-abandoned village sites of the Overhill Cherokees, places which some Cherokees viewed as sacred. Moreover, although TVA-sponsored archaeological digs had been going on in the Tellico region since the late 1960s, many recognized that filling the reservoir would mean the loss of valuable archaeological and historical data which the Cherokees recognized as part of their heritage. In the early 1970s there had been a phenomenal increase in concern (one might also say guilt) over the plight of the American Indians. The American Indian Movement commanded enormous media attention; Dee Brown's best-selling *Bury My Heart at Wounded Knee* (1970) and Thomas Berger's novel and later film *Little Big Man* (novel 1964, film 1971) became the rage among white Americans. Tortured by their ancestors' successes and excesses, many whites of the middle and upper-middle classes founded, supported, and often led movements to restore to America's Indians their heritage, their land, and their self-esteem.[63]

The problem for the opposition was that the Cherokees did not show much interest in getting involved in the Tellico controversy. In February 1972 the Cherokee Nation[64] in Oklahoma had protested TVA's plans for inundating important historical sites, including the birthplace of the revered Sequoyah. Though someone at TVA wrote "Ho! Ho!" in the margin of a xeroxed copy of that newspaper story, the agency responded quickly, inviting representatives of the Oklahoma Cherokees to Tellico to tour the site and to receive assurances that the archaeological work was being carried out in the most respectful manner possible. In spite of the fact that Chief W. W. Keeler, who would head the delegation, was the chairman of the Phillips Petroleum Company, the Oklahoma Cherokees replied that their limited financial resources would not permit them to make the trip. TVA paid their way, gave them a VIP tour and presentation, and then recommended that the Cherokee Nation "should not become involved in any way in the current controversies over the future development of the Little Tennessee River basin." Less than a month later the Oklahoma Cherokees announced that they would not join the opposition.[65]

In order to cement relations with the Oklahoma Cherokees, TVA wanted to hire one of their number as a consultant on Tellico and the proposed "model city" of Timberlake. After a good deal of searching, the agency chose Col. Martin Hagerstrand. A retired military man, Hagerstrand was managing the Cherokee Cultural Center in Oklahoma, had a good deal of administrative experience, and was "reasonable" on Tellico. Hagerstrand himself was not an Indian, but he had married a Cherokee woman, had often demonstrated his loyalty and value to his wife's people, and was respected and well-liked by important figures in the Cherokee Nation. He was hired. Over the next year, Hagerstrand was a valuable asset to TVA, solidifying the alliance between the agency and the Oklahoma Cherokees. He reported to TVA on the opposition of the North Carolina Cherokees (the "Eastern Band") and on one occasion recommended to TVA's general manager that Indians *not* be included in the planning for Timberlake.[66]

Given the Oklahoma Cherokees' passivity, Tellico opponents hoped that the Eastern Band would provide more fertile ground. Yet when they surveyed their chances, they couldn't be overly optimistic. Although only a couple of hours' driving time from Tellico, the North Carolina Cherokees had shown no interest in those "sacred sites" for years. Many residents of the Little Tennessee Valley had never seen a Cherokee visit the sites until TVA brought the Oklahoma delegation in April 1972. Moreover, the Eastern Band's tribal council traditionally had been very conservative, preferring to secure things quietly from the Bureau of Indian Affairs and save what political clout it had for crises closer to home. Too, TVA had worked with the Eastern Band for a number of years and thought it had amassed a good number of political credits that it could call in relatively at will. Indeed, the Eastern Band had seemed content to stand by while Cherokee, North Carolina, turned into one of the more garish tourist meccas in the East. The group appeared comparatively stoic in the face of high seasonal unemployment, growing alcoholism, problems with young people, and the denigrating if profitable tourist trade.

Still, some signs gave the Tellico opposition hope. The very problems mentioned above had caused a certain amount of restiveness among some of the Eastern Band. Those people, though a minority among the Cherokees, had been more receptive to the aggressive, confrontational style of the American Indian Movement. Winning some seats on the tribal council, they created a rift within that body that conservative Principal Chief Noah Powell found impossible to ignore and difficult to deal with.[67]

There is little doubt that members of the Association for the Preservation of the Little Tennessee River tried to bring the Eastern Band into the opposition. Their most telling point was their accusation that University of Tennessee archaeologists were "desecrating" the Cherokees' sacred burial places. This, many felt, was the issue that would stop the Tellico Project forever—if only the Cherokees could be brought into the fold.[68]

The archaeological problems were an almost continual nightmare for TVA. In 1967, Beverly Burbage of TVA's Office of General Counsel, an archaeology buff, unofficially informed University of Tennessee archaeologist Alfred "Ted" Guthe that TVA would begin filling the Tellico reservoir in eighteen or nineteen months and that "we had better get with it." Guthe, a rather low-key and self-effacing professor, for years had quietly presided over the university's small Anthropology Department and its modest McClung Museum. But Guthe recognized a good thing when he saw it. The son of an archaeologist (he had gone on his first dig while still in his early teens), Guthe had, in effect, spent a lifetime preparing himself for this project.[69]

Guthe had several contacts in the National Parks Service, so naturally he approached that organization first for financial support. The NPS was enthusiastic, providing Guthe with some money, and he began initial field work in summer 1967. But many within TVA were also genuinely concerned about archaeology (later one of Wagner's sons worked on a UT dig at Tellico), and the agency began adding financial support in 1968–69. Ultimately TVA would spend in excess of four million dollars on excavations at Tellico. By 1971 over 500,000 fragments of human remains and artifacts had been taken from two sites.[70]

To protect itself against anticipated criticisms, TVA suggested that Guthe give Cherokee young people summer employment as diggers. The youths were to receive room, board, and diggers' wages. "It was a political move," Guthe remembered, "and a damn good one." To make the contacts with Cherokee young people and act as one of the field supervisors, TVA recommended that Guthe hire Worth Greene, a one-time medical student and amateur archaeologist who claimed to be part Indian (one critic said, "The story is that he dyed his hair black"). Guthe was dubious of Greene's qualifications and said so. As Guthe remembered it, TVA responded, "We don't care whether Greene does a goddam thing or not as long as he keeps contact with the Cherokees." What Guthe did not know was that the self-styled "Dr." Greene was reporting to Corydon Bell of TVA on the internal politics and feuds of the Eastern Band. Ultimately, Greene's lack of archaeological expertise (UT's Jefferson Chapman classified his knowledge in that area as "demonstrably poor") forced Guthe to remove him.[71]

But the excavations themselves posed real problems. To begin with, Burbage's 1967 warning that the reservoir would be filled in eighteen or nineteen months struck archaeologists as "heartbreaking." Some sites had not even been located; some of the identified sites had not yet been acquired by TVA; and some landowners, such as Dr. Troy Bagwell, on whose property the Toqua site was known to be located, refused to allow any digging while they still owned the land. To some archaeologists, eighteen or nineteen months seemed just about the amount of time needed to put an archaeological team together. Just as bad, because of the Tellico Project's uncertain completion date, financial backing was always given on a year-by-year basis, which made long-range planning virtually impossible.

Although Guthe was normally a careful and deliberate scholar, obviously he had been called upon to preside over a "rush job." The task was to get as much out of the ground as possible before the water started pouring in. As an older "salvage archaeologist," Guthe was less concerned with collecting everything from a site to recreate an entire "cultural envelope," the favored technique of archaeology's "new breed," than with establishing chronological ordering of the material. For these and other reasons, the digging often was hurried, incomplete, and not fully documented. Burdened with other teaching and administrative duties, Guthe appeared only occasionally at the sites. Moreover, security at the sites was loose, and, according to Guthe, TVA never wanted to prosecute pilferers and "weekend hunters" because that would create a "bad image." In addition, Tellico opponents who visited the "digs" claimed (their assertions were much disputed by both Guthe and TVA) that the littering of soft drink cans, gum and candy wrappers, and other material was widespread. Littering was a trivial point to some, but to many Cherokees, it was a symbol of TVA's and UT's lack of concern for their sacred sites.[72]

Tellico opponents were quick to use what they called the "insensitive desecration" of Cherokee sites to appeal to the Eastern Band. Even conservative Cherokees grew concerned. In August 1972 the tribe requested a meeting with Governor Dunn to get the digging stopped. But Dunn, while already opposed to the dam and sympathetic to the Cherokee entreaties, could do nothing. Hence, with the more strident faction using the issue to increase its strength and with disturbing reports being brought in from the sites themselves, the Eastern Band Tribal Council, in a move both statesmanlike and political, went on record as formally opposing the Tellico Project. In October, after the National Congress of American Indians passed similar resolutions against the Tellico Project, CBS News began to awaken to the fact that there was a good story brewing.[73]

Tellico opponents, though elated, were not naive enough to believe that the Cherokees could not be "co-opted" by TVA. Indeed, at that very moment TVA's Mike Foster, in response to suggestions from Paul Evans and Flash Gray, was pulling together "a chronology of the assistance TVA had given to the Eastern Band." Obviously worried, Mark Harrison of the Association for the Preservation of the Little Tennessee River wrote to Principal Chief Noah Powell, urging Powell not to compromise with TVA and to postpone any meeting with Wagner "until after the November election," in which, Harrison obviously hoped, a Republican landslide would spell the end of Tellico appropriations. As a further incentive, Harrison informed Powell that the APLTR favored land for the Cherokees in the Tellico area.[74]

But Powell was temperamentally a compromiser. Sixty-four years old in 1972, he had worked in a Civilian Conservation Corps camp in the 1930s, driven a bus for the Bureau of Indian Affairs, served in the Army during World War II, and been on the tribal council since 1940. By 1972 he was the undisputed head of the conservative faction of the Eastern Band. Harrison and the Tellico opponents might plead and promise until they were blue in the face, but Powell had an almost unerring instinct for the realities of power. And, to Powell, power lay with the Tennessee Valley Authority.[75]

Probably the same day that Powell received Harrison's letter, the principal chief requested a meeting with Wagner, at the same time expressing genuine dismay over the archaeological work in progress. Evidently the meeting was a success from TVA's standpoint, for Powell later announced that "he personally and most of the members of the tribal council considered the [archaeological] work . . . as important and necessary to the understanding of the Cherokee heritage [and] . . . could see no reason for not continuing the work." Indeed, Powell even suggested that the Eastern Band itself might contribute to the financial backing of the archaeological project. Old and ill, Powell confessed to Worth Greene, who passed it along to TVA, "that he just did not have the strength to fight many battles anymore." With Powell and a majority of the tribal council apparently mollified, some within TVA must have felt that they were "out of the woods" on the whole Cherokee matter.[76]

But neither Powell, TVA, nor the opposition anticipated the appearance in the Cherokee debate of one of the most intriguing characters in the whole Tellico history. In mid-1972 a man who called himself Hawk Littlejohn began to attract widespread attention as a leading Cherokee opponent of Tellico. According to his "official" biography, Littlejohn had been born on the North Carolina reservation (officially called the "Qualla Boundary") in 1941 and had left in his early teens to "roam," taking jobs as a busboy and in a circus. Returning to the reservation in 1971, he had married a non-Indian woman who worked as a nurse at a Knoxville hospital and had gotten deeply involved in the anti-Tellico movement.[77]

TVA later disputed most of Littlejohn's "official" biography. For reasons that are not entirely clear (Paul Evans says that TVA's security people "overreacted"), TVA requested that the Federal Bureau of Investigation do a thorough investigation of Littlejohn. Although results of the investigation have never been made public (the FBI will not confirm or deny that such an investigation ever took place), they were leaked to almost everyone connected with the Tellico Project. Corydon Bell told Guthe; Paul Evans knew, probably from Wagner; Jeff Chapman heard the rumor, as did many, many others.

According to rumor, the FBI claimed that Littlejohn was not an Indian at all, that he had been born in Akron, Ohio; had not set foot on the Qualla reservation until 1971; and was "as phony as a three-dollar bill." Moreover, according to TVA files, Tennessee Highway Patrolman John Jones, who had had a bizarre conversation with Littlejohn, reported that Littlejohn had made thinly-veiled threats of violence against the dam and TVA people.[78]

Whatever Littlejohn was or was not, he had a talent for attracting publicity and focusing it on the Tellico Project. Therefore, briefly, he was the darling of the opposition and perhaps their last hope to use the Cherokees to gain national attention. In August 1972, Littlejohn made contact with Griscom Morgan, son of TVA's first director Arthur E. Morgan. The senior Morgan still lived in Antioch, Ohio, where he continued to rail against the "treachery" of Lilienthal and Harcourt Morgan nearly forty years before. Although still alert, he was ninety-eight years old, infirm, deaf as a stone (he kept an electrical "hearing aid" in the center of his floor and literally shouted back the answers to questions), and dependent on son Griscom and a private secretary.

At the time, Griscom was taken up with the idea of starting a moccasin factory to cure the foot and back problems of all America. So perhaps it was only natural that he and Littlejohn hit it off so well. Together they drafted a letter for the senior Morgan to sign, excoriating TVA for the Tellico Project and recommending that the project be "shelved." Littlejohn then deftly released copies of the letter to the press (the *Knoxville Journal* actually received its copy before TVA did). Privately Wagner was furious, referring to Morgan as an "old man." But outwardly the chairman was calm, sending a reasonable reply to Morgan. Morgan immediately apologized to Wagner and later retracted his anti-Tellico statements.[79]

Meanwhile, Noah Powell undercut Littlejohn's influence on the Qualla reservation. Worth Greene reported to Bell that Littlejohn "is no longer welcome in Cherokee and Noah Powell refutes any statements made by him." But by this time Littlejohn was virtually unstoppable. In early 1973 he and Fran Mashburn, a part-time student at the University of Tennessee who also claimed to be part Cherokee, presented the university with two trash bags of litter they asserted had been collected at the UT archaeological sites, implying that the excavations were being carried out in a slipshod manner lacking in respect for the sacred areas. Littlejohn himself showed up at UT's McClung Museum when some of the artifacts were put on display. University security forces were ready for anything, including violence, but Littlejohn showed up nearly alone (the media were never far from him) and offered dramatic but silent prayer before the artifacts. In February 1973 Littlejohn and Mashburn appeared on a local radio show. Mashburn was at her best, charging that Powell and the Eastern Band council were under the thumb of the Bureau of Indian Affairs. Littlejohn could easily top that rhetoric, claiming that those who supported the Tellico Dam were "guilty of genocide." Guthe's careful appearance on the program the following week to defend his archaeological work was, to say the least, an anticlimax. Little wonder the Memphis Commercial Appeal called Littlejohn "the most militant figure in the movement to stop TVA from building the Tellico Dam."[80]

Also little wonder that many Tellico opponents, while privately suspicious of Littlejohn, were delighted by the results. The Washington Post had run an article very critical of TVA over the whole Cherokee business, and other news organizations began to pick up the cry. In February 1973 Playboy Magazine published a letter to the editor which took the Cherokees' side. Not many at TVA shared the tongue-in-cheek attitude of Bevan Brown who, in a memo to Bell, exclaimed, "Corydon, We have hit the big time. We now have our knockers in Playboy! Bev."[81]

But just as Littlejohn's ascent had been swift, so was his fall. Although he was still in the area as late as 1974, he was no longer vocal, was completely without influence, and was almost universally ignored by the media and by the opposition. Perhaps Powell's actions against him within the Qualla Boundary eventually had their desired effect. Perhaps TVA's campaign to discredit him—the staff consistently referred to him as "the phony Indian"—gradually caught up with him. Perhaps he had to stop to mend the broken fences of his personal life. Perhaps he had just run out of tricks to amuse the fickle media. Or perhaps he had grown too frantic, too wild for his white defenders to stomach. Charismatic figure or media-made charlatan, messiah or fraud, Hawk Littlejohn passed from the scene almost as quickly and as mysteriously as he had come.

While busy beating down the brushfires of the opposition, TVA was simultaneously waging an all-out campaign to regain its momentum and get the Tellico Project started once again. Aided by TVA, local pro-dam boosters held rallies and circulated petitions to offset similar efforts by the opposition. The agency attempted to discredit the Phillips Report while at the same time bringing to Tellico three "expert" archaeologists to defend the quality of Guthe's work. In his typical self-effacing way, Guthe himself later admitted that TVA "loaded the deck" by bringing in three people the agency knew would find no fault with the digs. Wagner himself took to the stump against the environmentalists, arguing for a "reasonable balance" between the natural environment and the economic environment. Once the "reasonable balance" concept had occurred to him, the chairman had used it at every conceivable opportunity, backing his argument with plenty of statistics on projected jobs and economic growth. Following Wagner's lead, speakers fanned out across the land, carrying Wagner's message. The only snafu came in September 1972, when General Manager Lynn Seeber was a guest on a prerecorded Chattanooga television talk show sponsored by the city's Jaycees. One of the questioners got the Tellico and Duck River Projects confused, and, in trying to straighten things out, Seeber had made something of a mess of it. Reasonably, Seeber asked that both the question and the answer be deleted from the final tape. The Jaycees refused, offering to insert Seeber's answer which would be retaped. Seeber, back in Knoxville, demurred. Unfortunately Bevan Brown was not present to comment on the exchange.[82]

The March 1972 article in *Sports Afield* had been something of a setback, but TVA countered by asking its many contacts to send supporting letters to Washington to offset the flood of critical mail heading toward President Nixon. Director Don McBride called on his connections in Oklahoma with this blunt appeal:

> Help!!
> I need your help in reference to the Tellico Project that TVA had under construction, but which has been stopped by a court order on false environmental grounds. We have already expended over $40 million on this $69 million project. The preservationists and the "way out" wilderness buffs, including such as Trout Unlimited and Sierra Club, have launched a letter-writing campaign to the President. *We must counteract this effort.* You have my word that this is a good project. . . . Do me a personal favor and write Mr. Nixon. I am enclosing two fact sheets. . . . Time is running out.[83]

And the favorable mail did come, from development associations, electrical co-ops, chambers of commerce.

At the same time, the agency sought to build an impenetrable wall between itself and its enemies. Responding to a request by the University of Tennessee School of Architecture for maps to do a summer school project on Tellico, TVA, smelling another Phillips Report, refused because it judged that the request "was based on a different motive than mere academic study." That incident made the *Knoxville Journal*, but TVA stood firm.[84]

By the middle of 1973, TVA had withstood the worst of the national opposition, and the Tellico Project had not been irreparably damaged. True, construction had been stopped for the previous eighteen months, but other work had gone on, and property acquisition had continued. Nixon's presidential victory had not destroyed Tellico. Furthermore, the stubborn Dunn was nearing the end of his term as Tennessee's governor and, according to the state constitution, could not succeed himself. Although the Environmental Impact Statement had been badly mauled by the opposition as well as by a number of state and federal agencies, at least it was still intact, and TVA believed that the chances of having the temporary injunction lifted were good. Indeed, a 1972 U.S. District Court decision in Arkansas (*EDF v. Corps of Engineers*) had ruled that "The Court does not believe that the Congress intended that the National Environmental Policy Act be used as a vehicle for the continual delay and postponement of legislative and executive decisions." Obviously TVA expected that that decision would be helpful to Tellico. Admittedly, the opposition forces were still numerous, talented, and eager. But they lacked leadership and a good emotion-laden issue to arouse the populace and engulf the politicians. The "Cherokee problem" had been deftly neutralized, and Littlejohn was virtually out of the picture. Moreover, Jonathan Ed Taylor, a Cherokee who led the opposition to the dam inside the Qualla Boundary, for all his intelligence and tenacity, was hardly the equal of the headline-grabbing Littlejohn. In addition, a March 1973 flood which damaged parts of Chattanooga provided TVA with another arrow for its quiver—the claim that the Tellico Dam would increase the agency's flood control capacity. Indeed, while most at TVA were cautious, a bit of optimism returned to the beleaguered agency. From Justice Douglas to local property owners to economists to environmentalists to Hawk Littlejohn—TVA had met them all.[85]

And then, on August 12, 1973, zoologist David Etnier, who had testified against Tellico in the EDF suit and spoken to various groups on environmental issues, was snorkeling in the Little Tennessee River near Coytee Springs, about seven miles upriver from the Tellico Dam. Almost miraculously (so miraculously that many later were suspicious), Etnier spotted and scooped up a three-inch perch-like fish, a creature he had never seen before. Instantly Etnier realized the significance of his catch. For when a bystander asked him, "Wha'cha got?", Etnier immediately—and ebulliently—replied, "Mister, this is the fish that will stop Tellico Dam."[86]

7

THE MODEL CITY AS MICROCOSM:
THE TIMBERLAKE IMBROGLIO

In late June 1970 a group of top government officials, demographers, and urban planners, responding to a call by Daniel P. Moynihan, counselor to the President, attended a two-day closed meeting at the White House to discuss American population growth and the nation's cities. As later reported by *U.S. News and World Report*, President Richard Nixon "himself attended for about two hours and took an active part in the discussion."[1]

To all who attended the high-level meeting, it was clear that Nixon's view had changed considerably concerning federal support for construction of new "model" cities to alleviate demographic, economic, and social pressures on the nation's troubled megalopolises. Whereas the President had been enthusiastic about this "start from scratch" concept as late as the previous January, by June 1970 the Nixon administration had grown skeptical. As *U.S. News and World Report* commented, "There will be no vast program of building 'new towns' in America, if this administration has its way. A few new towns, mostly on the outskirts of the big cities, will probably get federal help on an experimental basis."[2]

In Knoxville, TVA officials read the report of the White House conference with consternation. Since early 1965, TVA had been planning the development of a model city, Timberlake, on the shores of the reservoir to be created by the Tellico Dam. But funding for this model city was dependent on large grants and guarantees from the Department of Housing and Urban Development (HUD), the Federal Housing Administration (FHA), and the Appalachian Regional Commission, as well as on the cooperation of a private developer. If the Nixon administration was calling for an end to the New Town movement, support from those federal agencies very likely would evaporate and the private developer would flee. Indeed, by mid-1970 the prospects for Timberlake looked bleak indeed.

In an effort to turn collective gloom into optimism, TVA's C. W. Blowers dashed off a quick memo to TVA regional planner "Flash" Gray. Responding to the *U.S. News and World Report* article, Blowers took the position that "Timberlake fits the 'new' policy as it could readily qualify as an experimental effort which is in an advanced stage of planning. It meets the satellite requirement [being approximately 40 miles from Knoxville, Tennessee, population 180,000], would need only modest funding and no new legislation."[3]

In fact Blowers was wrong on almost all counts. Despite TVA's insistence to the contrary, Timberlake was unlikely to be a bold national demonstration of community planning. Plans must flow from an overall project concept, a vision which gives both legitimacy and direction to the detailed plans themselves. The overall concept must reflect a general philosophy regarding the population for whom the planning is being done. Should the plans seek to alter individual and group behavior or reflect already-existing behavioral patterns? This question above all must be addressed and resolved before planning begins.

TVA never had either a planning philosophy or an overall concept for its model city. So each division within the agency developed its own concept of Timberlake. And, rather than impose one unifying version from the top, TVA's Board of Directors, headed by Aubrey Wagner, allowed Timberlake planning to embody *all* of these competing visions. The model city was to be a community to house workers in the industries TVA hoped to attract to the Tellico area, a commuting suburb of Knoxville, a second-home recreation community, a retirement center. It was to be multi-class and upper-middle-class. In short, Timberlake was to be all things to all people, a city in search of its own definition. As such, it could never be, as Blowers claimed, an "experimental effort."

In one sense the Timberlake concept was as old as TVA itself. While the agency was still in its infancy, Arthur Morgan and President Franklin D. Roosevelt had "agreed on the need for some kind of model community which would be more than just a construction camp." Those within TVA who shared Morgan's and Roosevelt's vision centered their attention on Norris, the temporary facility which housed those working on the Norris Dam. With the land already in hand, TVA believed that after the dam was finished the Norris site could be made into a "small-scale model for future [town] planning" in the Tennessee Valley.[4]

But that dream never took hold, not in Norris or anywhere else in the valley. By spring 1935, cost overruns made the town cost twice as much as had been originally budgeted. And as we have seen, Arthur Morgan's larger vision of TVA's role was ultimately defeated by his codirectors Harcourt Morgan and David Lilienthal. After World War II, when TVA sold much of its nonreservoir property, the town of Norris was sold at public auction to a Philadelphia developer. By the time the "model city" concept resurfaced in Timberlake, Norris had become a well-to-do suburb for current and former TVA officials.[5]

TVA staffers who embraced a broader mission for the agency readily admitted that Norris had been a failure. But in their eyes it was not the "model city" concept that had led to failure, but TVA's unwillingness to commit itself totally to the project. Thus the concept lay just below the surface at TVA, waiting for a propitious time to reemerge.

Credit for the revival of the "model city" concept can be given to Aubrey Wagner himself. As one former TVA planner recalled, "The whole thing was 'Red' Wagner's baby. It was his idea right from the start." Indeed, the notion of a "model city" on the shores of the reservoir fit in with all of the chairman's dreams: it could bring together the various concepts of mission that had been circulating throughout the agency since its founding; it could get the Tellico Project off the ground and thereby give new force to Wagner's "more dams" idea; it could save a benefit-cost ratio badly in need of salvation; it could make TVA once again the hero of the valley. Thus, while Wagner's intentions were a far cry from Arthur Morgan's goals at Norris, the "model city" form appeared to be the same.[6]

Early in 1965 TVA planner "Flash" Gray wrote a memorandum to Lawrence Durisch of the agency's governmental relations staff. "The possibility of developing a new town as a supplement of the Tellico project has been mentioned by Mr. Wagner and other staff people," Gray explained. Though Gray refused to commit himself to the idea, Durisch was more excited, enthusiastically noting to General Manager Van Mol even before he received Gray's memo that a planned town might have real possibilities as an Appalachian showcase project. In turn, Van Mol pressed Durisch: "Would you recommend that TVA undertake such a project?" Durisch replied in the affirmative.[7]

The decision to go forward with the model city at Tellico had, in effect, been made. Wagner had broached the idea, and Gray, Durisch, Van Mol, and others had followed it up. However, when Congressman Joe L. Evins shifted funds from Tellico to Tims Ford, TVA thought it politically wise to put the scheme on the back burner. Reporting to his staff, Van Mol explained that the board was "agreeable" to the idea of a planned new town at Tellico "but wants to defer until initial appropriation for Tellico is available."[8]

As TVA waited for a more propitious time to introduce its model city concept, national trends shifted to favor such a notion. American cities were in crisis, facing rapid suburbanization and "white flight," deteriorating tax bases, poverty, crime, and civil disorders. In such an atmosphere, new cities looked increasingly attractive. Moreover, in an effort to diversify their holdings and get in on what promised to be a very profitable trend in real estate development, some large corporations were beginning to consider "model city" development projects. Gulf Oil, Sunset International Petroleum, Humble Oil and Refining, Union Oil, and Goodyear Tire and Rubber were but a few of the mammoth companies attracted to the new town movement. In 1966 Gerald L. Phillipe of General Electric testified before a government operations subcommittee that GE was in the process of selecting sites for a number of planned cities.[9]

Almost immediately after President Lyndon Johnson's 1967 budget appeared in early 1966, Gray fired off a memo to Van Mol urging the general manager to act on the Timberlake Project as quickly as possible and to hire a private consultant to begin the planning. Apparently it had been decided that TVA would build the model city in cooperation with a private developer, largely because massive government appropriations to TVA for such a venture probably would not be forthcoming and also because, as a government agency, TVA could not apply for government grants such as those being awarded by HUD. Almost as soon as General Electric announced its plans, TVA's Peter Stern wrote excitedly to Van Mol, "Board should be interested. What a fine partnership it would be if we could lure GE's electrical city into the Tennessee Valley. Shouldn't we contact?" Directors Frank Smith and Don McBride agreed. Wagner met with GE representatives, who later assured the chairman that the corporation was "actively thinking about it." Smith tried to hire a GE employee as a consultant, obviously hoping to get some kind of commitment. But that commitment never came. Instead GE's G. T. Bogard told Wagner, "From a market growth standpoint it is difficult to identify a source of population or job expansion which will permit large-scale development in the rather short period of time of twenty years." Not only did GE decline the opportunity to develop a planned city at Tellico but also called the whole justification for the entire Tellico Project into serious question.[10]

Other early efforts to attract private developers met with similar lack of success. Frank Smith was the principal booster, writing to nearly every private corporation which might be interested in such an undertaking. He invited officers of Reynolds Aluminum on a TVA-sponsored trip to the area and tried to get David Rockefeller Associates to explore the project. Neither group was interested. Chief of Power George Wessenauer, never enthusiastic about Tellico, was asked to "mention" the new town plans to his friends and contacts in New York; Wagner himself made a formal presentation to Lehman Brothers at a luncheon meeting in New York City. But all responses proved disappointing. In an uncharacteristic understatement, Frank Smith noted that it "was not as much as I had hoped for."[11]

Responses from private developers cooled some of TVA's naive enthusiasm, but not its ardor for what was then being generally referred to as Tellico New Town. In a memorandum to the general manager, John Rozek criticized efforts to sell Tellico New Town to the private sector as premature: "I don't think a letter like this should be sent until we can demonstrate better an economic base for a proposed new town."[12]

In response to criticisms by private developers that the Tellico New Town site was too small, TVA began investigating the possibility of purchasing additional land for the project, as much as two thousand acres at a cost of one-and-a-half to two million dollars. Second, TVA hired the first of numerous private consultants to assist the agency with its Tellico New Town planning and help TVA keep the model city's costs down so as not to further damage the Tellico Project's benefit-cost ratio.[13]

That first consultant was Carl Feiss of Washington, D.C., an expert in town building and related legislation. At a lengthy meeting between Feiss and TVA officials, Pete Claussen of TVA's Office of Tributary Area Development boldly pointed out that the whole idea was to develop a town "with everybody's money—we want to make sure we can take advantage of anything incorporated in legislation." Feiss rejected out of hand the notion that new towns were a "resolution of central city problems" and made clear his antipathy to any new town which involved large numbers of economically disadvantaged people. If that idea was adopted, he reasoned, the myriad problems of the already-decaying cities would merely be shifted to the new towns. Feiss went on, making known his dislike of HUD legislation which provided lower mortgage interest rates for poor persons in large developments and his hope "that the design [of Tellico New Town] would be one in which the lower income housing was so scattered within the community that it wouldn't be noticed."[14]

At the same time Feiss urged the agency to pursue an aggressive real estate development approach. He clearly felt that TVA had passed its water project stage and should go on to "other goals." Feiss enthusiastically envisioned "a dozen site developments. . . . I wouldn't go at it with a feeling of fear." Moreover, he cautioned TVA against too much interaction with government officials who might make things more difficult.[15]

The Tellico New Town idea was drawing TVA further and further into a web of contradictions. One of the chief questions, raised earlier by G. T. Bogard of GE was who would live in the model city. The obvious answer was: people who worked in the industries to be attracted to the area after the dam was completed. But would industries be attracted if residential communities were not already there? And if housing for people of very modest incomes were not to be provided in the new community, as Feiss had strongly recommended, then who in fact would live there? Would the model city wind up another Norris, an enclave for retired TVA executives? And if that did happen— and there were those at TVA who wanted just that—would public opinion stand for it? If not handled carefully, Tellico New Town might hurt rather than help the Tellico Dam Project.[16]

Warning signs were abundant. Not one of the dozens of private developers contacted had shown much interest in the project. Indeed, GE had openly doubted both TVA's claims and the agency's accelerated timetable for Tellico New Town development. Moreover, plans to purchase additional land and the hiring of private consultants threatened to increase project costs. Finally, the dilemma of whether to begin with industries or housing had never really been addressed.

But TVA was determined to push ahead as rapidly as possible. In late 1967 Robert Gladstone and Associates from Washington, D.C., also experts in new town planning, were engaged to help with plans for housing and to work up a project justification to take to the TVA board. Gladstone's report, delivered on November 1, 1967, envisioned a community of roughly 30,000 people living in 8,500 houses. Construction was projected to begin in 1971, with housing starts gradually increasing over a twenty-five-year period. Like Feiss, Gladstone saw Tellico New Town as a middle- to upper-middle-class community, with some prospective residents commuting to Knoxville. In this way, Gladstone explained, Tellico New Town would be able to "crack the metropolitan market." TVA was delighted with Gladstone's report, and his consultantship was renewed for further "analysis of selected development factors."[17]

At the same time, Feiss was doing his best to prod TVA into taking a more aggressive planning and development role. He was critical of the fact that while the agency's power production had helped to urbanize the Tennessee Valley, TVA had done almost no planning for that urbanization. The result, said Feiss, had been "a slaphappy pattern of growth and development found everywhere else in the United States." Such a pattern, he continued, was strongly at variance with, and certainly far behind, the agency's engineering standards and achievements. "It would be regrettable," Feiss noted, "if the demonstration were to fail, as did Norris, in stimulating and influencing better urbanization and other well-planned new communities throughout the Valley." Feiss then strongly recommended the purchase of additional land to protect "Tellico views" and urged TVA to put pressure on the governments of Loudon and Monroe Counties to establish rigid zoning laws for the surrounding property not in TVA hands. Hence, although he denied it, what Feiss seemed to want was an enclave "walled off" from its surroundings. TVA planners were "very pleased" with Feiss's report and suggested that "it be left as it is."[18]

Encouraged by these reports, TVA put together a task force to carry on with planning for Tellico New Town. James Gober of Regional Studies chaired the group, which was heavily weighted with regional planners. Gray, Rozek, and Peter Stern of Regional Planning were key members, with James Bottorff, Architectural Design Branch; Charles Davidson, Health and Safety; F. D. Stansberry, Civil Design Branch; and Robert Roarks, Landscape Architecture, playing somewhat more secondary roles. Specifically, the task force was to coordinate consultants' reports and initiate more planning studies, investigate the possibility of securing additional federal money for the project, study other new town ventures, and court private developers. As Stern put it, "Tellico New Town will be a flop, no matter how brave the technology or how grand the architect's concept, if private investors don't see a way of making a project with a reasonable risk and within reasonable time."[19]

Cooperation between TVA and other federal agencies was viewed as crucial, both to pump money into the project and to attract a private developer. Hence the task force immediately set to work to woo those agencies. Initial contacts with FHA and HUD were encouraging. FHA was impressed by TVA's unique ability to use eminent domain to acquire large amounts of land and to "bank" such land without interest payments for a long period of time. Since FHA could not provide mortgage insurance to another federal agency, FHA and the task force worked out "a scheme under which TVA would transfer only the land needed for immediate development and supporting utilities to the private developer. . . . [who then could] be eligible for. . . . insured mortgage and could borrow money for infrastructure costs not to exceed a $25 million total." Thus, with FHA's imprimatur, a private developer could be assured of borrowing money at low interest rates for site preparation and for construction. For its part, HUD promised the task force that it could provide grants of up to fifty percent of the cost of putting in a water and sewage system, providing another tremendous saving to TVA and a private developer. Hence, although Gober was less sanguine about federal assistance ("I doubt that we can expect to compete with the Space Program for Tellico New Town funds"), most members of the task force were encouraged by their early forays into the complex world of federal guarantees and grants.[20]

Meanwhile, other members of the task force were hard at work on a fiscal viability study for the new community. Their goal was to secure the most revenue for running the town at the lowest possible tax rate to projected residents. This was because, they openly admitted, "the new town must compete to some extent with Knoxville for house sales." And their calculations, it must be said, were optimistic indeed. By 1995, assuming a Tellico New Town population of 21,120, they estimated that exceedingly low tax rates would still generate a surplus of over $1 million per year. These figures assumed that no schools would have to be built (county schools were expected to take New Town children), that HUD would absorb half of the cost of public utilities, and that TVA would build roads, parks and recreational services.[21]

Despite clear lack of interest on the part of private developers, less-than-firm assurances from FHA and HUD, and what some must have realized was an overly optimistic revenue and spending study, the task force was determined to take its "Tellico New Community: A Proposal" to the board in May 1968. With Wagner and Smith as principal boosters, the board was enthusiastic. Sitting with the directors, General Manager Van Mol was eager to set up Tellico New Town under Section 22 of the original TVA Act (which provided for demonstration projects), largely because it would be funded separately from the Tellico Dam. Joint funding, Van Mol stated, was risky because "it might bring up once again the changing relationships between costs and benefits of the original Tellico dam and reservoir project." Of course, this did not mean that TVA would not claim for the Tellico Project the land enhancement benefits generated by Tellico New Town.[22]

Van Mol also wanted TVA to move more aggressively to attract industry even as Tellico New Town efforts continued. He felt that the commitment of a major industry to locate at Tellico would provide "a captive market for home seekers." On the other hand, the general manager recognized that the new community itself might well attract the industries that TVA economist Mike Foster was trying so hard to lure to the Tellico area. "We would hope," noted Van Mol, "that preparatory land development in the new town would constitute an inducement toward industrial development and would help Mike Foster land his first big fish sooner."[23]

One is struck by the sense of unreality which seems to have permeated these deliberations. No industries had yet been attracted, and there were some people, albeit mainly opponents of the Tellico Project, who doubted if they ever would be. No private developers had expressed any interest in Tellico New Town, and some had questioned TVA's boldly optimistic projections for the new community. The war in Vietnam was siphoning off federal money at an alarming rate, and savage internecine battles within the Democratic Party offered at least the possibility that a revived Republican Party behind the once-eclipsed Richard Nixon would capture the White House, thereby endangering TVA funding. Yet the TVA board was determined to continue, hoping to bring plans for Tellico New Town before Congress in 1970 for inclusion in the 1971 federal budget. In the meantime, planning would continue, access roads to the Tellico New Town site would be built, a detailed budget drawn up, and a private developer sought. In the face of a host of warning signals, TVA chose to press forward.

Ever since the 1930s, when private power companies had fought TVA, there had been within the agency a strong distrust of private business interests. These interests, it was reasoned, put economic gain before community progress and betterment. TVA, many felt, was morally superior to private corporations and therefore had to guard both the valley and itself against those who thought only of profit. True, private enterprise had a place in the Tennessee Valley, but TVA's role had to remain dominant. Hence many within the agency had been uncomfortable with such calculations as benefit-cost ratios, which smacked too much of private-sector manipulation.

Fired by the vision of building a model city yet at the same time suspicious of private developers, some at TVA argued that the agency should go it alone in the New Town experiment. Fearing that private business interests would ruin TVA's idealistic plans for a truly "model" city, Porter Taylor composed a fiery memorandum calling on TVA to assume a bolder role at Tellico New Town. The whole idea, Taylor felt, was not for TVA "to support private enterprise in a demonstration. ... [of] how to build a fancy subdivision." Rather, the agency's role was to lead, "with the support of private enterprise." Exhibiting some of the passion which had motivated the old TVA (Taylor had come to the agency in 1934), Taylor urged TVA not to wait for a private developer but rather to make Tellico New Town "another example of TVA's innovative spirit."[24]

Taylor was not alone. Flash Gray, a key member of the task force, was opposed to turning the project over to a private developer. The time was ripe, Gray insisted, for TVA to obtain federal support, and waiting for a private developer might make it impossible to marshall the kind of support for a project "which would no longer qualify as a demonstration of national significance." Gray's superior, Mike Foster, agreed, explaining that most private interests viewed the Tellico New Town project as too small and too slow in developing to be immediately profitable. On the other hand, even if a private developer were found, Foster disliked the idea that TVA would do most of the work only to have some private business reap all the benefits.[25]

To their credit, many of those who urged TVA to take a dominant role in planning and building Tellico New Town were genuinely concerned that no private developer would provide sufficient housing for low- and moderate-income people. To them, Carl Feiss's vision of a model city for the well-to-do was exactly the wrong approach, a betrayal of TVA's commitment to the region's people. Responding to a gratuitous critique of Tellico New Town plans by Albert Mayer, an urban planner of considerable stature, Wagner himself wrote: "If our people remain here . . . we have a major responsibility to create a satisfying and stimulating living environment for them. . . . Here, uniquely I believe, we plan to start with the industrial base and its employment potential and then build the needed city." To Wagner, a "needed city" was one which provided sufficient housing for all income groups who would find employment in the protected industries on the reservoir.[26]

The commitment was there, but the problem remained a sticky one. Gober went to Phoenix to visit Litchfield Park, a planned community developed by the Goodyear Tire and Rubber Corporation, but returned disappointed. He explained that, while the developer had sold "a number of $40,000 town houses overlooking one of the golf courses," Litchfield "has been unable to attract many of the nearby industrial employees." Principally, Gober noted, this was because Litchfield had had trouble finding someone who would build a $15,000 house, a problem which "will be a major hurdle at Tellico as well."[27]

A solution to that problem was never found. TVA continued to search for ways to make Tellico New Town a model city for all socioeconomic groups, even going so far as to hire a team of sociologists from the University of North Carolina, Chapel Hill, to study how to incorporate blacks into the new community. But gradually idealistic notions gave way to economic realities. Deadlines had to be met, costs were mounting, and, as much as many within TVA regretted it, a private developer had to be found.[28]

By early 1969 TVA was ready to unveil its plans for Tellico New Town to the general public. In January Gober met with the residential subcommittee of the Tellico Regional Planning Council (as well as with representatives from the Tennessee State Planning Commission and the East Tennessee Economic Development District). But the group's reception must have been less than enthusiastic. Although the residential subcommittee felt confident that the council would give the project "its full support," Judge Harvey Sproul of Loudon County said the presentation should be done in stages in order to avoid what he termed the "opening shock which might result if the whole new town project was presented at once."[29]

Gober's report to Mike Foster on the meeting did not go into detail on what the local objections might be. But later TVA presentations, as well as correspondence between the agency and members of the Tellico Regional Planning Council, clarify what they were. To begin with, many local interests had supported the Tellico Project in hopes of stimulating prosperity, real estate development, and increased tax revenues. As noted earlier, area residents who strongly backed TVA were largely those people who wanted the region to change and who hoped that these changes would bring prosperity to themselves and their neighbors. But now, to some, it seemed as if TVA was undercutting that promise by funneling most of the prosperity into Tellico New Town. Local boosters from Vonore, Lenoir City, and other communities hopeful of cashing in on the Tellico Project might now withdraw their valuable support. If that happened, the New Town might actually hurt the Tellico Project rather than help it.

Furthermore, some county officials, including Sproul himself, were openly worried about what additional costs the counties might have to bear in providing services to the new community. Later Sproul complained to TVA that his own county's tax base was "stretched to the breaking point" and feared that the project would "leave a small, rather poor county to cope with the consequences."[30]

TVA did what it could to assuage local boosters' fears. Gober's presentation, which unveiled the project to the press and general public, was carefully screened by several people at TVA, including Director of Information Paul Evans and General Manager Van Mol. In a note attached to the announcement of the February 18, 1969, meeting, Van Mol had written, "Jim Gober in his remarks will make clear that a new town would in no way detract from Lenoir City." And in response to Sproul's and others' worries about taxes and services, TVA proposed a fifty percent subsidy for services such as solid waste disposal, basic road system, shoreline reshaping, industrial site development, recreation facilities, education, health, police and fire protection, and general government and operational expenses. Fears may have been calmed, but the price would be exceedingly high.[31]

Even as TVA was preparing to go before the Bureau of the Budget and Congress to seek funds for the new community, the project was caught in an almost self-destructive spiral largely of the agency's own making. Initial cost estimates for the rather modest venture had been disturbingly high, and many at TVA feared that congressional appropriations would not be forthcoming for so modest an effort. Hence, even though the agency was in the process of purchasing 3,500 additional acres for Tellico New Town, Mike Foster suggested that even more land might be needed so as to increase the scale of the project.[32]

Even more significantly, TVA quickly doubled the town's projected population size, from 25,000 to 50,000. Gladstone had been retained for $21,300 to do a market and economic analysis to see if such a population figure were feasible, and Gober was delighted when Gladstone reported that 50,000 people could be lured to the new city. At the same time, the estimated number of jobs that would be created by industrialization on the reservoir shoreline was jacked up to 10,000. Gober exuberantly told his task force on July 30 that "the 10,000 manufacturing goal is confirmed as supportable. Combined with other new town employment and Bob Gladstone's assumption concerning our ability to penetrate the greater Knoxville housing market, it looks as if a new town population of 50,000 is justified." Virtually overnight, TVA had doubled the size of its "model city."[33]

Having pumped the population of the planned city up to 50,000, TVA and Gladstone began to devise fairly detailed lives for the future inhabitants. The name Tellico New Town, altogether too futuristic and cumbersome, was changed to Timberlake. A larger community would need more of everything: another golf course, a McDonald's Restaurant, a hospital with outpatient nursing facilities, a junior college. Shoreline improvements to enhance the beauty of Timberlake and increase its recreational capacity were discussed, at an esti-mated cost of seven million dollars. Sewer, water, and police and fire protection all would have to be doubled. Possibly anticipating more wails from county officials like Sproul, Gober asked Lee Kribbs of the State Planning Commission, "Lee, do you know of anything that we could be doing that we're not at the local level to get control?"[34]

The TVA task force recognized that revising plans for twice as many people meant considerably more than just "doubling everything." Unlike smaller entities, cities of 50,000 or more people almost always are composed of several communities, clustered together to make up a city. These communities are shaped by ethnic, religious, socio-economic, or other common interests. During the late 1960s, the so-called "neighborhood school," occasionally a euphemism for an all-white school, was a focus of widespread, highly emotional attention. The neighborhood school concept evoked memories of a mythical golden age of youthful innocence, parental control of schools, and walking girls home from school with a stop for a soft drink on the way. To be sure, this image was romantic and outmoded, but for a rapidly urbanizing society with mammoth, seemingly impersonal schools which taught everything from abnormal psychology to sex education, it was powerful indeed.[35]

TVA planners recognized the attraction of the neighborhood school concept. Timberlake was to be divided into cluster communities, each with an elementary school within walking distance. In an interesting exchange, Pete Claussen of OTAD commented that a walk of three-quarters of a mile "would be good for them," to which Gober replied, "That's not going to kill any youngster." Of course, this meant that TVA was going to have to erect a number of elementary schools, the cost of which could be astronomical.[36]

Indeed, as the site of Timberlake and TVA's visions of it expanded, costs mounted precipitously. It was at this point that Chris Weeks of Gladstone and Associates tried to sound a note of warning. Cost per acre was running at $14,000. In November 1969 Weeks wrote, "At present we are faced with a problem equivalent to fitting a hippopotamus into a suitcase. At this end [Gladstone] we can definitely search for a larger piece of luggage, but I also think that we have to find some way to turn our hippopotamus into a lean and graceful antelope." Anticipating what TVA's reaction would be to that choice, Weeks signed his report, "Chris Weeks, senior hippopotamus packer."

TVA's Phil Ericson, assistant general manager and the man in charge of overseeing the Tellico Dam budget, was openly critical of the bloated new town project: "Tellico is a prerequisite to Timberlake. We can, I am afraid, sink Tellico by continuing to saddle it with Timberlake's cost." In other words, the project which was originally designed to enhance the Tellico Dam's benefit-cost ratio actually was threatening to drain strength and money from the Tellico Project.[37]

But TVA people involved in planning Timberlake appear to have been virtually deaf to criticisms and warnings. Like TVA's reaction to those who questioned the Tellico Project itself, the response to criticisms of Timberlake was to attack the critics. Weeks's and Ericson's comments were ignored, but those of Dennis Cody and Michael Byrnes of Bost and Associates, urban design consultants asked by TVA to evaluate the project, were not. Cody and Byrnes called Timberlake a "grim imitation of past errors" based on the "out-of-date" ideas of urban planner Clarence A. Perry. Gray was furious, writing in the margin of Best and Associates' report, "How righteous can you get? These 'urban designers' have finally discovered Clarence Perry vintage 1930." Under Gray's outburst another TVA official wrote, "I believe they are afraid of losing a job." And TVA Chief Architect Withers Adkins was even harsher: "It is almost inconceivable that this firm could, from a very cursory review of this project, give an accurate criticism. . . . I feel that we should take this criticism as that of what it is—lunch hour snoopers."[38]

The point is not whether Timberlake critics were correct or not, but rather that TVA chose to respond to them either by ignoring them or attacking them and their findings. Sometimes TVA hired consultants to study various aspects of Timberlake and then virtually ignored their reports. Such was the case with the analysis of Professor Richard E. Brown of the College of William and Mary's Department of Government. Responding to Brown's report, TVA's Charles Blowers, who was working with Gladstone, wrote, "Have decided to muddle through without direct examination of your Tellico working papers." Of course, consultants Feiss and Gladstone continued to be heeded within the agency. But Feiss was publicly calling for "a string of TVAs across the country," and Gladstone was working hard within the ground rules TVA had given him.[39]

By the end of 1969, as TVA was preparing to take its case for Timberlake to Congress, some estimates of projected total public expenditures for the project were running as high as $145 million. These outlays, the Timberlake task force explained, were "necessary to set the stage for private investment." Yet even then TVA was considering acquiring an additional 9,300 acres to enlarge Timberlake even further. Even some of the stanchest loyalists within TVA were shocked at the bloated Timberlake cost projections and at how much was being claimed for the new community. G. H. Kimmons, a tough-talking engineer who was TVA's manager of engineering design and construction, was aghast: "We would prefer to see qualified, or less positive, wording used throughout the [task force] report. . . . We suggest that the summary clearly point out that cost estimates are very preliminary, order of magnitude type figures and are based on 1969 costs."[40]

As TVA was about to go before Congress to request funds for Timberlake, the whole project was exceedingly vulnerable. Rising costs had gone hand in hand with expanding horizons, until they threatened the Tellico Project itself. Federal assistance from other agencies was conditional at best, and in some cases TVA was ineligible to receive it. The whole scheme rested on finding a private developer, but, in spite of concerted efforts, no progress had been made on that front. Finally, Richard Nixon's 1968 victory over Vice President Hubert Humphrey had signaled a change in the federal government's approach toward the nation's troubled cities. Nixon himself was openly skeptical of the "model cities" movement, characterizing the new cities built at a distance from large cities as "not a good bet." In this the president had been joined by an increasing number of urban planners who confessed that the benefits of the "model cities" movement of the 1960s had been meager. Instead, many planners had turned their attention to the "gentrification" movement, in which the well-to-do were "rediscovering" the central cities as desirable places to live and play. Taken together, these trends did not bode well for Timberlake.[41]

In his appearance before the Senate Appropriations Subcommittee on April 20, 1970, Wagner was confronted by his old nemesis, Senator Ellender. Wagner was asking for $473,000 for Timberlake, with $170,000 of it requested for fiscal 1971. A master of invective, innuendo, and outright bullying when it came to TVA, Ellender was immediately suspicious and began to prod Wagner, hoping to drive the restrained chairman into a corner. In response to Ellender's question of "what was involved," Wagner said that industries would soon be coming to the Tellico region and that without planned development the workers attracted to these industries would create an unwholesome "urban sprawl" in the area. Then, in what was more than a modest fabrication, Wagner testified that the area's community officials themselves wanted a planned town instead. As a sop to the conservative Ellender, Wagner added, "The actual building of the town, of course, would be done with private funds."[42]

But Ellender had Wagner on the spot and was not about to let him go. Did TVA intend to do anything beyond general planning of the new town? Wagner responded that the agency would buy the land; Ellender objected to this. Sensing that Wagner was not telling all he knew, the senator pressed harder:

ELLENDER: You do not contemplate building sidewalks and streets, sewer systems and so forth?
WAGNER: I think that that—the way they would be provided remains to be decided. [Almost all TVA studies had TVA doing these things, if not totally with TVA funds.] . . . You mentioned sewer systems. One of the things we hope to do is to make this town a town without

pollution. Now, I do not know whether by the time we get to building it, by the time it is built, technology . . .

ELLENDER: You said it right. That is in the back of your head. I can tell what you are planning to do by the way you are talking. . . .

WAGNER: It is not in my mind, Senator, it really is not.[43] [Italics added by the authors.]

Ellender said that he hoped TVA would "not go into the building of towns now because that is where you are headed for [sic]." To that Wagner replied, "We built the town of Norris, and we do not want to build another one."[44]

In spite of Ellender's harrassment of Wagner, the Timberlake appropriation did go through, bringing the project back into the spotlight. Probably because the project looked more like a certainty, in July 1970 the Boeing Corporation of Seattle expressed interest in becoming Timberlake's "master developer." Although Boeing had had no experience with planned communities, the aeronautical company was in the process of diversifying and saw the "model cities" movement as having considerable potential for profit with minimum investment and risk. A meeting between Gober and Blowers of TVA and three representatives of Boeing's Civil Systems Division of Huntsville, Alabama, was quickly arranged, after which Boeing loftily told TVA that "we view the Timberlake proposal as an opportunity to participate with an interesting industry-government structure in pursuit of model development having national significance." That was talk to thrill even the flintiest TVA employee.[45]

Yet, even after the Boeing nibble, TVA was willing to consider—and even to invite—proposals from other potential developers. The agency had announced with no little fanfare that it would choose the "master developer" for Timberlake from among a number of bids the agency had hoped to receive. As no one except Boeing had expressed interest, TVA actively began approaching corporations about submitting proposals to be the "master developer" of Timberlake. Among the dozens of corporations contacted were Westinghouse, Anheuser-Busch, North American Rockwell, Walt Disney Productions, U.S. Gypsum, U.S. Steel, Boise Cascade, and Standard Oil. When all this had been done, Boeing still remained the only company interested in the project. Most invitees cited heavy previous commitments or a general lack of interest, although Standard Oil did take time to point out that it had never been a community developer (Seeber's marginal comment was "Oops!"). The "competition" for a "master developer" had fallen flat.[46]

It became clear at the outset that Boeing was not prepared either to rush into the project or to take TVA's word on anything concerning Timberlake. In mid-December 1970, a series of meetings took place between TVA's Gober and F. B. "Red" Williams of Boeing. Boeing's continued interest, Williams said, depended on the industrialization of the Tellico region. Williams tried to press the agency for some firmer assurances of "whether or not industries will locate there." Nearly everyone at TVA was confident they would, and Gober said so. But Mike Foster still had not "land[ed] his first big fish," and no hard prospects were on the horizon. Though Boeing was far from satisfied with TVA's vague responses, more meetings were scheduled.[47]

By February 1971 Boeing was ready to present an eighteen-page outline detailing the "ground rules" on responsibilities for the project. The outline makes it clear that Boeing expected TVA, other federal agencies (mostly HUD), and yet-to-be-created local bodies to shoulder a significant portion of the costs. TVA was to act as a land bank, adjust its purchase boundary to accommodate Timberlake (TVA already had increased the town's size to 23,000 acres), build access highways, construct a multipurpose barge terminal, and secure funding for industrial waste disposal. HUD would finance sewers, water, and schools, and other bodies, through bond issues, would take care of solid waste disposal and recreation development. In addition, Williams told Gober that Boeing would accept a fifteen percent rate of return on the project.[48]

Unaccustomed to corporate negotiations, some at TVA must have blanched at this proposal. But with no other developers waiting in the wings, TVA had little choice but to go along. Boeing and TVA agreed to undertake an eight-month joint study of the Timberlake project. Boeing anticipated that it would spend approximately $200,000 on the study, while TVA, which claimed it had already spent $200,000 on previous work, would commit an additional $60,000. Boeing made it clear that, until the findings of the joint study were known, it was under no obligation to go forward.[49]

About the time Boeing was becoming interested in Timberlake, in August 1971, the EDF obtained its temporary restraining order against TVA's completion of the Tellico Project. Work on Tellico effectively stopped. Consultant Robert Gladstone commented acidly to Gober, "It seems that the waters of this river are spreading far and wide." Gober stiffly replied, "We are not discouraged by these challenges. I feel that we are on sound ground with Tellico and Timberlake."[50]

But if Gober was publicly confident, privately he and others at TVA must have feared the lawsuit's impact on both Boeing and HUD.

Gober quickly touched base with HUD which, according to him, was still interested, especially now that Boeing was in the picture. In fact, TVA seems to have been willing to mislead HUD to keep the federal agency in the picture. As Gober related to Gladstone, "They [HUD] were very receptive to the idea of a 'free standing' new community, although we both know that the greater Knoxville housing market is going to account for half of the Timberlake residents."[51]

Boeing appears to have been anxious both about the EDF lawsuit and about HUD's involvement. The injunction had temporarily frozen all federal funding for Tellico, and Boeing very likely feared that if the injunction were made permanent, government money for Timberlake would evaporate. Too, HUD money, if it came, would have strings attached. The most important of these, in Boeing's eyes, was HUD's requirement that the new community serve a wide variety of income and socioeconomic groups. This would catch Boeing in something of a bind: federal funds were crucial to keep Boeing's investment down and, under these circumstances, the freezing of federal money could be disastrous; yet guidelines accompanying federal funds might restrict Boeing's plans for Timberlake. As Gober reported to Gladstone, "Boeing is not prepared to admit that HUD's support will be essential to Timberlake feasibility." Gober, however, disagreed.[52]

There were clearly strains in the relationship between TVA and Boeing from the very start. Some of TVA's "old hands" were suspicious of all businessmen and continued to dream that TVA could develop Timberlake alone. This group's suspicion that Boeing was interested only in profits and not in the "experimental" nature of the project was confirmed by further meetings with Boeing. For example, on January 12, 1972, the newly-formed Timberlake Advising Board— composed of people from TVA, Boeing, various state agencies, and local government bodies—heard Boeing read a report on commercial recreation from a private consultant the company had hired. The report recommended that Boeing "develop all activities that yield a measurable return on investment while the public sector [TVA and the state] develop other unprofitable activities that are vital to the success of the project." Kristin Kercher of TVA's Regional Planning Staff reported in classic understatement, "This prompted much discussion and controversy."[53]

Boeing did not fully trust TVA, either. While the "joint study" continued, the corporation hired several private consulting firms to study various aspects of the project. Economic Research Associates prepared the commercial recreation study mentioned above. The Fantus Corporation was hired to check the validity of TVA's projections, and its report could hardly inspire Boeing's confidence. Fantus felt that TVA's figures were far too optimistic. The potential for jobs at Timberlake, Fantus argued, was closer to 4,000 than the 6,200 projected by TVA. And the thirty-seven industry types which TVA had identified as "possibles and probables" for moving to the area simply were not relocating at the time.[54]

Boeing had other reasons to wonder whether it had made a mistake getting involved with TVA and Timberlake. Relations between TVA and the State Department of Conservation and the State Planning Office had gone from bad to worse. With Tennessee Governor Winfield Dunn openly opposed to the Tellico Project, longtime opponents in these two state agencies now were free to voice their doubts and criticisms. Determined to get off to a good start with state government, Boeing decided to meet with the Department of Conservation officials with TVA not present—a clear sign that Boeing was beginning to lack confidence in the agency.[55]

By spring 1972 Boeing had begun to feel that it had been misled concerning HUD's commitment to the project, too. In a meeting with HUD, again without TVA present, the corporation discovered that HUD had some real doubts about the Timberlake plan. HUD had come to the conclusion that the Timberlake plan was too preconceived, inflexible, and lacking in the "social planning" component that many at HUD thought was necessary. More serious, Timberlake's projected population was too small, causing HUD to wonder if the project's social impact would be significant enough for HUD to become involved. What Boeing and TVA earlier had thought were assurances were beginning to evaporate.[56]

Boeing had been worried about the Fantus Corporation's report, which questioned TVA's analysis of the region's industrial potential. In an effort to determine whether TVA or Fantus was correct in its projectsion, Boeing hired the Chicago-based Real Estate Research Corporation (RERC) to study the problem. The findings, given to Boeing the month of the HUD meeting, called TVA's calculations further into question. TVA had argued that industries were attracted to sites of water recreation, but RERC disagreed, further noting that TVA had "conveniently omitted" the fact that wages in the greater Knoxville area were higher than the state average. Why, the consultants wondered, would industries move into high wage areas when they could go elsewhere? Pointing out that Timberlake would lie only ten miles from ALCOA and close to Union Carbide at Oak Ridge, RERC flatly stated that industries would not move into an area where they would have to compete for labor with these two giants. The report went on to criticize TVA's assumption that industries would move to areas having low construction costs and argued that the agency's survey of "potential locatees" was a "highly uncertain" business.[57]

It was beginning to seem that if Timberlake were going to develop at all, it would have to be as a community of vacation and retirement homes and as a bedroom suburb for Alcoa, Oak Ridge, and Knoxville. Therefore recreation facilities—and recreation profits—began to loom larger and larger in the deliberations of both Boeing and TVA. But state agencies (especially the Game and Fish Commission and the Department of Conservation), whose cooperation was crucial, openly objected to what they saw as the two giants' arrogant and highhanded tactics. State employees were miffed that Economic Research Associates (ERA) had not consulted them in doing its recreation analysis for Boeing in late 1971. The Department of Conservation's Walter Criley, a longtime Tellico opponent, was angry that ERA had claimed that it received the state's cooperation in putting together its report, "since no such assistance was requested or given." Criley and others questioned the constitutionality of having the State of Tennessee and TVA engage in recreation development so that Boeing could profit from it. If revenue was to be generated, Criley thought it reasonable that the state take its share. In a final blast in May 1972, Criley spoke for many in state government when he pointed out:

> Our joint studies with TVA that led to the current state plan for the development of Fort Loudoun State Park was the result of a long-standing state-TVA cooperative effort in developing parks or reservoirs. If the Tennessee Valley Authority believes that concepts developed to date for public recreation or reservoirs are no longer sufficient, and if TVA is in fact going to accept the concept presented in these [Economic Research Associates'] reports, then I believe it will be necessary that the state and TVA mutually develop new approaches to public recreation on these public reservoirs. It appears that we have been slowly drifting away from some established policies without really facing up to the fact that we are changing concepts.

TVA was paying the price for its "benign neglect" of the state agencies, and to some in state government it must have been sweet revenge to wound the Tellico Project by poking holes in Timberlake. One month later, Boeing drew up a "memorandum of understanding" between the corporation and the Department of Conservation, but the department refused to sign it.[58]

Even though the state was showing a rare independence of TVA, it looked as if recreation in the Timberlake area was going to be big business. That was enough to attract the Eastern Band of Cherokee Indians, who already had made Cherokee, North Carolina, a major tourist attraction and now saw a chance to do the same thing in East Tennessee. TVA was doing everything it could to mollify those Cherokees who opposed the Tellico Project, though their opposition had not been very aggressive to date. Hence the agency was willing to consider a series of proposals written by Martin Hagerstrand and collectively titled "Timberlake Project: Potential Educational, Economic, and Historic Opportunities for Indians."

Hagerstrand offered a number of ideas on how Cherokees could cash in on the recreation bonanza at Timberlake, each more tasteless than the last: an outdoor drama, an Indian-oriented theme park, an "Indian-style" summer camp, and a series of controlled archaeological digs in which amateurs could sate their mania for unearthing Indian treasures. Although C. M. Stephenson of TVA noted that "it would solve the problem of TVA relations with the Cherokee Indians and their non-Indian supporters and spokesmen," one TVA staffer labeled the amateur digs as "a disaster and an outrage" and another called the cost, which the Cherokees wanted TVA to shoulder, "rather staggering."

"The report," fumed TVA's William Wiley, "is somewhat insensitive to Cherokee interests because it does focus heavily on commercial exploitation of the Cherokee heritage." Of course, the Cherokees themselves had already made just such an exploitation in North Carolina, but TVA would have nothing to do with it in Tennessee.[59]

Earlier Jim Gober had written bleakly, "I am beginning to think that the obstacles to new community building are more numerous and of greater magnitude for the public sector than they are for the private." Even so, despite the host of problems and mounting mutual distrust, both Boeing and TVA appeared willing to press forward with Timberlake. On February 15, 1973, Boeing presented its *Proposal for Timberlake*. To sidestep the problems which would occur if TVA transferred the land for Timberlake directly to Boeing, the corporation recommended that TVA and Boeing jointly form the Timberlake Development Corporation, which would receive the land from TVA and then transfer it to Boeing. TVA and Boeing were to share the profits from land sales; grants would have to be forthcoming for utilities, educational facilities, and general development; and "Boeing proposes to accept a reasonable limit on its financial return . . . in exchange for an appropriate amount of risk limitation."[60]

To some at TVA, the Boeing proposal was very nearly an outrage. The agency would be obliged to provide or arrange to have provided roads, water systems, sewers, electrical facilities, open space corridors, shoreline alterations, public recreation areas and facilities, public buildings (including schools and a hospital), and public barge terminal facilities. In addition, TVA was to design performance standards, provide land use plans, and plans for energy and environmental research and be financially responsible, in an oddly vague phrase, "to the extent that private investors or public programs could not finance."[61]

To some it appeared that TVA would do the work and Boeing would take the profits. C. M. Stephenson complained to Gober, "It strikes me that Boeing, with joint control of the Timberlake Development Corporation, but in the position of administrator of TDC activities would be 'more equal' than TVA in the overall operation." Other doubts arose over the method of land transfer. TVA's Dwight Nichols reminded his colleagues that "the cost-benefit [sic] ratio was calculated based on a return to the treasury of money received from the sale of land." How, then, could Boeing be allowed to take some of that money?[62]

TVA was fully aware that the transfer of land to Boeing was potentially the most volatile issue. As William Wiley of the Regional Planning staff put it, "TVA must be very careful about what happens to project lands during the development period, since land purchases and disposal procedures have created a *delicate public relations problem* for the Tellico Project, and this question most likely will continue to arise during the Timberlake development period."

Larry Colaw was even more to the point: "I believe land is the most 'local' issue in the entire project from the standpoint of TVA investment and, therefore, potentially the most emotional. TVA bought the land from the 'locals' and it could well be that in the long run we could be criticized for letting Boeing make a 'big profit' out of the land that 'TVA took away from the locals.' "[63]

Another point of contention was the percentage of profit Boeing demanded. Even though Boeing expected TVA and other federal agencies to put in a combined sixty million dollars, greatly diminishing Boeing's risk, the corporation thought a *minimum* return of 15 percent on its own investment was justifiable, while some TVA people believed a *maximum* of 12 percent was more suitable. Moreover, Boeing wanted to share in revenues from the beginning, whereas TVA wanted to get its money back first. Negotiations of these points dragged on throughout the summer.[64]

With TVA dragging its feet, Boeing issued a kind of ultimatum. In June 1973, O. C. Boileau of Boeing told General Manager Lynn Seeber, "In our proposal of February 15, 1973, we did not place an expiration date on our proposal. Due to project schedule delays now apparent and in order to prevent a lengthy cost accumulation by our staff, we must now place a termination date on our proposal of September 1, 1973."[65]

Indeed, the whole relationship between TVA and Boeing seemed to be going sour. Negotiations continued, and some compromises were hammered out. Boeing did accept a 15 percent maximum rate of return, which some at TVA still thought was too high. But additional problems continued to pop up. Boeing demanded that TVA file a better Environmental Impact Statement, something the agency had consistently resisted doing. Ultimately TVA gave in, as much to save the Boeing deal as anything else.[66]

Meanwhile a curious drama was taking place in both the Tellico area and Nashville. As noted earlier, many of the business interests in the small towns surrounding the Tellico area had supported TVA and the dam, hoping that the project's benefits, especially industrialization, would turn their modest communities into boom towns. Even in the town of Vonore, which would be cut off by the impending reservoir from access to State Highway 72 and to the Louisville and Nashville Railroad, some businessmen supported the Tellico Project, reasoning that the projected benefits would far outweigh these liabilities. But Timberlake might well cost TVA all that support by arrogating all the benefits to itself.

As early as 1967 Walter Lambert, Governor Buford Ellington's director of the Office of Urban and Federal Affairs, had advised TVA to incorporate what was then Tellico New Town as quickly as possible, in order to avoid annexation thrusts by Lenoir City or Vonore. Under Tennessee law, cities and towns could annex unincorporated property without the consent of the affected population. When Boeing came on the scene in the early 1970s, the corporation was especially concerned about this problem and in 1972 had set up a meeting with Governor Winfield Dunn and members of his staff, including Ed Thackston and Niles Schoening, then director of the Tennessee State Planning Commission. "The tenor of the meeting," recalled Schoening, "was how to protect the Timberlake site from local zoning restrictions, annexation, and local [county] regulations." Governor Dunn, who initially had opposed the Tellico Project, was beginning to soften his stance, impressed by the amount of money Boeing's "Red" Williams told him the corporation was going to put into the project and, according to Schoening, by the persistent rumor that TVA was thinking of relocating some of its offices from Knoxville to Timberlake. As Schoening remembers it, the governor "wouldn't turn up his nose at that kind of talk.... Williams was a masterful salesman.... I saw him when he was pressing flesh in Nashville, where he made a great impression."[67]

Dunn immediately gave his staff the go-ahead to create legislation giving Boeing what it wanted. TVA already had drafted the general form of such legislation, and Boeing engaged the law firm of Baker, Worthington, Crossley, Stansberry, and Woolf (Sen. Howard Baker's firm) to work with Dunn's staff and TVA on the specific wording of the bill. Most of the work was done by Roger Kesley of the law firm and Schoening. "I didn't enjoy punching a hole in the state's planning and zoning enabling legislation," recalls Schoening. "But when Boeing came in I got the fever. Even I got the fever." Ultimately the bill, drafted by TVA, a private law firm, and state officials to protect the investments of a private corporation, became the New Community Development Act of 1973. From Boeing's point of view, the chief feature of the act was that new communities were immune from annexation and from county zoning restrictions.[68]

Before the legislation passed, however, some residents of Vonore began to speak angrily about annexing the Timberlake site. To counter this, a worried TVA promised the town that the agency would help Vonore get federal aid, firefighting equipment, and a new municipal park.[69]

But some Vonore citizens were not mollified. Early in 1974 TVA met with town officials, including the mayor, aldermen, and the chairman of the Vonore Planning Commission. TVA was accused of a callous disregard for Vonore, and angry talk rose again about annexing part of the Timberlake site to improve the town's tax base as well as to gain some choice developable land. But TVA reminded Vonore officials that the recently passed New Community Development Act prohibited the annexation of "certified" lands. At this point Carl Rasnic, who was taking minutes of the meeting for TVA, added the notation that "the Vonore people are unaware that the Timberlake project area has not been certified." No TVA person enlightened the gathering, and the Vonore officials' determination to engulf Timberlake simply collapsed.[70]

Later TVA's William Wiley summed up Boeing's position on the whole affair: "Uppermost in the mind of the private developer was the fear of annexation by a local government unsympathetic with the longrange planning goals for the community. A corollary of this concern was the fear that the new town might be incorporated by its [own] residents in a spirit of irritation with the developer and a desire to alter his plans." Boeing and TVA might talk about citizen involvement and planning participation by area residents, but in fact they wanted neither.[71]

But even as TVA and Boeing were sheltering Timberlake with state legislation, the project was rapidly coming unravelled. In February 1974 HUD informed Boeing and TVA that it would not approve any grant applications which Boeing would submit. HUD had twice extended deadlines when Boeing couldn't meet the agency's dates. Now HUD's Albert Trevino indicated that "substantive problems and obstacles" still remained: the parent Tellico Project was embroiled in legal difficulties, a satisfactory Environmental Impact Statement had yet to be filed, and the project was too "dependent on congressional authority for $30 or $40 million in grants to proceed with this development." Reporting to Mike Foster, TVA's Larry Colaw noted disconsolately that "their decision was made *before we arrived.*"[72]

From this point on, it was as if both parties wanted to end the relationship but didn't know how. In late August 1974, Boeing insisted that TVA reimburse it for $149,091 which the corporation claimed it had spent on "evaluation and planning" from mid-1971 to mid-1972. Colaw was openly contemptuous, stating that Boeing had spent most of its time and money reviewing TVA's work, identifying and contracting with "consultants" to perform various services, "buying good will in the Knoxville area," establishing a "Boeing presence" in the Southeast, and educating itself about new communities. Colaw added that in spite of all this, Boeing had not brought much "original thinking" to the project and scoffed at Boeing's claim that it had put 3,000 labor hours into the effort. Colaw recommended that TVA pay one-third of what Boeing wanted.[73]

Eight negotiating sessions were held during the summer and fall of 1974 to resolve differences, but each collapsed amid accusations, arguments, and recriminations. TVA attacked what it considered to be Boeing's inordinate greed, to which Boeing replied that TVA showed a determined "unwillingness to protect us from loss." TVA insisted that Timberlake provide low-income housing; Boeing labeled the goal an "arbitrary housing-type mix." Boeing charged that TVA was not performing on schedule and that the agency was wedded to "a cumbersome land transfer procedure." Anyone attending these heated sessions could see the handwriting on the wall.[74]

Meanwhile the "model cities" movement was collapsing throughout the nation. Eight of the fifteen HUD-sponsored new town projects were near financial ruin, and the rest were in varying degrees of trouble. HUD was pulling out of the whole enterprise, admitting that it had been an enormous failure. Private corporations were looking for other areas in which to diversify their holdings; those who could get out did so.[75]

On March 3, 1975, Boeing formally withdrew from the Timberlake Project, citing "current general economic conditions, the nation's economic and energy priorities, the prospect of . . . [no] federal funding of Timberlake, and the relationship of those factors with our own 1975 investment planning." Boeing did say that it might be interested at a later date, "when funds have been allocated for the Timberlake Project," but almost everyone knew that the breach between Boeing and TVA would never be mended.[76]

Not uncharacteristically, TVA continued to maintain that Timberlake was not dead. Pollyanna Mike Foster claimed that "we're going to have a Timberlake, but the rate of residential and second home building may not be as great as it would have been with Boeing with us." Parroting Foster, Wagner insisted that Boeing's withdrawal "will not eliminate Timberlake, but the initial rate of development will not be as rapid nor the beginning scale as broad as if with their participation." Instead, TVA would "alter the pace of development," concentrating first on recreation and letting industrial development set the pace for residential growth. In what must have been fervent wishful thinking, Wagner publicly stated that Boeing's withdrawal would have "no great impact" on Timberlake.[77]

Wagner and Foster were wrong, of course. For all practical purposes, TVA's "model city" of Timberlake was dead. For over forty years TVA had dreamed of building a completely new city in East Tennessee. Norris had been a crushing disappointment, but the Tellico Project had offered a second chance. But, unlike Norris, Timberlake lacked both goals and ideals. Created as a justification for the parent Tellico Project, Timberlake was inexorably tied to the completion of the Tellico Dam. Cooperation with a private developer had only made matters worse, confirming TVA's preconceived opinion of private enterprise. However, TVA was still unwilling to relinquish its dream of a "model city" once and for all.

8

THE SNAIL DARTER, THE COURTS
AND THE CHANGING OF THE GUARD AT TVA

Not long after he became a TVA director in 1977, David Freeman was
called into Chairman Wagner's office on the twelfth floor of TVA's
East Tower in Knoxville. By the time Freeman arrived, General
Manager Lynn Seeber was already there.

In his patient, grandfatherly way, Wagner laid out the problem:
Freeman's directorial style. In an effort to become more familiar with
the giant agency, the new director had bombarded every division
with requests for information and had shown open irritation when he
found the responses inadequate. Brusque, abrasive memos and verbal dressings-down had angered men and women long used to Wagner's
courteous, easygoing, even courtly style. Worse, Freeman was travel-
ing throughout the Tennessee Valley listening to the opinions of
private citizens, businesspeople, state and local government officials,
and even TVA critics. All of this, Wagner and Seeber explained, had
created a morale problem among TVA people, some of whom had
served the agency for over thirty years.

Freeman's reply was characteristic: "I said, 'Well, then the
problem is *your people*. They've been here too damn long,' and then
I got up and walked out."[1]

The incident in Wagner's office suggests the character of the
metamorphosis that took place in TVA as it neared its fiftieth
anniversary. Although TVA had always remained a major force for
change in the Tennessee Valley, simultaneously it had become an
entrenched conservative establishment, unable to entertain alternatives to the course it had charted for itself and the region. Moreover,
the agency seemed to have lost its ability to listen to "outsiders" or
even to engage in internal dialogue and self-criticism. This extremely
defensive posture had precipitated the rise in power of the Office of
General Counsel, whose advice and opinions were actually dictating

agency policy by the mid-1970s. Although Freeman's opinion that TVA was "moribund" when he arrived in 1977 was overdrawn, clearly TVA had taken on many of the trappings of the stereotypical government bureaucracy, where past successes were endlessly replicated, where life was comfortable, and where "outsiders" and iconoclasts were viewed with suspicion and alarm.[2]

The Tellico Project was both a cause and a reflection of the changes within TVA. The agency had responded to punishing criticism from farmers, sportsmen, environmentalists, and Indians with a "wagons-in-a-circle" mentality. Amid lawsuits, the agency's embarrassing shifts on the snail darter and the Endangered Species Act, heightened scrutiny by national news media and the General Accounting Office, the Sixth Circuit Court of Appeals' injunction halting the Tellico Project, and the growing restlessness of the Eastern Band of the Cherokees, David Freeman arrived at TVA. The occasion reminded one TVA staff member of Daniel Webster's comments on the arrival of Andrew Jackson at the latter's first inauguration as president of the United States: "Gen. J. will be here about 15 Feb—Nobody knows what he will do when he comes. My opinion is that when he comes he will bring a breeze with him. Which way it will blow, I cannot tell. My fear is stronger than my hope." The Tellico opponents' hope was greater than their fear. But they had no way of knowing the road that lay ahead. Neither did Freeman.

Immediately prior to Etnier's discovery of the snail darter in August 1973, the opposition, while large, seemed disorganized and dispirited. True, construction work had been temporarily halted by court order. But few expected the stoppage to last, since it was well known that TVA was preparing to return to Taylor's court in the fall with a beefed-up case.

Opponents were convinced that Tellico's Environmental Impact Statement was a farce. Etnier, an opposition witness in the 1972 lawsuit, had drafted a carefully-worded three-and-one-half-page document in that same year which argued against the dam *on the chance* that some unspecified species would be endangered or destroyed. The Environmental Protection Agency apparently agreed, criticizing the Tellico statement for its lack of sufficient environmental information. Others, among them the Appalachian Regional Commission and the Tennessee Office of Urban and Federal Affairs, warned darkly that TVA's prediction of new jobs might be inflated. The earlier Phillips Report had said essentially the same thing. But TVA had stoutly defended its Environmental Impact Statement (while continuing to maintain that no such statement was required). The agency, in short, seemed to be waiting out its statement's critics.[3]

Another opposition tactic which seemed to be going nowhere was the effort to have the old Cherokee village sites listed on the National Register of Historic Places. This attempt probably was given impetus by longtime Tellico foe Mack Prichard, now Tennessee's state archaeologist. Listing on the National Register would enable opponents to call on support from historic preservation enthusiasts on the national level. An irritated Corydon Bell, assistant to the director of TVA's Water Control Planning Division, wrote, "Mack Pritchard [sic] is given credit for the photographs [in the nomination] and 'his heavy hand' is evident throughout. . . ."[4]

TVA tried to get the nomination "shelved" until after the upcoming court hearing on removal of the injunction. Apparently some officials in the National Parks Service, which supervised the National Register of Historic Places, were agreeable to that, realizing, as Bell put it, "that they and the Register were being used as pawns in the lawsuit." But the Tennessee Historical Commission, the State Department of Conservation, and Governor Dunn's office kept pressure on the National Parks Service until, in September 1973, the Chota and Tanasi sites were added to the National Register. TVA's initial response was to attempt to invalidate the nomination by picking at procedural and technical errors in the nomination form. Then the agency decided to "stonewall." As Bell noted in an October 1973 memo, "We will not conform to [Executive Order] 11593 or [Section] 106 [of Public Law 89-665] at this time." In the meantime, the agency used a new court decision to assert that it had complied fully with the National Historic Preservation Act.[5]

Indeed, wherever the opposition probed, TVA seemed ready. Members of the Eastern Band of Cherokees tried to excite the general public with protests concerning the University of Tennessee's archaeological procedures and its refusal to turn over Indian remains from Tellico that were stored in boxes in the basement of UT's McClung Museum awaiting research and analysis. But TVA trotted out its Board of Archaeological Consultants to praise the high quality of UT's work and brought in a National Parks Service archaeologist who claimed the UT work demonstrated "excellent field technique." To the general public, then, the weight of evidence in favor of TVA was overwhelming.[6]

Nor did the shrinking group of immovable local landowners elicit more than passing interest. In July 1973 an angry Thomas Burel Moser had written to his congressman, Knoxville attorney John Duncan, who supported the project, to accuse TVA of giving favored treatment to its employees who owned land inside the Tellico taking line. Moser named four TVA employees (one of them his brother-in-law) who, he claimed, had been paid unusually high prices for their holdings. Potentially more serious, Moser thought, was the fact that two additional TVA employees had purchased river property in 1964; Moser asserted that they had used prior knowledge of the project to engage in land speculation. These accusations bore the earmarks of a real scandal, and TVA moved quickly to blast them. In a long letter to Duncan, Don McBride angrily denied Moser's charges, and then by implication characterized Moser as a "hardhead," since he was one of only 2.4 percent of property owners whose property had had to be acquired by condemnation. And although rumors of preferential treatment, prior knowledge, and land speculation by TVA employees continued to be circulated in the Little Tennessee Valley, proof was lacking. And without it the news media and most of the general public refused to take the charges seriously.[7]

So confident was the agency that it could overcome its opposition, win its court battle to have the injunction lifted, and resume Tellico construction that it even showed signs of overconfidence. In March 1972, Lamar Alexander, a lawyer originally from Maryville who was making a name for himself as the possible Republican successor to Governor Dunn, wrote a letter to Wagner that began, "Dear Red." Alexander and some Maryville physicians had purchased property for development in the Tellico area which was inside the taking line but would not be inundated by the reservoir. Wouldn't TVA, Alexander requested, be willing to make an exception in this case and redraw the taking line so as to place their property outside it? After consulting with Taylor, Claussen, and Elliot, Foster and Gray were dispatched to Alexander's Nashville law office to explain to Alexander that this was not possible. Alexander and the consortium were obliged to sell their holdings to TVA at virtually no profit. While several meetings took place to discuss this matter, TVA was confident enough of its position to refuse to budge, even for a man who eventually would become governor of Tennessee.[8]

Reflecting TVA's self-confidence, the handling of public requests for information on Tellico changed. Originally the agency had believed that if the general public was provided with all the facts, then public approval of the project would be overwhelming. But this had not proved to be the case, and gradually TVA began to refuse potential critics access to information. Phillips had "difficulty," and the UT Architecture School's requests for information were rebuffed. But the treatment of actual or potential critics had been extremely courteous. Now that the agency was confident of victory, some of that civility ebbed away. When Bill Owen, a former UT student who had run once for the state legislature and was planning his second campaign, used his position at the Oak Ridge National Laboratories to request interviews with TVA people on "Tellico and decision making," he was coldly turned down. In a memo to Seeber, Information Director Paul Evans complained, "Frankly, I am both puzzled and irritated by this seemingly gratuitous involvement of ORNL in the Tellico Project." TVA remained firm, and Owen got nothing.[9]

At last, in October 1973, Judge Robert Taylor ruled that the agency had complied with NEPA, the Environmental Protection Act, and the National Historic Preservation Act. He therefore removed the injunction. Taylor's decision was upheld on appeal, and construction resumed on November 12. TVA was triumphant, its enemies beaten and scattered.[10]

But there was still Etnier's fish, the unusual one he had found near Coytee Springs on August 12, 1973. Etnier and others had long warned of the possibility that the dam might endanger or eradicate some species not yet found or classified. But here, Etnier sensed immediately, was the real thing, a rare species of fish literally right in his hand.

Yet it was some time before frustrated dam opponents realized the significance of Etnier's find. For one thing, the UT zoologist was remarkably tight-lipped about his discovery. Not wanting to make a premature public disclosure until more research had been done on the fish, Etnier shared the information with only a few graduate students. One was Wayne Starnes, who ultimately, with financial assistance from TVA, did his master's thesis on the life cycle and habits of Etnier's fish.

But the discovery could not be kept under wraps for long. Word leaked out to Sam Venable, outdoor writer for the *Knoxville News-Sentinel*, who called Louis Gwin at TVA's Information Office for confirmation. "That was the first time," recalled Gwin, "that TVA had even heard of what was to become the snail darter." Venable ran the story weeks before the national media recognized the significance of Etnier's find.[11]

Opponents were also slow to grasp the potential importance of Etnier's discovery because the Endangered Species Acts of 1966 and 1969, while they were steps in the right direction, had weak enforcement provisions. The 1973 Endangered Species Act, passed four months *after* Etnier's discovery, put enforcement of the act in the hands of the federal judiciary. The 1973 act made it possible for citizens to sue federal agencies which threatened any species listed by the Secretary of the Interior as endangered.[12]

Etnier's discovery, once its import dawned, presented environmentalists and other opponents of the Tellico Dam with a real dilemma. Since the 1960s the opposition had argued that Tellico was simply a bad project. Could they afford to abandon their old arguments and pin their hopes on a fish that had not even been classified, much less listed as endangered? And how many of their supporters would see the fish as cheapening their own arguments and fall away? Finally, would using the fish to stop this expensive, nearly completed federal project expose the Endangered Species Act itself to ridicule and tempt an inconstant Congress to water it down or to gut it entirely? In sum, opponents would have preferred a weapon like a bald eagle or a bear or a buffalo. But what they had was Etnier's fish.[13]

While opponents were considering their less-than-attractive options, a new figure appeared on the battlefield, a person who would galvanize the opposition with the force of his personality, his intelligence, and his near-obsession with bringing the Tellico Project to a stop. In fall 1973, Zygmunt "Zyg" Plater arrived in Knoxville to take up his duties as assistant professor in the UT College of Law. Plater came from a cultured, cosmopolitan, but not especially wealthy background. His father had been a Polish aristocrat and diplomat stranded in the U.S. when World War II broke out, and his mother had been born in England to a moderately well-to-do American family. Born in the U.S. in 1943, Plater had received, partly on scholarship, the best education that America could offer—boarding school, Princeton, Yale, and the University of Michigan. Before coming to Knoxville, in 1973, he had spent three years teaching in Ethiopia. Intelligent, energetic, handsome, Plater was usually at the center of activity, whether at a cocktail party or in a worthy cause. Almost as soon as he arrived in East Tennessee, he became caught up in the Tellico controversy.[14]

In some ways Zyg Plater embodied the upheaval within the American legal profession in the late 1960s and early 1970s. In those years a new breed of young attorney had emerged, advocating the causes of the dispossessed—the poor, the blacks, the "victims" of society—against powerful entrenched interests. Not since the early days of Louis Brandeis and sociological jurisprudence had a group of lawyers argued so vociferously and articulately against "the Establishment" from which many of them had come. And in the late 1960s that "Establishment" seemed to be losing ground. Blamed for growing poverty in the midst of plenty; for the strangulation of America's cities in pollution, congestion, and want; for the "failures" of the civil rights movements; and for the escalating war in Vietnam, the "military-industrial complex" was roundly condemned and their "rules of the game" openly flouted or simply ignored. Brash and irreverent, openly contemptuous of their elders, shrewd and naive, worldly-wise and innocent, these new professionals were playing a new game with new rules.[15]

This is not to say that Plater was a wild-eyed radical. He wasn't. Elders and professional associates describe him as "charming," "forceful," "charismatic," and "a nice guy." TVA's Louis Gwin refers to him as "smooth, very impressive." But some faculty colleagues thought he and his wife were "trendy;" he shocked some law school professors by admitting that he had stopped for some fishing on the way to his job interview; occasionally he wore a tweed coat, turtleneck sweater, and blue jeans to class; he had as many friends among the student body as among the faculty; and once he and his wife (they had no children) took some time off for an around-the-world tour.[16]

Although he had done some consulting work for TVA on land use planning, by early 1974 Plater was a dedicated opponent of the Tellico Project. His legal specialty was environmental law, a relatively new field. He was a devoted fisherman and hiker and loved the outdoors. These factors, as well as his intelligence and forensic talents, made Zyg Plater a natural leader of the opposition. He became deeply and emotionally involved in the Tellico controversy.[17]

Etnier and Plater made a fearsome combination. Etnier himself was not particularly interested in the legal and political ramifications of his study; in fact, he maintained a certain distance from Plater and other hardcore dam opponents. But by October 1974, research on the fish that became known as the snail darter (for its somewhat peculiar diet) had established it as a distinct species and its situation as "endangered." Plater, "excited" as soon as he heard of Etnier's discovery, recalled that he "originally made an attempt to urge its discoverers to name the species something more mediagenic, like 'Tennessee Darter' ('Old Glory Darter'? 'Motherhood Darter'?)." He quickly set the wheels in motion to get the snail darter officially declared endangered and the Tellico Project stopped.[18]

Not only did Plater's magnetism galvanize the dispirited opposition, but it also attracted to the fold a new group of talented people who would be important in the future. Perhaps the most important of these was Peter Alliman. A Knoxvillian since the age of twelve, Alliman had served in the United States Army, then earned his undergraduate degree in history at the University of Denver and returned to Knoxville to attend law school at UT. A hiker and outdoorsman, he was sensitive to environmental concerns. "I was very much a child of the sixties," Alliman said later. When he read a newspaper article on Tellico, he marched into Plater's office at the UT College of Law and offered his services. He was simultaneously seeking a second degree in the UT School of Planning, which made him extremely valuable to Plater. In addition, Alliman was "courting" Sharon Lee, a female law student from the Tellico area, which gave him an access to the community that most of the dam opponents did not have. Though it made for some tense situations (Alliman's future father-in-law was a wealthy real estate broker who owned land in the Tellico area and would surely become rich if the dam went through), Alliman was able to campaign against the dam and still maintain good relations in the area.[19]

Plater also tried to breath new life into those who had been fighting the Tellico Project for years. From the earliest days, Judge Sue Hicks, Alice Milton, and the Fort Loudoun Association had been skeptical of TVA's plan and had been on record as opposing it. TVA genuinely had tried both to mollify the association and to preserve the historic fort. Originally the agency had hoped that a series of dikes could be built around the site, but core drillings showed that such a scheme would not work. Finally TVA proposed raising the entire fort seventeen feet, with fill underneath, to save the historic edifice. The association resisted. Losing patience, TVA's Jack Rountree wrote in an internal TVA memorandum that if the association opposed the "raise and fill" scheme, then TVA should condemn the property and raise the fort anyway in spite of the association. Plater tried valiantly, but TVA's determination took the association out of the picture, though Alice Milton (TVA staff dubbed her "Crazy Alice") would remain a thorn in the agency's side.[20]

Gradually Plater realized that, in spite of all the arguments and tactics he could have used against the Tellico Dam, the snail darter was virtually his only weapon. Though Keith Phillips and others believed that the use of the right of eminent domain to acquire land for resale was unconstitutional, Plater and Alliman realized that the courts had consistently refused to question what a government agency had called the "public good." "That argument," Alliman mused later, "wouldn't have lasted twenty days." Moreover, the Cherokees once again seemed frustratingly uninterested, and benefit-cost arguments had become mired in unprovable claims and counterclaims. If the Tellico Dam was to be stopped, the snail darter would have to do it.[21]

While applications were being drawn up that would place the snail darter on the Endangered Species List, Plater and his allies began drumming up interest in a lawsuit which would prevent TVA from working on the Tellico Project while the petition was being considered. At an October 1974 meeting of opponents at Fort Loudoun, local landowner Asa McCall probably summed up the general feeling when he exclaimed, "We've never heard of this little fish before, but if it can save our farms, our rivers and valley, I say let's try it." The twenty-three dollars that was collected in McCall's hat that evening became the beginning of a litigation fund.[22]

Meanwhile, fearing that TVA was trying to destroy what snail darters there were before the lawsuit by cutting timber and silting the fish's ecologically fragile habitat, Plater and others requested that the agency stop these operations until the issue was settled in court. At the same time, dam opponents tried to reach Tennessee's new governor, Ray Blanton, who had earlier suggested the Tellico Project might be a "bureaucratic blunder," to get him to waive the ninety-day waiting period before a petition to add a species to the Endangered Species List could be published in the *Federal Register*.[23]

Although technically legal, TVA's reaction to this new turn of events was, in the eyes of the opposition, shockingly callous. In a reply to Bruce Foster, who had asked TVA to stop its logging to protect the snail darter, General Manager Lynn Seeber gruffly refused, denying that siltation was hurting the fish and adding that the snail darter was not yet listed as endangered anyway. Meanwhile, work on the Tellico Project actually *increased*. The agency apparently reasoned that the farther along the project was, the less likely the courts would be to stop it. At the same time, TVA Fish and Wildlife staff members were ordered to search the lakes and streams of Tennessee for snail darters and, if none were found, to begin transplanting snail darters into the Hiwassee River. Former Cornell University ichthyologist Dr. Edward S. Raney, who had occasionally done consulting work for TVA, was brought in to offer his opinion that it was "extremely likely" that snail darters would be found elsewhere. He ended his statement with a carefully-worded but hardly subtle reference to Etnier as a "young icthyologist" with limited resources who "postulates on limited evidence" that the snail darter was rare. Seeber took a similar swipe at Etnier in an August 20, 1975, letter to the editor of the *Knoxville News-Sentinel*. All the while TVA was officially referring to the fish as the "so-called snail darter" and continuing to maintain that the Tellico Project was not subject to the 1973 Endangered Species Act.[24]

Even more cynical, thought the opposition, was TVA's use of political leverage to move the project along. Initially opponents had hoped Governor Blanton would be sympathetic. But Blanton's 1972 statement had been made for political reasons. Soon after he became governor, Blanton had received a letter from powerful Tennessee Congressman Joe L. Evins, who had become a warm supporter of

Tellico once TVA got the Tims Ford Project under way. The congressman boldly wrote:

> TVA has asked that I intercede with you and request you as Governor
> of Tennessee to decline and refuse to waive the 90 day waiting period
> for publication in the Federal Register required before any action can
> be taken [on placing the snail darter on the Endangered Species List].
> TVA would like for you as Governor to affirm your support for the
> project and refuse to waive the 90 day waiting period, assuring contin-
> ued construction for at least 3 months.

In a 1981 interview, Walter Criley, director of planning and development for the Tennessee Department of Conservation, remembered:

> Early during the Blanton administration, half a dozen of us were
> called in for a meeting with him. We thought we'd go in there and help
> develop the policy [toward Tellico] but Joe Evins was already there.
> You could tell that he had gotten to Blanton. Blanton then informed us
> that he'd already taken a position, that he accepted TVA's claims and
> would give TVA 100% support.

Meanwhile Congressman John Duncan from Knoxville attacked efforts
by the Department of the Interior (DOI) to list the snail darter as
endangered. In a letter to DOI's Nathaniel Reed, Duncan complained,
"In my opinion this is 'bureaucratic' meddling at its worst. There was
never any intent for this act to be retroactive on projects already
under way or partially completed." In essence Duncan was serving
notice that he would take the lead in helping TVA find a way around
any environmental roadblock. Finally, in September 1975 TVA, according to CBS News, announced that it would keep going on the Tellico
Project even if the snail darter were placed on the Endangered
Species List. How arrogant, opponents asked, could this immensely
powerful government agency become?[25]

Few were surprised when Judge Robert Taylor threw out the
opponents' petition for an injunction to stop construction while the
snail darter was being considered for inclusion on the Endangered
Species List. But this time, spearheaded by the forceful Plater, the
opposition did not fold. Instead, it helped to form the Tennessee
Endangered Species Committee, prepared for more legal battles when
the snail darter was finally listed as endangered, and continued to
harass TVA concerning its logging operations and work speedup. But
until DOI acted, there was little the opponents could do.

In 1974 TVA officials had to address the issue of open board
meetings. Dubious machinations in the Johnson and Nixon administrations had increased public calls for an end to secrecy in government.
Of TVA's division heads, only one or two supported opening board

meetings to the public; General Manager Lynn Seeber was also strongly against it. However, Information Director Paul Evans, while personally opposing open meetings, argued that they were inevitable, that TVA ought voluntarily to move in that direction before the agency was required to do so, and that to resist would make TVA look bad.[26]

At that time the board was composed of Aubrey Wagner, Bill Jenkins, and Don McBride, who was to retire in 1975. Jenkins had operated under the Tennessee "sunshine law" while in the state legislature and "knew that we could live with it." Wagner apparently was not committed to either side. However, McBride was adamantly opposed, largely, according to Evans, because he was hard of hearing, needed comparative quiet to hear what was going on, and feared that open board meetings would be too noisy. Wagner needed McBride's support on the board on other issues and so voted to keep the meetings closed.[27]

In the 1960s TVA had contracted with Alberta and Carson Brewer to write a history of the Little Tennessee Valley, hoping in this way to preserve, at least in print, much that was being lost or destroyed by modernization. Carson Brewer was a popular columnist for the *Knoxville News-Sentinel*, whose work included an entertaining mixture of local history, folklore, unusual local happenings, and whimsy. Both Alberta and Carson were outdoorspeople who loved nature and often wrote pieces favoring its preservation. Hence TVA was not surprised when the Brewers announced that they planned to devote one chapter to the Tellico Dam controversy. In a prospectus submitted to TVA, they wrote: "This chapter would give background of the Tellico project and the controversy over it. It would end with the view of Timberlake, the town of the future, and its impact on the region."

TVA had no objections, and the Brewers began their research. The "final draft," which Alberta Brewer dated July 1, 1971, included such a chapter, titled "Battleground." Although brief, their account of the controversy was balanced, fair, and so uncontroversial that it could irritate almost no one. TVA approved, the Brewers' fee was paid, and the East Tennessee Historical Society was approached about publishing the book.[28]

The officers of the historical society balked, however. Tellico was a highly controversial topic, and, despite the fact that the chapter on Tellico was bland, the society agreed to publish the book only if that chapter were deleted. The Brewers apparently voiced no objection, and *Valley So Wild* added nothing to the controversy, save for rumors, absolutely untrue, that the Brewers had been "bought off" by TVA.[29]

In fall 1975, the tenured members of the UT law faculty met to consider Zyg Plater for tenure.[30] According to some, Plater had been spending so much time on Tellico that he had let some of his duties slide. A few faculty members who had visited his classes felt that he was not well prepared, that he was somewhat disorganized, and that he wasn't "giving the students much." Moreover, there was "some disappointment" as to the amount of research Plater had published.

The discussion lasted over two hours. Plater's defenders (and some critics) believed that he could be a fine faculty member and that Tellico "had distracted him." Once that was over, it was implied, things would be all right.

At the end of the discussion, Law School Dean Kenneth Penegar called for the usual secret ballot, appointing one person to count the votes and another to record them. It was an established tradition at the UT Law School that five negative votes (roughly one-third of the tenured faculty) was enough to deny a person tenure. Plater received more negative votes than that. It was Dean Penegar's unhappy task to tell Plater, who was understandably "hurt" and "upset."[31]

Not one shred of evidence exists that Plater was dismissed because of his vocal and highly publicized position against TVA. Nor is there any evidence that TVA made any overt or covert moves to influence the decision. "We wouldn't have stood for that," one faculty member stated indignantly.[32] But without doubt the tenure decision on Zyg Plater shook and embittered the Tellico opposition. If he wanted to continue teaching, Plater would have to leave Knoxville. Ultimately he went to Wayne State University in Detroit and then to Boston College. Although Plater had no intention of dropping the Tellico fight (it had become a personal crusade by 1975), when he left Knoxville some organizational restructuring was required to keep the opposition together and maintain the legal battles. It was agreed that Plater would continue as the central figure by making frequent visits to Knoxville. The local "chief of staff" was to be Plater's friend and protege Peter Alliman. Another law student, Hank Hill, would also be a key figure, as were Doris Gove of the UT zoology department and UT German professor C. J. "Jeff" Mellor (the latter two as officers of the Tennessee Endangered Species Committee). Although Plater's absence would be inconvenient, it would not hurt the opposition as much as anti-Tellico people at first feared.[33]

The opposition's next move was to have the U.S. Fish and Wildlife Service (USFWS) declare the Little Tennessee River the "critical habitat" for the endangered snail darter. That designation would have identified the Little Tennessee as the only remaining place where snail darters could exist and breed. The listing of the snail darter as endangered and the Little Tennessee River as its "critical habitat" would give the opposition a strong basis for suing the agency for violation of the 1973 Endangered Species Act.

TVA's reaction was characteristic of much of its behavior toward the end of the Wagner era. The agency vigorously resisted efforts both to list the snail darter as endangered and to designate the Little Tennessee River as its critical habitat. Around-the-clock work on the project continued, in the night under floodlights, prompting the Tennessee Endangered Species Committee later to charge — correctly — that over $40 million had been spent on Tellico after TVA knew it might be violation of the Endangered Species Act. Meanwhile, the search for the snail darter continued in other lakes and rivers, occasionally with unsolicited volunteer help. Louis Gwin received a fish in the mail with the query, "Is this a snail darter?" It wasn't. A teenager brought a bottle to his office with what the boy mistakenly thought was a snail darter inside. Private citizens claimed to have spotted snail darters outside the Little Tennessee, and TVA dutifully checked out all these reports. At the same time the agency, as required by law, was consulting with the USFWS about the snail darter. However, Kenneth Black of the USFWS complained that TVA was dragging its feet, was continuing work on the project, was not really consulting at all, was not in full compliance with the 1973 Endangered Species Act, and in essence had been unwilling to talk about any alternatives except finishing the Tellico Project as planned. TVA had spent nearly $100,000 on transplantation of the snail darter when the USFWS put a stop to it while consultations were going on. TVA also was spending around $25,000 on a scheme to incubate and hatch roughly fifteen hundred snail darter eggs for future transplantings (under ten percent actually hatched and almost all of them died within twenty-four hours). The evidence clearly showed TVA's determination to resist any course other than the one already chosen.[34]

Inevitably the conflict once again wound up in the federal courts. In February 1976 Plater and the opposition petitioned Judge Taylor of the Sixth Circuit Court to issue a temporary injunction while the "critical habitat" status for the snail darter was being considered. Taylor refused, making it clear that he didn't want to hear anything about cost overruns, benefit-cost ratios, or anything else. When the USFWS Service did declare the Little Tennessee as the snail darter's critical habitat the following May, Plater and his allies were back in Taylor's court to argue for a permanent injunction. True to form, Taylor refused that petition, too, and the opposition announced that it would appeal to the Sixth Circuit Court of Appeals in Cincinnati, presided over at the time by Judge Anthony Celebrezze. According to Plater, he was later told by a confidential source within TVA that TVA officials in summer 1976 boasted at a meeting that "by the time Plater stands up to argue in Cincinnati, there won't be a tree standing in the reservoir area." If Plater's source and Plater himself quoted correctly, TVA had no intention of abiding by the spirit of the law when it came to the Tellico Dam.[35]

The opposition had good cause to think it had a good chance to win on appeal. Celebrezze was considerably more sympathetic than Taylor, making it clear on the opening day of the hearing that he feared the precedent that would be set if the snail darter was allowed to become extinct. What would that do, he mused, to the Endangered Species Act? Moreover, Plater and his allies had lined up a solid array of witnesses, including the ubiquitous Etnier. Etnier asserted that it would take up to five years before TVA's transplanting of snail darters could be judged successful (TVA was already claiming success). TVA responded with a multifaceted defense of the project. Since the national concern with the energy shortage, the agency had increasingly emphasized power benefits, but it set forth a good case for other benefits as well. Moreover, TVA lawyers argued that so much money had been spent on the project, it was unreasonable not to allow its completion, especially since the lake could, they said, be impounded immediately. Too, the agency continued to argue that the 1973 Endangered Species Act was not retroactive and hence could not be applied to the Tellico Project, begun in 1967.[36]

The Court of Appeals was unmoved by TVA's arguments. On January 31, 1977, Celebrezze ruled that the Tellico Project could not be completed unless Congress specifically exempted the project from the Endangered Species Act. Celebrezze went on to state that it was immaterial to the court how much money had been spent or how far along the project was. The key issue was the status of the snail darter. Therefore, Celebrezze ordered Judge Robert Taylor to issue a permanent injunction.[37]

Press reaction to Celebrezze's decision was mixed. The *Washington Star*, the *Birmingham News*, the *Nashville Banner*, and the *Memphis Commercial Appeal* saw the ruling as ludicrous. That a project whose costs had recently been put at $116 million could be stopped "by a fish that wasn't even known to exist until seven years after Congress funded the project" seemed to them the height of absurdity. As the *Nashville Banner* put it, "The TVA had its critics, yet this is one instance of a project being too far along to stop. The snail darter might stand a chance of thriving elsewhere, but what do you do with a nearly finished dam? At $116 million, how does having the world's costliest aquarium grab you?"[38] Few newspapers (the *Knoxville Journal* was one) came to the defense of the court.[39]

More important than the press reaction to Celebrezze's decision was TVA's reaction. Almost immediately local newspapers began to receive complaints that TVA was continuing work on the canal to connect the planned Tellico Reservoir with Fort Loudoun Lake. When asked about this, TVA Information Director John Van Mol (son of Louis Van Mol) replied that the Sixth District Court had not formally issued its injunction yet and so the work was not illegal. "It is not our place," Van Mol stated loftily, "to interpret the Court of Appeals' decision." Technically, of course, Van Mol was correct. Yet clearly the spirit of Celebrezze's decision was being flouted.[40]

As soon as Taylor issued the permanent injunction, TVA petitioned the USFWS to "delist" the Little Tennessee River as the critical habitat of the snail darter. The agency's reasoning was ingenious. TVA argued that the area could no longer sustain a viable, natural population of snail darters because the agency's construction of cofferdams in 1975 prevented the snail darters from reaching their nursery area below the dam. In other words, TVA should be allowed to complete the Tellico Project because it had effectively, if accidentally, made sure that the snail darter population would not be able to reproduce. This must have struck the USFWS as a strange argument, and TVA's petition was denied.[41]

At the same time that the agency was preparing to appeal Celebrezze's ruling to the Supreme Court, it was lobbying Congress to grant an exemption from the Endangered Species Act for the Tellico Dam. In March 1977, members of Congress John Duncan of Knoxville and Marilyn Lloyd of Chattanooga introduced House Resolution 4557 which proposed just that. HR 4557 was promptly sent to the Committee on Merchant Marine and Fisheries, whose chairman was John M. Murphy. Almost immediately Wagner informed Murphy, "We support HR 4557." Meanwhile, Tennessee Senator Howard Baker, a strong supporter of the project, said that the Supreme Court should reverse Celebrezze and that if it did not, he personally would introduce legislation to weaken the Endangered Species Act and make it "reasonable." Robin Beard, a congressman from Middle Tennessee, introduced a bill which would exempt *all* public works projects from the Endangered Species Act if they had been started before the act was passed.[42]

Though probably not very hopeful, the opposition too made a concerted lobbying effort. The Tennessee Endangered Species Committee newsletter of March 14, 1977, appealed to its members to write letters. A "research package" was sent to all congressmen, "outlining and documenting the benefits of preserving the Little 'T'." Dan Burgner of Greeneville, Tennessee wrote to his congressman,

James H. "Jimmy" Quillen at the committee's urging. Quillen's reply was highly noncommittal and eventually found its way into the committee's files, where someone had added about Quillen, "So much for this S.O.B.!"[43]

Opposition strategists were more optimistic about influencing the executive branch. Recently-inaugurated President Jimmy Carter seemed sympathetic to environmental causes, had hinted in February that he would oppose any weakening of the Endangered Species Act, and had appointed environmentalists to key posts in the Department of the Interior. Carter had won the presidency partly because of voter disgust with the Nixon Administration and partly because he portrayed himself as a plain, decent man with simple tastes and a penchant for telling the truth. Initially his ideas on the environment, energy conservation, and foreign policy were idealistic, deeply felt, and forthrightly stated.

Tellico opponents began sending appeals to Carter, Secretary of the Interior Cecil Andrus, and others in the executive branch. But while TVA and its allies were lobbying Congress for Tellico's exemption and preparing the agency's Supreme Court appeal, they were not ignoring the Carter people, either. In late March 1977, retired Congressman Joe L. Evins wrote to Director of the Office of Management and Budget Bert Lance, a Carter confidant, trying to win Lance to Tellico's side. Moreover, some Tellico opponents who wrote to Carter experienced some rather unusual repercussions. One, Professor Charles Johnson of the University of Tennessee history department, had written to the president to voice his opposition to Tellico. An avid trout fisherman, Johnson could barely conceal his anger at TVA: "Let me urge, very strongly, that the Tellico Dam-Little Tennessee River project be stopped right now. Those arrogant bureaucrats of TVA are like the proverbial mule: to get their attention you have to hit them between the eyes with a 2×4. That would be a lovely 2×4."

What Johnson did not realize was that copies of all letters sent to the president or Congress regarding TVA are returned to TVA's headquarters in Knoxville. Someone at TVA's Office of General Counsel saw a copy of Johnson's letter, written on Department of History stationery, and passed it along to UT's General Counsel Beauchamp Brogan, a former TVA lawyer. Attached to the letter was a query whether Johnson was speaking for the history department in an official capacity. Brogan dashed off a letter to President Carter assuring him that Johnson did not speak for the university (Carter had never asked). Then, using University stationery, Brogan went on to "personally . . . disagree very strongly with Dr. Johnson's statements" and offer his opinion that the Tellico Project did not come under the Endangered

Species Act (a position Brogan had helped formulate while at TVA). Copies of Brogan's letter were sent to five TVA people (Wagner, Seeber, John Van Mol, George Kimmons, and OGC's Herbert Sanger) and to Johnson. "It was all a little frightening," Johnson said six years later.[44]

By mid-1977, however, the opposition was receiving support from three unexpected areas. Prompted by the Senate, the General Accounting Office (GAO) conducted a study of the Tellico Project, a draft of which was sent to TVA in July. The findings could not have been more unfavorable. The GAO took the agency to task for its overstating of benefits, its parlous benefit-cost ratio, its double-counting, and its subsequent increasing of benefits with no apparent foundation. In a twenty-page response, TVA said that the GAO's report was "unbalanced" and that the agency would not undertake a new benefit-cost study, since the present claims were accurate. Moreover, in an effort to disparage the Tennessee Endangered Species Committee, which had provided information to the GAO, TVA characterized the group as "a Knoxville area organization of about 100 current members, mainly students or recent graduates of the University of Tennessee, who oppose the Tellico Project." Unmoved, the GAO recommended to Congress in fall 1977 that no more money be appropriated for the project until a more defensible benefit-cost analysis was made.[45]

TVA plainly was worried about the benefit-cost problem. The ratio for Tellico had always been shaky, but with costs rising above $116 million and with no complete review of the problem since 1971, who could tell how bad the ratio would look now? Yet the GAO wanted a review. At almost the same moment that Wagner was telling a Senate subcommittee that TVA would not do a new one, Seeber was quietly ordering a new study. Yet, to soften the numbers, TVA chose to set projected benefits against *remaining* costs, *not* costs from the beginning of the Project. That ploy put the benefit-cost ratio at 7 to 1, a figure which few but the most committed took seriously. It even embarrassed some TVA people.[46]

The second unexpected boon to the opposition was hearings held on the project in July 1977 by the Senate Subcommittee on Resource Protection, chaired by John C. Culver. Although caught off guard, the opposition quickly marshaled witnesses to counter TVA's claims. One was University of Tennessee anthropologist Dr. Jefferson Chapman, who had replaced Guthe as the director of archaeological work at Tellico. The older Guthe had been more willing to work under TVA-imposed limitations than was Chapman, who told the Culver subcommittee that the Tellico reservoir should not be flooded until considerably more archaeological work was done. In an effort to cast a shadow on Chapman's expertise and testimony, TVA induced Guthe to write a letter to Culver undercutting Chapman. Yet Chapman's testimony doubtless carried a good deal of weight.[47]

The third unexpected source of aid to the opposition came from the Eastern Band of the Cherokees. Since the days of Hawk Littlejohn, the Cherokees had assiduously tried to stay out of the limelight and let others do their fighting for them. According to Alliman, however, the opposition strategy was to bring over the Interior Department completely to their side. If the Cherokee opposition could be revived, the Bureau of Indian Affairs, powerful within the Interior Department, might just be enough to tip the balance. Before TVA people had gotten wind of it, Plater, Alliman and others had convinced the Cherokees to join the opposition. On June 10, 1977, a meeting of opposition leaders with the Cherokees was held at the Cherokee, North Carolina, Holiday Inn. Plater, Alliman, Hank Hill, Prichard, and Jeff Mellor were all there. Together these opposition leaders forged an alliance with the Cherokees that was to last—albeit shakily—until the very end.[48]

TVA quickly learned that it had been outflanked on the Cherokee front. John Van Mol suggested to Seeber that something be done quickly. Seeber in turn wrote to Ed Lesesne (Kilbourne's successor at OTAD) asking, "Do you think that your people could do any good with the Eastern Band?" Lesesne hurriedly set up a meeting with Cherokee leaders at which the agency pledged to rebuild a replica of the council house and other buildings near the Chota site and may even have encouraged the Cherokees to use those buildings as the nucleus of a tourist attraction not unlike Cherokee, North Carolina. Surprisingly, however, the meeting ended with the Cherokees still opposed, although they did say they would accept TVA's offers if their opposition ultimately failed. Little wonder the Tellico opposition was never quite sure of the Cherokees' constancy.[49]

By mid-1977, things were pretty much at a standstill. The project had been halted by the Court of Appeals, but TVA would appeal that decision to the Supreme Court. The House of Representatives had seemed willing to go along with TVA, but the Interior Department, the GAO, and the Senate were considerably more skeptical. In spite of all that TVA had done, the opposition had held together fairly well and actually had outmaneuvered the agency with the Cherokees. TVA's efforts to delist the Little Tennessee River as a critical habitat and to overturn the snail darter's status as endangered had been turned back. Project boosters in the Tellico area were furious, but there appeared to be little they could do. According to the opposition, TVA had spent approximately $900,000 in public relations and lobbying efforts, and yet the project was still going nowhere. President Carter seemed to be opposed, although some within his official family—notably Attorney General Griffin Bell—supported TVA's position. Howard Baker was warning that he would punch holes in the Endangered Species Act if the Supreme Court did not find for TVA. In sum, after over twenty years of planning, roughly fifteen years of opposition, ten years of on-and-off construction, the purchase of 692 pieces of land, the condemnation of 58 others, and the expenditure of over $116 million, the Tellico Dam lay in the countryside like a beached whale, unable to move forward and unable to return to the sea.[50]

This was roughly the situation when S. David Freeman was confirmed as TVA's third director in August 1977. His appointment marked a significant shift for an agency about to celebrate its fiftieth anniversary.

Freeman was the son of a Russian Jewish immigrant who had found his way to Chattanooga, opened a modestly successful umbrella repair shop, and become known locally as an almost indefatigable writer of letters to newspapers, in which he extolled his adopted country. David Freeman graduated from Georgia Tech in engineering and was immediately hired by TVA in 1948. He spent the next thirteen years working for the likes of J. Porter Taylor, a demanding supervisor. "He was a damn good engineer," Taylor recalled, even after his differences with Freeman.

In the late 1950s, Freeman enrolled at the University of Tennessee to pursue a law degree. TVA encouraged him by giving him a part-time job in Mattern's Projects Planning Branch. When he graduated from UT Law School with record marks, TVA's Office of General Counsel (OGC) snapped him up.

But TVA was too confining for Freeman. Following a colleague who had left TVA's OGC, he moved to Washington in 1961. Soon he had secured a $4 million Ford Foundation fellowship to study governmental energy policies. In 1970 he published a report, *Electric Power and the Environment* (August 1970), which called for a "broader view" of "environmental damage" which could result from the unquestioning pursuit of more power. The whole report was a manifesto for energy conservation and for alternatives to the demands for more and more power. Obviously, when President Carter recommended Freeman to the Senate, he had in mind the idea that Freeman would be able "to turn the agency around."[51]

Although Freeman had worked for TVA for many years prior to his appointment as director, he was clearly viewed as a threat to the entrenched agency. For one thing, Freeman represented a new managerial style which TVA was not used to. Managers in the Wagner era had been loyal to the institutions they served (many had seen TVA as their life's mission). The post-1970 managerial style required that a person establish a personal power base and then move on through the "network" to another agency. Institutional loyalty was sacrificed to upward mobility and to ideas, wherever these ideas were applied.

Clearly Freeman's philosophy and style would clash with those dominant at TVA. Traditionally it had been expected that the two other directors would take a back seat to the chairman. Moreover, it was expected that a director would believe that "what was good for TVA was good for the Valley, and vice versa." But Freeman simply would not conform. His well-publicized trips throughout the Tennessee Valley, even meeting with TVA critics, often stole the limelight from Chairman Wagner, something that "just wasn't done." Freeman prided himself on being an impeccable dresser, whereas Wagner always looked as though he would have been more comfortable in overalls and an open-collared shirt. Nor did Freeman try to hide the fact that he thought TVA was out of step with the times ("they were completely wrapped up in building power plants").[52] His impatience with subordinates, many of whom had been at TVA for over forty years, his biting memos, and his off-the-cuff remarks ("this is the most inbred place I've ever seen") made him the object of fear and, in some cases, outright hatred.

In style, manner, and personal habits, Freeman and Wagner were very different. Yet both were men of vision—even if Wagner's was a vision of power plants, factories, and progress, whereas Freeman's was a vision more attuned to the energy-conscious 1970s. Both were impatient and hot-tempered, although years of practice had taught Wagner to control or hide those impulses. Both were stubborn men, although usually Wagner was able to disguise his inflexibility, and the younger Freeman was not. Finally, both were able to command the personal loyalties of others. Wagner was a champion of TVA's older employees who constituted a vast majority of the staff, while Freeman was more popular with the younger people, many of whom expected him to root out some of the older, entrenched employees. And, indeed, many people either retired or left TVA rather than remain during the Freeman years. Freeman's abrasive style and his penchant for spewing forth new ideas faster than people could absorb or implement them helped create a real morale problem among many employees who stayed. When TVA staffers, from file clerks to division heads, used the sobriquet "S. David," it was often in derision. One retired TVA person spoke for many of the Agency's first generation when he remarked that "if I was young enough, I'd go up to his office right now and kick his ass."[53]

It was almost inevitable that Wagner and Freeman should clash. True, they remained civil, even courteous, to one another in public. At one meeting of employees, Freeman received perfunctory applause, whereas the greeting of Wagner was thunderous. In a gesture of self-deprecation, Freeman bowed grandly to Wagner and stepped backwards to let the chairman upstage him. But Freeman made no secret of his belief that TVA "had practically gone to sleep" and that President Carter's motive in appointing him had been "to get TVA back to its roots." Such statements were barely disguised criticisms of Wagner and those who shared Wagner's vision. The meeting at which Freeman walked out on the chairman's rebuke of his management style was the first open recognition of the rift. After that, Wagner all but ignored Freeman whenever he could, while the new director continued to keep the agency in a state of mild panic. It was left to Seeber to try to put out the resulting brushfires.[54]

Peter Alliman was in Washington for the Culver subcommittee hearings when he heard of Freeman's appointment as TVA director. He telephoned the director designate and engaged Freeman in a five-minute conversation about the Tellico Project. "He was very noncommittal and told me that he was making no commitments. But I sensed a more open attitude. The other board members up to that time had been confrontational."[55]

Indeed, Tellico did not fit into Freeman's notion of what TVA should be doing, which was to provide for others a model of environmental sensitivity and energy conservation (his residential weatherization program ultimately aided some 700,000 homes) and to give farmers and agriculture a role in the agency as large as the one they had formerly enjoyed. Privately Freeman loathed the whole Tellico Project, for it represented everything about TVA that he would attempt to change. Later he said, "I would have been inclined to stop the whole thing." Publicly, however, Freeman cleverly maintained that, as a new board member, he had not yet made up his mind about Tellico and would have to study the situation further. He ordered that all correspondence emanating from TVA concerning Tellico should contain the following sentence: "Our new Board member, S. David Freeman, has not as yet formed an opinion on the Tellico project." To many within TVA, including Wagner, the statement was galling. But they honored Freeman's wishes, and the sentence was typed at the bottom of hundreds of letters. Usually it was up to Wagner to sign them, just below the statement.[56]

Yet Freeman left clues nearly everywhere as to his true feelings. For example, when TVA was about to contest the GAO's October 1977 report to Congress, Freeman wrote a short memo to Seeber which, in full, said: "Lynn: I'd like the reply [to the GAO] to reflect my non-involvement in the matter and possibly a belief that we should know the future benefits and costs of Tellico and *alternatives thereto* [italics ours] before proceeding. Dave." A few were convinced that this was Freeman's way of dissociating himself from a potentially damaging issue.[57] But until the Supreme Court heard TVA's appeal, Freeman was off the hook.

Meanwhile, early in 1978, the conflict shifted to the Carter administration. The president himself was temperamentally inclined to oppose the project, as was Secretary of the Interior Cecil Andrus. But Attorney General Griffin Bell argued vehemently for the project's completion. Moreover, Bell maintained that, as head of the Department of Justice, he should file a brief with the Supreme Court supporting TVA and that that brief should constitute the government's official position. Andrus was aghast, and, backed by a nearly-united Interior Department, opposed Bell forcefully. The result, according to political scientists Rechichar and Fitzgerald, was "some of the most bitter infighting of the Carter era."[58]

Carter was in a quandry. His own cabinet members were becoming increasingly strident on the Tellico issue, and other government agencies, especially the Council on Environmental Quality, were getting involved in the fray. In an attempt to restore peace within his own official family, the president personally pressured Bell to switch sides. The attorney general remained adamant, and in January 1978 he filed a brief with the Supreme Court supporting TVA.[59]

Unsuccessful with Bell, Carter tried another tack which he hoped would prevent members of his administration from trading verbal punches in front of the Supreme Court. Would Interior be interested, Carter asked Andrus, in sitting down with TVA and reaching some sort of compromise? Andrus was receptive. Later a White House press officer denied that any pressure had been brought to bear on the secretary. But now Wagner dug in his heels. In a March 31, 1978, letter to Andrus, the chairman stated coldly, "We are unwilling to discuss the alternatives mentioned." For Wagner, the reservoir was the only alternative.[60]

This gave Freeman the opportunity he had been waiting for. Perhaps his move came at Carter's bidding, perhaps in response to his own reading of the president's "signals," or perhaps out of his own determination to "turn TVA around." In any case, over Wagner's outraged objections, on April 6, 1978, Freeman undercut Wagner's position by writing to Andrus suggesting that a compromise was possible and that he disagreed with his fellow board members Wagner and Jenkins on alternatives to filling the reservoir. "Industrialization," he said, "can take place without another lake." Then Freeman released his letter to the press. At last the break was out in the open.[61]

Freeman's release of the letter struck TVA as tantamount to treason. The Supreme Court was scheduled to hear TVA's appeal in less than two weeks. Moreover, making public the fissure within the board would only embolden the opposition and embarrass the agency. Freeman's public stance might well prompt those TVA people who had been suspicious of the project but afraid to voice their doubts to "come out of the closet," deepening the chasm already existing within the agency. For example, Roger Woodworth, an agricultural economist at TVA, wrote in support of farmers' interests. After Wagner retired and Freeman became chairman of the board, TVA biologist W. Douglas Harned wrote a blistering critique of the project to Freeman, starting out, "With you as chairman a letter such as mine is now possible." In contrast, Wagner loyalist Mike Foster added a margin comment to a copy of one of Freeman's letters: "Clever!"[62]

The uproar which greeted Freeman's "defection" could be heard nearly to Washington. Congressman Duncan reported to Freeman that a straw poll he had taken recently showed that over 80 percent of his respondents favored completing the project as originally planned—a reminder that Freeman's position was opposed by a vast majority of Tennessee's Second Congressional District. Meanwhile, letters from lawyers, bankers, real estate brokers, developers, and private citizens from the Tellico area flooded into Freeman's office. Some were correct, though cold, while others were frankly abusive. Tellico Plains Mayor Charles Hall complained that Freeman was flouting the will of Congress. Moreover, when Freeman attended a public meeting on June 20 in Madisonville to explain his position, the pro-dam people, who had conducted a massive telephone campaign to pack the meeting, were unruly and rude, shouting the director down when he tried to speak. In a mixture of understatement and pun, Freeman told Maryville department store executive Harwell Proffitt that he had had a "close encounter" at Madisonville.[63]

Though they were not as numerous, Freeman also had his supporters. Attorney William Vines III saluted the Director's position, "not because it is for or against the Dam, but because it cries out for compromise of a difficult issue." Ben Plant of the Tennessee Wildlife Resources Agency told Freeman it was "refreshing to hear your new and realistic approach." Congratulations also arrived from the Tennessee Conservation League and the Tennessee Citizens for Wilderness Planning. Those groups must have been even more pleased when the director testified before the House Subcommittee on Fisheries and Wildlife Conservation that "perhaps the Tellico Project was the best possible project when it was designed decades ago, . . . but there have been dramatic changes in land values and values of society since this project was planned."[64]

As the Supreme Court began hearing the agency's appeal of Celebrezze's 1977 ruling, an era was coming to an end at TVA. Aubrey Wagner, a TVA employee since 1934 and chairman of its board since 1962, was about to retire. For the past two decades, Wagner had institutionalized his vision of what the Tennessee Valley ought to be and of TVA's role in its transformation. Many credited him with having "saved TVA" during the Vogel years and having brought the mammoth agency into the nuclear age. "Red" to his contemporaries and "Mr. Wagner" to younger employees, he was loved and respected within the agency. "Something went out of TVA when he left," an unabashed admirer declared years later. On Wagner's last day, May 18, 1978, an era ended.

But Wagner had also contributed to the difficulties TVA was now experiencing. His absolute intransigence on the issue of Tellico had created much bitterness both inside and outside TVA, had left the agency bitterly divided, and had caused an increasing number of people to see TVA as a callous and insensitive bureaucracy, deaf to cries of protest and blind to environmental damage.[65]

Wagner's retirement could not have come at a worse time. Earlier that month, outraged at what he saw as interference in TVA business by other government agencies, especially the Interior Department, Director Bill Jenkins had suddenly announced his resignation. After May 1978, then, Freeman would be the lone director, free to undo much of what Wagner had done. Since May 10, TVA, under pressure from the Carter administration and from Freeman, had been meeting with the Department of the Interior to consider possible alternatives to completing the Tellico Project. Wagner had until now been able to block any compromise. It was *immoral*, he insisted, to buy land from people for a stated purpose and then not follow through. But once Wagner retired, Freeman and the Interior Department would be free to consider other alternatives. Indeed, by this time Freeman was already being referred to as "Dry Dam Freeman" by critics who sensed his intention.[66]

On June 15 the Supreme Court announced its decision in TVA v. Hill. Attorney General Bell had argued that the Court should reverse the lower court's ruling on the grounds that the 1973 Endangered Species Act had been passed more than six years after Tellico Dam construction began and that transplantations of the snail darter meant that the fish was no longer endangered. Chief Justice Warren Burger asked Plater whether he would abandon his opposition if the snail darter transplants proved successful, to which the young attorney replied that no such proof existed, and that even if it did, the project was so lacking in merit that even the federal government (the Departments of Justice and the Interior) had presented diametrically opposed briefs to the Court. That fact had more than irritated Justice Lewis Powell.[67]

In the end, by a six-to-three vote (Powell, Blackman, and Rehnquist dissented), the Supreme Court upheld the lower court ruling and continued to block completion of the Tellico Project. Burger's majority opinion barely touched the question of the project's merits. Rather, the chief justice focused on whether the courts could interpret Congress's intent in passing the Endangered Species Act. Did the legislative branch mean for the act to apply to all projects and all species retroactively? Following the principle of judicial self-restraint, Burger stated, "Once a law has been passed, courts should not tinker with it unless a constitutional issue is raised. Congress has made it abundantly clear that the balance has been struck in favor of . . . endangered species. It is not for the courts to appraise 'the wisdom or unwisdom of a particular course.' " Powell's dissent argued for "some modicum of common sense. . . . Nor can I believe that Congress could have intended . . . the absurd result in this case." The opposition had bested TVA. If the Tellico Project were to be completed as originally planned, Congress would have to act.[68]

Senator Howard Baker immediately moved to accomplish just that. While congressional allies were trying once again to push an exemption for Tellico through Congress, Baker himself, in cooperation with Culver in the Senate and Robin Beard in the House, secured passage of an amendment to the Endangered Species Act which would set up an Endangered Species Committee with the power to grant exemptions to some federal projects where on balance the project itself was more valuable than the species endangered. Because of its power of life or death over certain species, the new committee was immediately nicknamed the "God Committee." The panel initially included Andrus (chairman), Agriculture Secretary Bob Bergland, Army Secretary Clifford Alexander, Charles Schultze of the Council of Economic Advisers, Environmental Protection Agency administrator Douglas Costle and National Oceanic and Atmospheric Administration head Richard Franch. Immediately the committee announced that it would hold hearings on Tellico both in Knoxville and in Washington.[69]

But before the "God Committee" could begin its work, Baker had other business to take care of. As the lone director, David Freeman had been examining the possibility that development could take place without closing the dam and thereby filling the reservoir. Freeman and Robert Herbst of the Department of the Interior were confidentially circulating for comments the draft of an "Alternatives Report" which mentioned the "dry dam" as a distinct option.[70]

On June 29 Baker wrote to Freeman:

> It has come to my attention through a number of newspaper accounts that you have begun an inquiry into alternatives to the Tellico Dam Project as presently constituted. I believe that any significant alteration in the original project would constitute a major change in policy by TVA requiring a decision by a quorum of the Board. Almost certainly the Tellico Project will be an item for careful inquiry when the two new TVA Directors are submitted to the Senate for confirmation.
>
> I hope, as well, that TVA would take account of the fact that the Congress is presently engaged in the reauthorization of the Endangered Species Act and that amendments and changes in the act appear, at least, possible. In view of those facts, I would very much hope that no final action in respect to changes in the TVA project would be undertaken until at least one of the vacancies is filled on the Board, and until the Congress has had a reasonable opportunity to act on the reauthorization of the Endangered Species Act.
>
> Sincerely,
> Howard H. Baker, Jr.

That same day Baker released the letter to the press.[71]

Much later, Freeman remembered the letter very well, but denied that it was a veiled threat from Baker to back off. "Howard was running for re-election against [Democratic candidate] Jane Eskind and she was trying to get to his political right. The whole thing was a political move to show the people of Tellico he was with them. I never took it seriously."[72]

However, the available evidence contradicts Freeman. His reply to Baker did not go out until July 7. Before that, it had gone through several drafts, after suggestions from John Van Mol, Louis Gwin, and Jane Parker. Gwin wrote the final draft and later recalled that Baker's letter was taken extremely seriously. After all, even though President Carter had named Freeman chairman of the board, it was in Baker's power to put him in the minority once again.[73]

Freeman sent a soothing reply to Baker, but he had no intention of altering his course. Two weeks later Freeman told ABC News that it was possible that the snail darter "performed a very useful service" for taxpayers by forcing TVA to take a hard look at the project. And in August drafts of the "Alternatives Report" were distributed widely to supporters and opponents alike for their comments. Ultimately over a hundred copies were sent out, and over three thousand reactions received. And while the "Alternatives Report" was not all the opposition could have wished for, it did raise the possibility that all the benefits claimed for the Project could be realized without the reservoir. Although Plater had some serious reservations, he praised the report in general, as did Joe Congleton of Trout Unlimited and other Tellico foes. The final report, essentially a minor revision of the draft, was published and released in December.[74]

Not everyone, of course, was pleased with the "Alternatives Report." A host of supporters, many of them retired TVA employees who had worked on the project, protested the report's "slant" toward the "dry dam" option. Don Mattern and Reed Elliot wrote detailed, carefully reasoned critiques, Mattern lamenting "that the competence and professionalism of the TVA staff is largely ignored." But Herbert Vogel could not contain himself:

> "What in the name of Heaven is a dam for but to contain water for power generation, flood control, navigation or some other service to man? . . . As I read your joint report I come to feel that its contributors must be living in a dream world. Certainly I have never read such fantasies since "Alice in Wonderland" and "Through the Looking Glass." Must we now add a snail darter to the White Rabbit, Humpty Dumpty and the other weird creations of Lewis Carroll [?] . . . To say that the report shocks me is to put it too mildly.[75]

In addition, some people inside TVA were determined to undercut Freeman, the "Alternatives Report," and the "dry dam" option. Indeed, many Wagner loyalists still within the agency disliked Freeman and strongly supported Tellico. In Mike Foster's Division of Navigation Development and Regional Studies, the OTAD staff devised a plan "for citizen participation in the Tellico Project" which would marshal public opinion in the Tellico area for one big push for congressional exemption. A step-by-step strategy was laid out which included disseminating highlights of the "Alternatives Report," holding a series of public meetings, getting a letter-writing campaign started, pumping OTAD money into the area for youth employment (through the Young Adult Conservation Corps) and sponsoring bus trips to Duffield, Virginia, and the Tims Ford Dam to highlight the benefits of those projects. In all, as one unidentified memo writer put it, it would be a well-coordinated effort to "use the OTAD organizational approach to marshal citizen participation."[76]

The campaign generated numerous letters to Freeman from

Tellico area citizens hostile to any alternative but the original project. Madisonville real estate broker Norman Lee (Alliman's father-in-law) charged that TVA was "only listening to the environmentalists;" Monroe County banker Brent Heiskell charged that the report was "biased" and that David Freeman was at the core of that bias. Madisonville attorney Clifford Wilson noted sarcastically that "While peasants in their straw hats laboring in the fields may be quaint and cute to the tourists, it is not the sort of employment which is sought by the high school graduates in our county. . . ." Freeman appeared unmoved. Indeed, the decision was out of his hands.[77]

It was in the hands of the "God Committee." After a brief round of hearings in Knoxville and Washington, the committee met on January 23, 1979, to consider exempting the Tellico Project from the Endangered Species Act. After a bit of foot-shuffling, committee member Charles Schultze spoke:

> Well, somebody has to start. Unlike my eminent colleague on the prior question, I have not prepared a resolution. However, I think the sense of it would be clear. It seems to me the examination of the staff report, which I thought was excellently done, would indicate that it is very difficult, it would be very difficult, as far as I'm concerned I can't see how it [exemption] could be done, to say there are no reasonable and prudent alternatives to the project.
>
> The interesting phenomenon is that here is a project that is 95 percent complete, and if one takes just the cost of finishing it against the total benefits and does it properly, it doesn't pay, which says something about the original design!

After a few more remarks from Schultze, a vote was taken, and the petition to exempt was unanimously defeated. Afterward Andrus commented, "Frankly, I hate to see the snail darter get the credit for stopping a project that was ill-conceived and uneconomical in the first place."[78]

Senator Baker was furious. Reacting angrily to the decision, he promised, "If that's all the good the committee process can do, to put us right back where we started from, we might as well save the time and expense. I will introduce legislation to abolish the committee and exempt the Tellico Dam from the provisions of the [Endangered Species] act."[79]

Now Freeman was in a quandary. If the reservoir was not to be filled, what was to be done with all the land TVA had purchased or acquired through condemnation? Earlier, state legislator F. "Benny" Stafford had warned Freeman that if the project were not completed as planned, then TVA "took the land under false pretenses" and ought to allow the original holders to repurchase it at no increase in price. An Association of Dispossessed Landowners had been formed in the Tellico area and was saying the same thing. On the other hand, many former landowners did not want their land back, and a good number already had used the money to purchase other property. Moreover, in the years since TVA began buying property, it had been leasing that land to farmers, some of whom had not been the original owners of the tracts they were leasing. Bankers complained that if TVA started selling the land back, local banks would be virtually drained of capital that they were already in the process of investing elsewhere. Meetings were held with former landowners, but the whole problem remained mired in confusion.[80]

Then, almost anticlimactically, it was over. Late in the afternoon on June 18, 1979, in a sleepy and near-empty House of Representatives, with the clerk laboring through the complete text of the 1980 public works appropriations bill, Congressman John Duncan rose to offer an amendment. The clerk received the amendment and began droning his way through it, when Duncan moved that the reading be waived. Tom Bevill of Alabama rose to say that he had seen it and would approve it without its being read, and John Myers echoed Bevill's sentiments. A voice vote was taken, and in forty-two seconds a drowsy fragment of the House membership had exempted Tellico from the Endangered Species Act.[81]

It was a tawdry move, one which violated any definition of fairness and, it might be added, the House rules as well. And Duncan was subjected to a punishing attack by outraged opponents and neutrals alike. The editor of the *Kingsport (Tenn.) Times* spoke for many when he wrote, "Duncan . . . , who has always been a bit muddled about economics in the Tennessee Valley, has now shown he also is confused about how democracy is supposed to work." Yet surely the ploy had not been Duncan's idea. Opposition leaders believed that the strategy had been put together by Richard Herod, a staff assistant to Baker. Duncan took most of the fire, though he was probably the wrong target.[82]

Now Baker came out in the open to force the rider through the Senate. He failed the first time, on July 17, by a vote of 45–53. On the second try, on September 11, with Baker pulling out all the political stops, the amendment passed by four votes.[83]

The Tellico Project, which had passed through so many areas of the government, was back in Carter's lap. The president had tried to exert pressure behind the scenes in the Senate to have the rider killed, but Carter and his "Georgia Mafia" had never had much of a talent for dealing with Congress. Now he had to choose: sign or veto.[84]

Frantic letters poured in from the opposition; Andrus begged his chief to veto; and TVA, which had been caught completely by surprise by the whole scheme, remained silent. In the end, on September 25, 1979, Carter signed, bartering that signature for help with other legislation he wanted. He did so reluctantly, with (as he put it) "mixed reactions." The night before, he had telephoned Plater from Air Force One to tell the opposition leader beforehand of his decision to sign. Plater tried to talk him out of it. But, after forty years, the battle over the Tellico Project had ended almost anticlimactially, far from the Tellico area, in the cloakrooms of Congress.[85]

Confused and bitter, the project's opponents tried to make sense of the long conflict and its sudden ending. For some, the outcome confirmed their fears that government had become impervious to the will of the people. Others concluded that Americans almost always chose shortrun profits over longrange benefits. But most realized that the meaning of Tellico would only emerge later, gradually. For them, this chapter in their lives was over.

CONCLUSION: TVA AND THE TELLICO DAM

Nationally the public reaction to the backdoor approval of the Tellico Project was a mixture of bitterness, anger, and outrage. Newspapers as diverse as the *Los Angeles Times*, the *Boston Globe*, the *Chicago Tribune*, the *New York Times*, the *Philadelphia Inquirer*, and the *Christian Science Monitor* attacked the outcome, the last calling Tellico "a boondoggle if ever there was one." Members of Congress and the president were deluged with mail not unlike Philip Ryan's letter to Congressman Hamilton Fish, Jr., in which he called the Baker-Duncan strategy "incredibly cruel, venal, stupid, or possibly all three." Many refused to believe that TVA had not had a hand in the matter.[1]

Legal battles droned on, but most knew the fight was finished. In refusing still another petition, Judge Robert Taylor said that he was "sick of hearing about the snail darter" and that dam opponents were harassing TVA. The Eastern Band of Cherokees finally filed suit on its own, arguing that inundation would be a violation of the Native Americans Religious Freedom Act. But that charge came too late. On October 19–21, 1979, a "campout" of dam opponents was held at Chota (TVA provided portable toilets). A few felt like fighting on, but the vast majority realized that this party was to bid farewell to the valley and to each other. Although the weekend was mostly one of guitar-playing, remembrances, and tears, it was marred by nails on the road and a bomb threat (police did find some dynamite).[2]

On November 13 federal marshals arrived to remove the last three holdouts from their land—Thomas Burel Moser, Jean Ritchie, and Nellie McCall. Freeman was furious with his predecessors for not removing these people immediately after condemnation in 1972. "They didn't have the guts to take Nellie McCall off her land, and I had to do the dirty work. . . . what a bunch of gutless wonders."[3]

A CBS camera crew had arrived from Atlanta to film the removal for the evening news. They were interviewing Nellie McCall in her parlor when the marshals showed up. "They didn't know we were there when they arrived," recounted Clarence Gibbons of the CBS crew. "When they saw us they refused to identify themselves, and then they finally asked us to leave."[4]

That evening on the CBS Evening News, with Ed Rabel narrating from the scene, Thomas Burel Moser told the nation that it was "a hell of a country, ain't it?" His house was bulldozed into a giant hole which had been dug for the occasion. Jean Ritchie called what TVA had done "stealin'." And when Rabel asked eighty-four-year-old Nellie McCall where she would go now, tears came to her eyes, and she could not answer. Back in New York, Walter Cronkite, the veteran newsman who had seen everything since the Battle of Britain, was visibly moved and had trouble getting into the next story. So did more than a few viewers.[5]

On November 29, 1979, TVA construction workers began manually closing the sluice gates on the Tellico Dam. The retired Wagner arrived for a short ceremony; victorious at last, he was in good spirits. The agency agreed to pay Plater and his legal allies $20,000 in "full and final settlement of all claims for attorneys' fees or costs." The Fort Loudoun Historic Area, including the fort, now perched on an artificial bluff, was soon to be given to the State of Tennessee for use as a park, and TVA had agreed to construct a modern visitors' center at its own expense. The agency promised to help the Cherokees with their plans "to reinter Cherokee remains removed during archaeological investigations." All crops on leased land had been harvested. And in approximately three or four weeks the Tellico Lake would be filled to its normal winter level.

As the reservoir began to fill, interest quickly died, as it had with so many reservoirs in the past. Few came by to watch the water rise, and TVA security guards were left pretty much to themselves. Indeed, by the time the water began spilling over the dam, only Tellico Plains Mayor Charles Hall was there to record the historic moment with his camera.[6]

Upstream, Charlotte Hughes and her grandson watched as the Tellico Lake put an end to the valley which she and her people had inhabited for so long. And suddenly, although she had remained composed for years, tears began to trickle down her cheeks. Noticing this, her grandson begged her, "Don't cry, Grandma. They haven't killed the river. We just can't see it anymore."[7]

David Freeman once remarked that Tellico was a "relatively insignificant" episode in TVA's recent past. But Louis Gwin was more correct when he characterized Tellico as "a microcosm of all that had gone on in TVA in its fifty-year history."[8]

TVA had begun as a bold social experiment in a proud, independent, but impoverished valley. Bit by bit, that initial vision had been eroded, in the almost savage criticisms of the agency as "socialistic," in the bitter fratricidal war among the first three directors, and in the resulting, wrenching shift of mission dictated by the triumphant Lilienthal and Harcourt Morgan. And so it continued through the crucial years of World War II, in which the agency made a significant contribution to the nation's ultimate nuclear victory.

By the 1950s, however, it had become clear to TVA that cheap power alone, while it might have helped to win a war, was not in and of itself revitalizing the Tennessee Valley. And so the agency shifted to a more aggressive posture, including a return to a large land-taking policy and active promotion of industrial development. Weathering the difficult but not terribly damaging Eisenhower years, TVA reemerged in the "New Frontier" era of John Kennedy as a dynamic advocate of the principle that the end (industrial progress) justifies the means. By then TVA employees, many of whom had been with the agency since its inception, had learned to keep their heads down, wait for signals from the board (and especially from the chairman), and to "follow the company line." Attempting to repeat the successes of an earlier era, TVA came to believe that projects meant progress, that dams—even on the tributaries—would bring prosperity. If citizen participation was needed, TVA would create it; if detailed justifications (in the form of benefit-cost analyses) were required, TVA would create them, too.

But in the 1960s and 1970s the agency's vision was called into question. To many who had watched Vietnam, Watergate, and accelerating environmental destruction, TVA appeared to be less a vision for the valley than just another arrogant government bureaucracy, capable of lying, deceit, venality, and "dirty tricks." And, in its intransigence, TVA seemed to play into the hands of its own enemies. To some, even within TVA, the agency's desires seemed almost insatiable and its leaders increasingly insensitive to the realities of the world outside. In the end, TVA won the Tellico Dam controversy. But that victory was so pyrrhic as to be almost a defeat. Many within TVA would like very much to put Tellico behind them and get on with the search for the agency's new mission. But they cannot.

By the time the Tellico Project was completed, the chasm between TVA and its critics was unbridgeable. Most Tellico opponents had come to believe that TVA had conspired to foist a bad project on the public for the purpose, they argued, of providing continued employment for the personnel of a bureaucratic agency that had outlived its usefulness. It was, in short, the "conspiracy theory of history" in modern dress. Indeed, propelled by their own bitterness, TVA critics widened the scope of their attack, charging that the agency had never helped the Tennessee Valley and that what progress and prosperity had come to the valley had been caused by other factors.[9]

In many ways those accusations were both untrue and unfair. While TVA has not been the only causal factor in the economic and social progress of the Tennessee Valley, clearly the agency has contributed to that progress in a number of ways. Moreover, to ask how much progress would have been attained *without* TVA is a counterfactual exercise that can never be conclusive. Although counterfactual econometric analysis is a useful tool which can help historians understand the contributions of certain policies, innovations, or institutions, in this case no amount of econometric modeling can demonstrate with certainty whether TVA, in its first fifty years, has been a "good" or a "bad" thing for the valley. Instead, anyone who would assess the contributions of TVA must do so on an era-by-era, project-by-project basis.[10]

As one step in such an assessment, this book has analyzed TVA's Tellico Project. It is clear that the agency sometimes treated critics of the project unfairly and harshly; that it was dilatory in its compliance with NEPA and the Endangered Species guidelines; that many of the project benefits claimed were illusory; and that, until criticism had brought the project to its knees, the agency never seriously considered an alternative to constructing the dam.

But the authors do not feel that a "bad" project makes an agency bad. In its long history, TVA has affected the region in numerous ways, many of them not amenable to quantitative analysis. TVA's test demonstration farm program has been a model of its type; its reforestation and afforestation programs have been widely emulated; its engineering and design offices have set civil engineering standards in functionality and aesthetics alike, and the mere presence of its numerous employees has served as a human leaven in the region. Many more achievements could be cited. The point is that there are many TVAs—too many to condemn them together on the basis of either a single econometric model or the "failure" of a single project.

The Tellico story must be understood in the context of a certain phase of TVA's development. TVA adopted the Tellico Project as part of its search for a "new mission", only to find that its "new" mission was outdated from its inception. In its failure to adjust to the changing realities and concerns of post-industrial America, TVA found itself and its project increasingly under attack.

Tellico constituted a turning point for TVA. Pre-Tellico TVA and post-Tellico TVA were radically different entities. Tellico was, for TVA, a pivot, a catalyst, and a watershed. The same agency that had received bad marks for noncompliance with NEPA standards under Aubrey Wagner would, under S. David Freeman's direction, become an environmental bellwether. The agency that had refused to drop Tellico on the grounds that it contained too much in sunk costs would later scrap a nuclear program whose sunk costs could have paid for dozens of Tellicos. The agency that had begun as a planning beacon for millions of Americans lapsed into a kind of mediocre commercialism.

The Tellico Project encompassed the end of Wagner's TVA leadership and the beginning of Freeman's. But it fell to Freeman to point TVA in new directions, as it fell to Wagner, who ironically thought himself a trailblazer, to maintain the old way. History was on Freeman's side. And all institutions undergo historical change. Whether they do so willingly or not does not, in the long run, matter. What matters is the way they are affected. TVA in the future will have to work decisively and publicly to identify its goals. At its inception TVA reflected the idealistic dreams of a depression-wracked nation and region. Symbolically it embodied the faith that planning and concerted effort could improve the lives of a region's people. But by the time of Tellico, the region and the nation were desperate for new ideals and new symbols, and TVA was like a vast mountain torrent which runs into an equally vast desert. It had dried up.

It is too soon to tell what will happen to the Little Tennessee Valley or, for that matter, to TVA itself. In April 1982 the Tennessee legislature created the Tellico Reservoir Development Agency (TRDA), a reincarnation of the Tri-Counties Development Association. This body was given the authority to develop the land inside TVA's taking line. Not surprisingly, Charles Hall was elected as TRDA's first president. On August 25, 1982, the TVA board approved a contract which would transfer roughly eleven thousand acres to TRDA (TVA retained an equal amount for recreational and historical uses); TVA would provide money for promotion of industrial and residential development.[11]

Immediately TRDA hired a fulltime executive director (a former TVA man) to begin aggressively promoting the project. An attempt to interest the Coors Brewing Company failed (many local residents, good Southern Baptists, were opposed to that anyway), but other industries were vigorously pursued, and one boat manufacturing company did sign a letter of intent. Plans went forward for a recreational marina complex, with TVA staffers continuing to provide technical advice. But TRDA was always short of funds and often had to go back to TVA, hat in hand, for more help.[12]

Meanwhile Fort Loudoun became an attractive, tasteful state historic site, although controversy continued between TVA and state officials as to what each had promised to do. Reenactments of the Battle of Fort Loudoun were held each spring, attracting an increasing number of tourists and "reenactment buffs." Most of the time the fort and visitors' center are quiet, with an occasional school bus breaking the stillness.

So far the fishing has been disappointing. The bass population apparently has not been breeding in its normal cycles, and other fish are not yet abundant. TVA, which went to a lot of trouble calculating anticipated benefits of visits per day by fishermen, ignores this fact. "We don't know how many fishermen or tourists have visited the lake," one TVA official remarked, "and we have no intention of finding out."[13]

Some insist that all will come in time—industry, tourists, jobs, money—and that TVA's vision for the Little Tennessee Valley will be realized. And it may indeed. The national recession, economic uncertainty, and high interest rates made the early 1980s a bad time for TRDA or anyone else to go out in search of industry. Critics crow that it will never happen. Almost every time local farmer Wade Swafford sees Charles Hall on the street, he asks him "where all the industries are" and tries to goad him into betting a steak dinner on the eventual outcome (Hall took the bet). Vonore grocer Ben Snider is embittered; he feels that he and his neighbors were "taken down the road, sold a bill of goods" for a prosperity that hasn't materialized. Many local residents share his frustration and believe that TVA "has built their dam and now has abandoned us." But all that, perhaps, is premature.[14]

For now, however, the Tellico area remains a quiet place, off the beaten track. The town of Vonore appears to have stagnated, its old main street sagging and neglected. Tellico Plains is busier and more prosperous looking, but the improvement is only relative. Lives have been altered by the Tellico Dam, but in many ways things are much the same.

Whether TVA has lived too long is a question we cannot answer. Curiously, many retired TVA people think precisely that—that the agency has accomplished all it can to bring the Tennessee Valley into the modern, post-industrial age. Others, naturally, disagree. They maintain that this enormous collection of talented men and women still has much to offer the Tennessee Valley, a region which in many ways is just beginning to benefit from the hard work of an earlier day. But, in its uncertain future, whichever of many roads the Tennessee Valley Authority chooses to travel, the ghosts of Tellico will always go with it.

APPENDIX: LAND PURCHASES BY TVA AT TELLICO, 1967–76

As with all previous projects, TVA created and retained voluminous records of the agency's land purchases at Tellico. Information on each tract contains the names of owners; the price TVA paid for the tract; the amounts of any mortgages, liens, or back taxes on the property; a general description and map of the tract; and a complete history of the tract's ownership. As of 1984, these files were housed in TVA's Division of Property and Services in Chattanooga. The authors transferred some of the variables onto computer tape so that we could analyze more easily the trends in property ownership and purchase.

During the first half of the twentieth century, property transfers in the Tellico area were almost nonexistent (an average of 2.06 transactions per year in 1901–44); those that did occur almost invariably were between relatives. Beginning in the 1950s, however, real estate transfers increased markedly (to an average of 21.5 sales per year in 1957–69), a large number of them to unrelated individuals, corporations, and a few development companies. Indeed, of the 635 tracts purchased by TVA at Tellico, less than 20 percent (123 tracts) had been in the same families since 1900, a figure considerably lower than it would have been but two decades earlier.

According to the recollections of TVA people, the agency's purchase policy was to acquire the land at the planned dam site and diversionary canal first. This may have been the case. It also appears, however, that TVA sought to acquire as much of the cheapest land and the small holdings as it could, regardless of the location of those holdings in the Tellico area. In 1967–69, 60 percent of the poorest (least expensive per acre) land was purchased. By 1970, therefore, the agency had acquired over half (53.42 percent) of all the land it would purchase or condemn at Tellico. The peak transaction year was 1969, when 113 transactions took place.

Next to be acquired were the larger holdings. With the exception of large tracts (over 100 acres) at the dam site and canal, the majority of these tracts (64.0 percent) were purchased or condemned in 1971–74. Holdings in excess of 100 acres were owned by 158 persons, 33 of whom received over $100,000 for their land. Private farms in this category ranged from that of Knoxville attorney Ray Jenkins, who received $173,650, to that of Dr. Troy Bagwell, who got $496,616. The Northumberland Land Company received $701,220 for its multiple holdings. Not surprisingly, these large tracts contained much of the best land, and TVA was obliged to pay more per acre for them. Land purchase price by quartiles is as follows: First Quartile, $2,995; Median, $9,328; Third, $26,208; and Maximum, $701,220. The mean purchase price was $27,287 (standard deviation, $56,546). Over the whole purchase, the average per acre price was $356. The authors' figures do not include tracts of less than one acre on condemnations and do represent purchase price after liens and taxes.

Table 3   **Acreage Purchased by TVA for the Tellico Project**

Year	Total Acres Sold to TVA	Total Amount of Sales in $ (Less Liens & Taxes)	N
1967	4,341	2,071,381	60
1968	5,062	2,263,790	97
1969	5,663	1,837,121	113
1970	7,762	2,169,877	80
1971	6,596	1,305,227	46
1972	7,411	2,231,936	61
1973	312	131,355	13
1974	4,596	2,516,606	55
1975	1,095	670,079	11
1976	161	128,250	2
Totals	42,999	15,325,622	538

Note: Sales rounded to nearest dollar; sale of only one farm in 1966 not included in figures. TVA's figures, including taxes, liens, condemnations, and less than one-acre tracts, came to $22,158,000. Source: TVA Land Branch Records, Chattanooga. Figures compiled by authors.

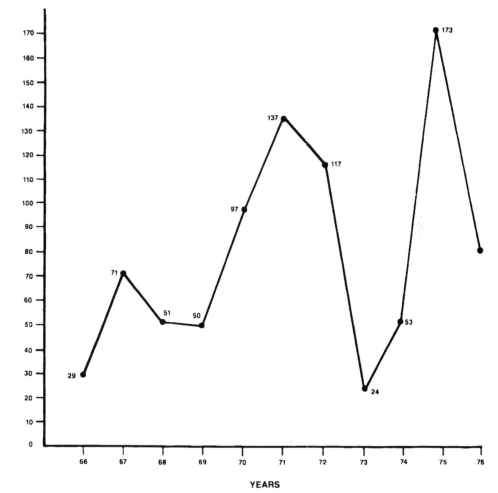

Chart 2. Tellico Land Purchases, Mean Farm Size Per Year 1966–76.

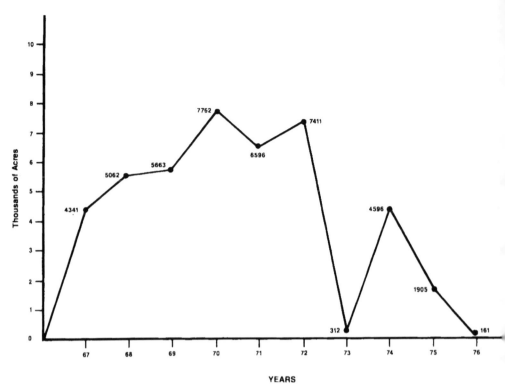

Chart 3. Tellico Land Purchases, Acres Purchased by Year, 1967-76.

KEY TO TVA DOCUMENT COLLECTIONS

AF — Administrative Files, Federal Records Center, East Point, Ga.

CF — Central Files

COF — Chairman's Office Files

DPS — Division of Property and Services, Chattanooga, Tenn.

DRPF — Division of Reservoir Property Files

DWM — Division of Water Management

FUDAR — Future Dams and Reservoirs Committee Files

GMF — General Manager's Files

NDRSF — Navigation Development and Regional Studies Files, now housed in Office for Economic And Community Development, Knoxville, Tenn.

OACD — Office of Agricultural and Chemical Development, Muscle Shoals, Ala.

OAF — Old Administrative Files, Federal Records Center, East Point, Ga.

OCENTF — Old Central Files, Federal Records Center, East Point, Ga.

OCOMF — Old Commerce Department Files, Federal Records Center, East Point, Ga.

OECD — Office for Economic and Community Development Files, Knoxville, Tenn.

OTAD — Office of Tributary Area Development Files, now housed in Office for Economic and Community Development, Knoxville, Tenn.

PPBF — Project Planning Branch Files, in Water Control Planning Files, Knoxville, Tenn.

TL — Technical Library

WCPF — Water Control Planning Files, Knoxville, Tenn.

NOTES

PREFACE

1. Samuel P. Hays, *Conservation and the Gospel of Efficiency: The Progressive Conservation Movement, 1890–1920* (Cambridge, Mass., 1959).

CHAPTER 1

1. Wagner to L. G. Allbaugh, Feb. 5, 1959, FUDAR File, OACD.
2. For a list of the twenty-six who were invited, see ibid.
3. For Wagner's opening remarks, see Wagner to Allbaugh, Mar. 2, 1959, FUDAR File, OACD.
4. For a brief analysis of the directors and their differences, as well as for Harcourt Morgan's dismissal of Arthur Morgan's broader vision for TVA, see Michael J. McDonald and John Muldowny, *TVA and the Dispossessed: The Resettlement of Population in the Norris Dam Area* (Knoxville, 1982), 14–23. For a fuller treatment see Thomas K. McCraw, *Morgan Versus Lilienthal: A Feud Within the TVA* (Chicago, 1970).
5. See Draper to Board, Mar. 26, 1934, COF. For Draper's ideological closeness to Arthur Morgan, see McDonald and Muldowny, *TVA and the Dispossessed*, 165. The evolution of the board of directors is historically significant. Initially there was a good deal of confusion about the roles of the various directors. The original organizational plan had each director responsible for one broad area: Engineering, Natural Resource Development, and Power. This almost inevitably led each director to champion his own area. After the removal of Arthur Morgan in 1937, the board was reorganized into a unitary body, with all directors committed to act together on all policy matters concerning all areas. At that time, the office of general manager was created to handle all general administrative duties, leaving the board to decide policy.
6. Lilienthal to Morgan and Morgan, Mar. 26, 1934, COF.
7. Ibid. See also Draper to Board, n.d. (soon after Mar. 26, 1934), COF. For the eventual dropping of regional planning, see Roscoe Martin, "Retrospect and Prospect," in *TVA, The First Twenty Years: A Staff Report*, Roscoe Martin, ed., (University, Ala., and Knoxville, 1956), 265.
8. David E. Lilienthal, *TVA: Democracy on the March*, new ed. (New York, 1945), 204–205.
9. Interview with George Palo, Apr. 7, 1983; interview with J. Porter Taylor, Apr. 11, 1983.

10. Interview with Paul Evans, Apr. 7, 1983.

11. One might also add that the goal of industrial development of the Tennessee Valley was not realistic during the Great Depression of the 1930s. As economist John V. Krutilla recalled, "In the face of a depressed national economy . . . the prospects for expanding regional employment in manufacturing did not look at all promising." See Krutilla, "Economic Development," in Martin, ed., TVA, The First Twenty Years, 220. For Menhinick's caution, see Menhinick to Those Listed Below, Aug. 7, 1944, OCOMF. For the reports, see "Forecast of Physical Resources of the Tennessee Valley—1945," Step I Working Papers (Knoxville, May 1944); "Opportunities for Postwar Development of Tennessee Valley Resources," Step II Working Papers (Knoxville, May 1944); "Development of Tennessee Valley Resources—Program Suggestions," Step III Working Papers (Knoxville, Nov. 1945). On the fear that the agency would have to give up many of its functions and dismantle some of its divisions, see [?] to Gant, Nov. 7, 1947, AF.

12. On the national political situation, see Richard Couto, "New Seeds at the Grass Roots: The Politics of the TVA Power Program Since World War II" in TVA: Fifty Years of Grass Roots Bureaucracy, Erwin C. Hargrove and Paul K. Conkin, eds., (Urbana, 1983), 233–34. On the cutting of the appropriation for the New Johnsonville steam plant, see ibid. For Wessenauer's opinion of steam versus hydroelectric power, see interviews with J. Porter Taylor and George Palo, Apr. 7 and 11, 1983. For Clapp's comment, see Gordon R. Clapp, The TVA: An Approach to the Development of a Region (Chicago, 1955), 3. For the move to steam and congressional reaction, see Aaron Wildavsky, Dixon-Yates: A Study in Power Politics (New Haven, 1962), 10–13.

13. On Wessenauer, see interviews with J. Porter Taylor and George Palo, Apr. 7 and 11, 1983. For a list of coal-fired steam plants begun by TVA between 1949 and 1952, see Harry Wiersema, "The River Control System," in Martin, ed., TVA, The First Twenty Years, 89. For breaking of congressional logjam see Wildavsky, Dixon-Yates, 12–15.

14. On early land purchase policy, see McDonald and Muldowny, TVA and the Dispossessed, Ch. 4; Claude W. Nash, "Reservoir Land Management," in Martin, ed., TVA, The First Twenty Years, 138–39. For Harcourt Morgan's comment on tourism, see interview with Robert Howes, Apr. 19, 1983.

15. For the "Jandrey Report," see A. S. Jandrey, "Review of Reservoir Land Policy for the Kentucky Project" (unpubl. TVA report, July 1944); see esp. 3, 46 for conclusions. For the fear that others would force TVA to dispose of surplus property, see interview with Robert Howes, Apr. 19, 1983. For the progress of property sales to 1960, see L. J. Van Mol to L. F. Goyette (Bureau of the Budget), Jan. 28, 1960, and Herbert Vogel to Congressman Clifford Davis, Aug. 5, 1960, DRPF. For overview of the issue, see Nash, "Reservoir Land Management," in Martin, ed., TVA, The First Twenty Years, 138–39.

16. Martin, "Retrospect and Prospect," in Martin, ed., TVA, The First Twenty Years, 226.

17. For Eisenhower's post-inaugural statement, see Herbert S. Parmet, *Eisenhower and the American Crusades* (New York, 1972), 174.

18. On Ike's budget cuts, see Wildavsky, *Dixon-Yates*, 20. On the "creeping socialism" remark, see *New York Times*, June 12, 1953, and Wilmon H. Droze, "The Tennessee Valley Authority, 1945–1980: The Power Program" (paper delivered at Vanderbilt Institute for Public Policy Studies Conference, Dec. 3–5, 1981), 14. For remark at cabinet meeting, see Marguerite Owen, *The Tennessee Valley Authority* (New York, 1973), 99.

19. The best treatment of the whole complicated affair is Wildavsky, *Dixon-Yates*. Without it, the explanation by Marguerite Owen (*The Tennessee Valley Authority*, 103–13) is barely comprehensible.

20. On the effort to get Clapp to resign before his term was up, see Sherman Adams to Joseph Dodge, May 20, 1953, cited in Parmet, *Eisenhower and the American Crusades*, 467. For Eisenhower's opinion of Clapp see Owen, *The Tennessee Valley Authority*, 113.

21. For news conference, see *New York Times*, May 20, 1954. For Adams's statement, see Adams to Sheila Turney, Jan. 26, 1954, cited in Parmet, *Eisenhower and the American Crusades*, 467. For number of employees, see Clapp, *The TVA*, v.

22. On the Carbaugh choice and abandonment, see Parmet, *Eisenhower and the American Crusades*, 467.

23. On the formation of interest groups, see Couto, "New Seeds at the Grass Roots," in Hargrove and Conkin, eds., *TVA*, 230–60. Similar groups formed after 1954 included the Associated Tennessee Valley Chamber of Commerce (1958) and the Tennessee Rural Electric Cooperative Association (1961).

24. On petition see Parmet, *Eisenhower and the American Crusades*, 467. For Vogel's appearance before the Senate committee and his confirmation, see *New York Times*, Aug. 10–11, 1954; *Congressional Record*, 100, 13988; Wildavsky, *Dixon-Yates*, 144. For more on Vogel and on the belief that he was sent to dismantle TVA, see Erwin C. Hargrove, "The Task of Leadership: The Board Chairman of the TVA" in Hargrove and Conkin, eds., *TVA* 105–106.

25. The best treatment is Wilmon H. Droze, "The Tennessee Valley Authority, 1945–1980: The Power Program," in Hargrove and Conkin, eds., *TVA*, 73–77. See also George E. Rawson, "The Process of Program Development: The Case of TVA's Power Program" (Ph.D. diss., Univ. of Tennessee, Knoxville, 1978), esp. 67.

26. See Droze, "The Tennessee Valley Authority, 1945–1980," 17–20. See also Rawson, "The Process of Program Development," 67.

27. North Callahan, *TVA: Bridge Over Troubled Waters* (South Brunswick, N.J., 1980), 203–205. Interview with A. J. Gray, Mar. 9, 1982. For the views of nuclear power, see *TVA Annual Report*, 1959. See also Droze, "The Tennessee Valley Authority, 1945–1980," 23.

28. *Public Papers of the Presidents of the U.S.; Dwight D. Eisenhower 1953, Jan. 20–Dec. 31, 1953*, #153, pp. 528–33.

29. It is fairly evident that TVA engineers had little patience for cooperat-

ing with anyone outside the agency. See interviews with George Palo, Apr. 7, 1983, and J. Porter Taylor, Apr. 11, 1983.

30. On Wessenauer's shift, see G. O. Wessenauer, "A Short History of TVA" (remarks to Power Distributors, Electrical Development Personnel Training Institute, Jun. 18, 1962), TL; see also interview with A. J. Gray, Mar. 9, 1982. The decline of Agriculture within TVA mirrored agriculture's decline throughout the Tennessee Valley. Agriculture's share of the valley's combined income had declined from 23 percent in 1929 to 18 percent in 1946 and would reach 11 percent by 1953. The percentage of valley residents employed in agriculture also deteriorated, from 56.4 percent in 1929 to 29.1 percent in 1953.

31. The above and some of the following relies on a biographical sketch by Callahan, TVA: Bridge Over Troubled Waters, 246–47. Other material has been gleaned from interviews of Wagner's longtime associates within TVA, esp. George Palo, J. Porter Taylor, Don Mattern, Paul Evans, Robert Howes, and an interview with Wagner himself.

32. Much of the above is speculation based on clues offered in interviews with Palo, Taylor, Evans, Mattern, and Howes.

33. On Wagner's early strict constructionist ideas, see interview with Howes, Apr. 19, 1983. On the Lutheran connection, see interview with Dr. Kerry Schell, Apr. 5, 1983.

34. On Wagner's change of philosophy, see interview with Howes, Apr. 19, 1983. For good examples of Wagner's later thinking on TVA's mission, see Aubrey J. Wagner, "The Future of TVA," in The Economic Impact of TVA, John R. Moore, ed. (Knoxville, 1967), esp. 151–59; see also Knoxville Journal, Feb. 12, 1966, and Hargrove, "The Task of Leadership," in Hargrove and Conkin, eds., TVA, 108–12. On Vogel, see ibid., 105–108. Many talk of Wagner's influence on Vogel, and in interviews Evans, Taylor, and Howes all mentioned it.

35. For TVA's early interest in tributaries, see Donald T. Wells, The TVA Tributary Development Program (University, Ala., 1964), 1–3. For Wagner's ideas, see interview with Wagner, Mar. 9, 1982; interview with Gray, Mar. 9, 1982; Hargrove, "The Task of Leadership," in Hargrove and Conkin, eds., TVA, 109. On the committee, see Wagner, "The Future of TVA," 151. See also interview with Kilbourne, Feb. 12, 1982. For "factories in fields" statement, see interview with Gray, Mar. 9, 1982.

36. Todd to Wagner, Aug. 4, 1958, GMF.

37. Wagner to Todd, Aug. 7, 1958, GMF. For suggestions to Wagner, see J. Ed Campbell to Wagner, Oct. 26, 1959, DRPF. For board's decision, see L. J. Van Mol to L. F. Goyette, Jan. 28, 1960, DRPF. See also Nashville Tennessean, Apr. 17, 1960, and Knoxville News-Sentinel, Apr. 6, 1960.

38. For failure to acknowledge an important committee report which displeased Wagner, see interview with Howes, Apr. 19, 1983, and below, ch. 3.

CHAPTER 2

1. For creation of FUDAR, see Wagner to Those Listed, Apr. 2, 1959, in FUDAR Files, Agriculture Division, Muscle Shoals. For Future Hydro Projects Committee, see Palo to Wagner, July 22, 1959, in Administrative Files, OTAD. For pressure on Elliot, see interview with Don Mattern, former chief of Project Planning Branch, Apr. 18, 1983.

2. The July 7 meeting was recalled by George Palo in an interview on Apr. 7, 1983. To get a benefit-cost ratio, the benefits of a proposed project are calculated in dollar terms and then set against the project's anticipated cost to produce a ratio, such as 1.5 (benefits):1 (costs). In his meeting with Palo, Elliot was speaking in a common shorthand. When he spoke of a "benefit-cost ratio of .25," he was actually saying a benefit-cost ratio of .25:1.0, a ratio which augured ill for a project. Normally a project would not be considered seriously unless it could show a benefit-cost ratio of at least 1:1. The use of benefit-cost ratios became prevalent at TVA, the Corps of Engineers, and elsewhere in the 1950s.

The spelling of Loudoun (or Loudon) merits a word of explanation. The correct spelling of the person's name for whom the fort, county, town, and dam were named was Loudoun. However, when the county was organized in 1870, apparently no one bothered to look up the correct spelling, and the county's name was spelled Loudon. The town of Loudon followed the county spelling. When the fort was reconstructed in the 1930s, it returned to the correct spelling, Fort Loudoun. TVA followed suit, naming its dam, completed in 1941, Fort Loudoun Dam. We have used the respective spellings that the people themselves adopted, Loudon for the county and town, and Loudoun for the fort, dam, and extension project.

3. Palo to Elliot, July 8, 1959; Robert E. Frierson to Elliot, July 10, 1959; and Elliot to Frierson and Mattern, July 13, 1959; all WCPF.

4. Wagner to Palo, Aug. 9, 1960, PPBF. For name change, see Evans to Wagner, June 22, 1960, GMF.

5. Wagner to Palo and Howes, Aug. 9, 1960, PPBF. This memo was the result of a conversation among Wagner, Palo, and Elliot. Interview with Palo, Apr. 7, 1983.

6. Until 1940 what was to become the Fort Loudoun Dam was known as the Coutler Shoals Project. To simplify matters, we have used the modern term. Many groups, including the Tennessee Society, Daughters of the American Colonists, petitioned TVA to name the dam Fort Loudoun Dam. See Jennie W. (Mrs. R. J.) Yearwood to Clapp, Mar. 9, 1940, AF. For official name change, see TVA Office of the General Manager, Administrative Memorandum No. 109-0 (May 8, 1940), AF. Other groups, who wanted the dam named for Admiral David Farragut, an East Tennessee native who served the Union cause in the Civil War, lost out.

7. Interview with Taylor, Apr. 11, 1983. See also Taylor to C. T. Barker, May 13, 1936, OCENTF. Early in his career, J. Porter Taylor called himself James P. Taylor.

8. C. T. Barker to James S. Bowman, Aug. 29, 1939, TVA Commerce Dept. File. For one later study, complete with maps, see John B. Blandford, Jr. (general manager), to Forrest Allen, Nov. 23, 1937, OCENTF. For decision to build a second dam and a canal, see Bowman to Dana Wood, Sept. 1, 1939, OCENTF.

9. For Harcourt Morgan's concern, see W. J. Hayes to John B. Blandford, Jr., Feb. 11, 1938; and Carl A. Bock to Blandford, Mar. 26, 1938; OAF. The two options are best summarized in Howard K. Menhinick to John I. Snyder, Apr. 15, 1941, and Neil Bass, Howard K. Menhinick, and John I. Snyder to Gordon Clapp, Apr. 11, 1941, OAF.

10. For various board decisions on land purchase policies for the Fort Loudoun Project, see A. Fletcher Percefull to Snyder, June 20, 1941; Gordon Clapp to Snyder, Aug. 26, 1941; Percefull to Snyder, Sept. 5, 1941; Clapp to Board, Nov. 5, 1941; Clapp to Snyder, Nov. 7, 1941; all OCENTF. For local protests at Fort Loudoun and Guntersville, see Snyder to Clapp, Oct. 3, 1941, OCENTF. For early indications of what the ultimate policy would be, see Paul Carringer and E. N. Torbert to E. S. Draper, May 19, 1938, in AF. But the question did not end there. See Clapp to Snyder, Oct. 3, 1941; Clapp to Board, Oct. 10, 1941; Morgan to Clapp, Oct. 14, 1941; Clapp to Snyder, Nov. 7, 1941; Clapp to Board, Nov. 5, 1941; all OCENTF.

11. Nearly every former landowner contacted by the authors believed that TVA intended to purchase only that land which would be inundated. See esp. interviews with Alfred David, Paul and Charlotte Hughes, Louise Stamey, Liona Cain, Margaret Sexton, Nellie McCall, Mar. 16, 1983.

12. For the intra-agency discussion of the fourth generating unit see Dana M. Wood to James S. Bowman, Sept. 1, 1939, OCENTF; J. A. King to T. B. Parker, Nov. 14, 1940; Parker to Clapp, Dec. 27, 1940; Parker to Krug, Dec. 27, 1940; Parker to G. R. Rich, Dec. 27, 1940; Krug to Parker, Dec. 20, 1940, all AF; Paul W. Ager to Clapp, Jan. 23, 1941, and Clapp to Board, Jan. 28, 1941, both in TVA David Lilienthal Records, microfilm roll 48. For request for authorization and approval see TVA, "Request for Authorization," Serial No. 200.2, Jan. 9, 1941, AF.

13. Parker to Clapp, Jan. 13, 1941; Parker to A. L. Pauls (chief construction engineer), Jan. 31, 1941; Clapp to Parker, Feb. 6, 1941, all AF.

14. For TVA's version of the McKellar episode, see Owen, Tennessee Valley Authority, 86–93. For one example of McKellar's wrath (quoted above), see McKellar to Lilienthal, Oct. 2, 1941, OCENTF.

15. TVA Water Control Planning Department [W. L. Voorduin], The Fort Loudoun Extension on the Little Tennessee River, Report 10–81 (Knoxville, Apr. 1942); Parker to Clapp, Apr. 27, 1942, CF; TVA Activity Authorization, Serial No. 200.6 (Apr. 28, 1942), and Paul W. Ager to Clapp, June 29, 1942, AF. For board approval, see A. Fletcher Percefull to Ager, July 10, 1942, AF. For submission to WPB, see Clapp to Ager, July 30, 1942, AF.

16. For WPB delay, see C. E. Nichols to Parker, July 25, 1942, DWM, Project General—Tellico; C. H. Garity to Herbert S. Marks (WPB), n.d., AF. For reasons and subsequent withdrawal, see A. S. Jandrey to Files, Aug. 7, 1942, AF.

17. For 1945 TVA priority lists see Clapp to Baird Snyder (Federal Works Agency), Aug. 4, 1945, with attachments, OCENTF. Glover was pressing Clapp because the Milk Producers Association was trying to increase dairy farming in the area but was hesitant because of the uncertainty surrounding the Fort Loudoun Extension Project. For that and Clapp's reply, see Clapp to Files, Apr. 11, 1946, AF. For Oliver's statement on costs, see Oliver to Kefauver, Oct. 16, 1951, WCPF, Project General—Expressions of Public Interest—Tellico. For similar letters from TVA, see Clapp to Kefauver, Nov. 7, 1951, and Marguerite Owen (TVA's Washington representative) to McKellar, Jun 23, 1952, AF.

18. For Wagner's standard response see his letter to Kefauver, Mar. 21, 1957, WCPF—Public Interest.

19. Interview with A. J. Gray, Mar. 9, 1982. For Wagner's statement, see *Knoxville Journal*, Feb. 12, 1966. For a fuller discussion of the subject by Wagner, see his "The Future of TVA," in Moore, ed., *The Economic Impact of TVA*, 151–54. See also interview with Wagner, Mar. 9, 1982.

20. U.S., *Statutes at Large*, LXVIII, 666–668; U.S., *Executive Order 10913* ("Rules and Regulations Relating to the Administration of the Watershed Protection and Flood Prevention Act"), Dec. 18, 1954. For a brief discussion of the national attention being paid to tributary area development, see *The Mid-Century Conference on Resources for the Future* (Washington, 1953), 123 and cited in Ralph F. Garn's very helpful "Tributary Area Development in Tennessee: TVA's Changing Perspective" (Ph.D. diss., Univ. of Tennessee, Knoxville, 1974), 52.

21. For a good discussion of the origins of the concept at TVA, see Wells, *The TVA Tributary Development Program*, 1–3.

22. Interviews with Kilbourne, Feb. 12, 1982, and Wagner, Mar. 9, 1982; Garn, "Tributary Area Development in Tennessee," 68–72.

23. Robert E. Lowry, "Beech River Experiment" (unpub. TVA report, OTAD), 4; interview with Kilbourne, Feb. 12, 1982.

24. Interview with Ralph Dimmick, a professor in the Univ. of Tennessee College of Agriculture, Mar. 29, 1983.

25. Interviews with Kilbourne, Feb. 12, 1982, and Gray, Mar. 9, 1982.

26. The above comes from conversations with a number of present and former TVA employees, some of whom had known Wagner from the time he joined the agency in 1934. Examples of all the above characteristics will appear in later chapters.

27. TVA [Robert E. Lowery], *Working with Areas of Special Need, With Examples from the Beech River Watershed* (Knoxville, 1953), 6–12; interview with Kilbourne, Feb. 12, 1982.

28. TVA, *Working with Areas of Special Need*, 6–12; interview with Kilbourne, Feb. 12, 1982; interview with Wagner, Mar. 9, 1982; Garn, "Tributary Area Development in Tennessee," 72–79.

29. Vogel to Joe Sir (Pres., Elk River Development Assn.), July 9, 1959, Administrative Files, OTAD; Garn, "Tributary Area Development in Tennessee," 98–102.

30. OTAD *Annual Reports* (Knoxville, 1963-71); interview with Kilbourne, Feb. 12, 1982.

31. For creation of FUDAR, see Wagner to Those Listed, Apr. 2, 1959, FUDAR File, OACD.

32. Interview with Howes, Apr. 19, 1983; Howes to Wagner (on first meeting), June 19, 1959, in Elliot's FUDAR File, WCPF.

33. Interview with Howes, Apr. 19, 1983. For the long efforts of Howes and the "recreation people" to appraise recreation benefits, see Memorandum, [?] to Williamson, McCarley, and Parrott, Aug. 7, 1952, in FUDAR File, OACD.

34. For Wagner's original hope for a July 1959 progress report, see Wagner to Those Listed, Apr. 2, 1959, in FUDAR File, OACD. For problems, see Perry to Howes, July 12, 1960; Howes to Wagner, June 19, 1959; and Taylor to Kilbourne, Mar. 22, 1960; all in Elliot's FUDAR File, WCPF.

35. For Wagner's early familiarity with the site, see Wagner to Arthur Schweier, Mar. 22, 1938, OCOMF.

36. Wagner to Palo and Howes, Aug. 9, 1960, PPBF; Palo to Those Listed, Aug. 18, 1960, GMF.

37. For Wagner's opinion of benefit-cost ratios, see interview with Howes, Apr. 19, 1983. For Palo's opinion, see interview with Palo, Apr. 7, 1983.

38. Agenda, FUDAR Committee Meeting, Sept. 29, 1960, in FUDAR Files, Agriculture; Ashford Todd, Jr., to Howes, Sept. 23, 1959, DRPF; Allbaugh to Palo and Howes, Sept. 23, 1960, DRPF; Howes to Wagner, Oct. 14, 1960, and E. L. Baum to Jack Knetsch, Aug. 10, 1960, FUDAR Files, OACD.

39. Howes to Those Listed [FUDAR members], Jan. 6, 1961, FUDAR Files, OACD. See also Howes to Wagner, Jan. 5, 1961, FUDAR Files, OACD.

40. Frierson to Elliot, Jan. 18, 1961, and Rutter to Elliot, Jan. 17, 1961, in Elliot's FUDAR File, WCPF.

41. For the final report, see Howes to Van Mol, Apr. 7, 1961, with attachment, FUDAR Files, OACD. For the failure to acknowledge and Howes's conversation with Shelley, see interview with Howes, Apr. 19, 1983.

42. The work from 1961 to 1966 makes an interesting, if frustrating, story, too long to be told here. See Allbaugh to Gerald Williams, Oct. 31, 1961, FUDAR Files, OACD; Howes to Van Mol, Dec. 18, 1961, GMF; and E. S. Brosell to Files, May 10, 1962, PPBF. Also see the following in GMF: E. P. Ericson to Van Mol, Mar. 20, 1963; Van Mol to Howes, Apr. 4, 1963; Wagner to Van Mol (in which he said, "It's almost frightening to read that we have studied since 1948 and still don't know"), Apr. 22, 1963; Ericson to Van Mol, Oct. 21, 1963; Van Mol to Todd, Oct. 29, 1963; C. W. Nash to Van Mol, June 12 and Sept. 18, 1964, and Apr. 23, 1965; and Nash to Van Mol, Aug. 13, 1965. For Shelley's marginalia, see his comments attached to Nash to Van Mol, May 19, 1965, and cover of FUDAR study, June 1, 1966, both GMF. On McGruder, see interview with Mattern, May 5, 1983. On Knetsch, see Howes to Shelley, Jan. 12, 1962; Howes to B. M. Haskin, Jan. 18, 1962; Howes to Van Mol, Jan. 18, 1962; Shelley to Howes, Jan. 22, 1962; Howes to Van Mol, July 20, 1962; Howes to Van Mol, Aug. 7, 1962; Howes to Van Mol, Sept. 19, 1962;

Howes to Howes to Van Mol, Aug. 7, 1962; Howes to Van Mol, Sept. 19, 1962; Howes to Robert Coker, Oct. 18, 1962; June Williams to Dorothy Morgan, Jan. 15, 1963; June Williams to Edith Spradlin, Feb. 13, 1963; Howes to Van Mol, Mar. 12, Apr. 10, and Nov. 26, 1963; and Ericson to Van Mol, Aug. 24, 1965, all GMF.

43. For the studies of the Tellico area, see Memorandum, Jan. 15, 1960; Berlen C. Moneymaker to Robert J. Coker, Jan. 18, 1960; W. F. Emmons to Elliot, Feb. 10, 1960; T. C. McCarley to Coker, Feb. 11, 1960; Taylor to Mattern, Mar. 30, 1960; all PPBF. For 1:1 benefit-cost ratio, see Mattern to Elliot, Mar. 31, 1960, PPBF. For plan, elevation and sections drawings see Wm. E. Loose, "Dam, Powerhouse & Spillway—Plan, Elevation and Sections, Tellico Project," Apr. 21, 1960, PPBF. For Wagner's order, see Wagner to Palo and Howes, Aug. 9, 1960, PPBF.

44. The identification of Kennedy as a neo-New Dealer was not universally held. See Theodore C. Sorensen, *Kennedy* (New York, 1965), 86, 118, 123. For Kennedy's 1963 statement at Muscle Shoals see Moore, ed., *Economic Impact of TVA*, 159. For the TVA planner's statement, see Stephen J. Rechichar and Michael R. Fitzgerald, "The Big Dam and the Little Fish: TVA's Tellico Dream in an Era of Intragovernmental Regulation" (paper prepared for April 1983 ASPA National Conference on Public Administration, New York), 7. For TVA's identification of Kennedy as sympathetic to the agency see interview with Kilbourne, Feb. 12, 1982.

45. Taylor to Palo, Sept. 15, 1960; Kilbourne to Palo, Sept. 19, 1960; Howes to Palo, Sept. 23, 1960; Gartrell to Palo and Howes, Sept. 23, 1960; Hamilton to Project Planning Branch Files, all PPBF. For Wessenauer's opinion, see Wessenauer to Palo, Sept. 28, 1960, PPBF. For shaving of costs in Palo's statement, see Palo to Wagner, Feb. 10, 1961, GMF.

46. *Knoxville News-Sentinel*, Jan. 1, 1961.

47. Ibid., Jan. 1 and 13, 1961; *Knoxville Journal*, Jan. 13, 1961.

48. *News-Sentinel*, Jan. 15, 1961. See also *Knoxville Journal*, Feb. 15, 1961.

49. For the interesting story of reintering the dead at Norris, see McDonald and Muldowny, *TVA and the Dispossessed*, Ch. 6. For some early endorsements, see Madisonville Jaycees to Vogel, Apr. 1, 1961, and Lenoir City Regional Planning Commission to Wagner, Apr. 8, 1961, in TVA Newspaper File, TL. Though more will be said later about the Knoxville Chamber of Commerce, for now see interview with Evans, Apr. 7, 1983.

50. Appleby to Kennedy, Feb. 23, 1961; Appleby to Kefauver, Mar. 15, 1961; Gore to Vogel, Mar. 29, 1961; and Jones to Gore, Apr. 7, 1961; all PPBF. For other examples of this posture, see Vogel to Kefauver, Apr. 13, 1961, PPBF; Reed Elliot to Alice W. Milton, Jan. 19, 1961, in Archaeological Exploration File: Old Fort Loudoun, WCPF; Van Mol to Loudon Regional Planning Commission, Apr. 19, 1963, OTAD Files.

51. Zygmunt Plater was an assistant professor in the Univ. of Tennessee's College of Law and one of the leaders of anti-Tellico forces in the 1970s. His role will be discussed later.

52. For a preliminary land value enhancement study, see Robert Coker to Howes, Dec. 29, 1960, PPBF. See also Elliot to Frierson, Jan. 3, 1961, PPBF;

E. P. Ericson to Van Mol, Mar. 6 and 9, 1962, GMF; Palo to Van Mol, Mar. 19, 1962, PPBF. See also TVA Division of Water Control Planning, Project Planning Branch, "Tellico Project: Economic Appraisal," report submitted Oct. 1962. For ten-year waiting period, see ibid. For Palo's later opinion, see Palo to S. David Freeman, Jan. 31, 1979, COF. On maps, see Loose to Gangwer, Feb. 1961, and Mattern to Coker, Sept. 21, 1962, in PPBF. For residents' confusion about benchmarks, see interview with Alfred Davis, Mar. 16, 1983, and interview with J. Porter Taylor, Apr. 11, 1983.

53. For meeting with Bowaters, see Kilbourne to GM Files, Feb. 6, 1961, PPBF. On TVA's long history of trying to accomodate the Fort Loudoun Association, see H. L. Callan to H. A. Morgan, July 31, 1942; George F. Gant to Ray H. Jenkins, June 10, 1947; Vogel to Arthur M. Fowler, Dec. 22, 1960; Elliot to Alice W. Milton, Jan. 19, 1961; all Land Use, Archaeological Explorations Folder—Tellico, WCPF.

54. For early meetings of LTRVDA see *Maryville-Alcoa Times,* Apr. 18, 1963, and *Loudon County Herald,* Apr. 18, 1963. A short discussion of LTRVDA can be found in Rechichar and Fitzgerald, "The Big Dam and the Little Fish," 12–13. Subsequent activities, division, and dissolution will be discussed in detail later.

55. On assumption, see TVA Project Planning Branch, "Tellico Project: Economic Appraisal" (Oct. 1962), 7. For other examples of work on benefit-cost ratio, see Howes to Elliot, Feb. 14, 1962; Elliot to Palo, Mar. 13, 1962, and attached "Economic Appraisal" (Mar. 1962); Palo to Van Mol, Mar. 19, 1962; Coker to Mattern, June 1, 1962; Palo to Van Mol, July 5, 1962; Mattern to Coker, Sept. 21, 1962; Palo to Those Listed, Jan. 10, 1963; Kenneth Seigworth to Palo, Jan. 25, 1963; Ericson to Palo, Jan. 25, 1963; Coker to Todd, Jan. 28, 1963; Taylor to Palo, Jan. 29, 1963; Todd to Palo, Jan. 30, 1963; Palo to Van Mol, Feb. 7, 1963; all PPBF.

56. 87th Congress, 2nd Session, Senate, "Policies, Standards, and Procedures in the Formulation, Evaluation, and Review of Plans for Use and Development of Water and Related Land Resources" [Senate Document 97] (May 29, 1962). On varying ratios, see Elliot to Palo, Mar. 13, 1962, and Palo to Van Mol, Feb. 7, 1963, both PPBF.

57. On Wessenauer and Seigworth, see interviews with Kerry Schell, Mike Pelton, and Ralph Dimmick, Mar. 29, 1963. On A. R. Jones, in a Feb. 1965 memo to E. A. Shelley on land purchases, Jones quipped sarcastically, "There must be some land in Tennessee that we will not need!" See Jones to Shelley, Feb. 9, 1965, GMF. On board approval, see Van Mol to Palo, May 7, 1963, OTAD. On Tellico and TVA's new mission, see Kilbourne to John C. Mitchell, May 8, 1963, OTAD.

58. Gray to Files, May 28, 1963, PPBF; *Knoxville Journal, Clarksville Leaf-Chronicle, News-Sentinel,* and *Oak Ridger,* Sept. 4, 1963.

59. For a judicious evaluation of TVA's first project, see McDonald and Muldowny, *TVA and the Dispossessed,* esp. Ch. 9.

CHAPTER 3

1. The portraits below are composites of impressions drawn from observation by the authors and from interviews with numerous valley residents (see following notes for those who allowed us to use their names).

2. "Prospect" from *The New Harp of Columbia* (Nashville, 1919), 15. As far as we can tell, "sacred harp" singing, though once popular, is no longer practiced in the area.

3. The description that follows is of the Little Tennessee River Valley prior to the creation of the Tellico Reservoir in 1979.

4. A good source on the very early history is Inez E. Burns, *History of Blount County, Tennessee: From War Trail to Landing Strip, 1795-1955* (Nashville, 1957), 14-38; Loudon County Citizens Advisory Council, "Loudon County" (unpubl., n.d.), copy in NDRSF. On early settlements in Knox County, see Mary U. Rothrock, ed., *The French Broad-Holston Country: A History of Knox County, Tennessee* (Knoxville, 1946), 22-33; 326-54. Blount County was carved out of Knox County in 1795. Burns, *History of Blount County*, 31.

5. On the difference between northern and southern migration patterns, see Peter D. McClelland and Richard Zeckhauser, *Demographic Dimensions of the New Republic: American Interregional Migration, Vital Statistics, and Manumissions, 1800-1860* (Cambridge, 1982), 7.

6. On Morganton, see Sarah G. Cox Sands, *History of Monroe County, Tennessee: From the Western Frontier Days to the Space Age* (Baltimore, 1982), vol. I, pt. 2, p. 283. On other towns, see ibid., 1, 277, 283. On early families, see Land Purchase Records—Tellico, DPS.

7. On the grants for education and churches, see Land Purchase Records—Tellico, (tracts 22-101, TELR 3907, TELR 5209, TELR 5308), DPS.

8. On the temperance petition, see Sands, *History of Monroe County*, I, pt. 2, 59-60.

9. See TVA, Division of Water Management, *Cemetery Relocations for the Tellico Project* (Chattanooga, Mar. 1977), 1-35. TVA discovered six slave cemeteries.

10. On Union sentiment, see Alberta and Carson Brewer, *Valley So Wild: A Folk History* (Knoxville, 1975), 159; Burns, *History of Blount County*, 59. On Knoxville engagements, such as they were, see Maury Klein, "The Knoxville Campaign," *Civil War Times Illustrated*, X (Oct. 1971), 4-10 and 42.

11. Interviews with Louise Stamey and Thomas Burel Moser, Mar. 16 and 31, 1983, and with James "Buster" Carver, Mar. 30, 1983.

12. Ibid. Also interviews with Charlotte and Paul Hughes, Mar. 16, 1983, and Reece Nicholson and Louise Kyker, Mar. 31, 1983.

13. Ibid. On McLemore and similar family names, see Land Purchase Records—Tellico, DPS. For family rumor, see interview with Billy Wolfe, Mar. 16, 1983.

14. Maynard, Gray, and Kennedy material (and voluminous other examples)

are in Land Purchase Records—Tellico, DPS. Tract numbers are Maynard (TELR 4511), Gray (TELR 3621) and Kennedy (TELR 3513). For a general discussion of inheritance patterns in Appalachia, see F. Carlene Bryant, *We're All Kin: A Cultural Study of a Mountain Neighborhood* (Knoxville, 1981), 40, 68–74. Bryant also studies changes in those patterns, and for this her work is particularly valuable; see esp. 67.

15. On J. C. Anderson, see material on tract TELR 1919 in Land Purchase Records—Tellico, DPS. For other material, see interview with Charlotte and Paul Hughes, Mar. 16, 1983.

16. Economic disparities drawn from Land Purchase Records—Tellico, DPS. Those records trace the ownership of each tract which TVA either purchased or condemned back to the late nineteenth or early twentieth century. Other material comes from sitting with a male conversation group on the morning of Mar. 31, 1983, at the Teddy Bear Restaurant in Vonore.

17. Interview with Louise Stamey, Mar. 31, 1983.

18. Interview with Charlotte Hughes, Mar. 16, 1983. The frequency of mortgaged property was taken from Land Purchase Records—Tellico, DPS.

19. Interviews with Louise Stamey, Mar. 31, 1983, and Louise Kyker, Mar. 31, 1983.

20. Ibid. See also interviews with James "Buster" Carver, Mar. 30, 1983, and Charles Hall, Mar. 31, 1983. The Coors incident was related to us by Louise Stamey. Incidentally, according to her, she was the sole person who did not stand up. "Many who put money in the collection plate worked for ALCOA where they made the cans [beer cans]."

21. All the interviewees who provided the material above were quick to point out that all the problems they mentioned "are worse now." One man, who refused to be identified, made it clear that he believed society was in worse shape now, and then listed a host of offenses he remembers having been committed long ago in his community, including unpremeditated murder.

22. *U.S. Census: Agriculture*, 1880, 1900, 1910. The yield statistics are for Monroe County.

23. On population increase, see Tennessee State Planning Commission, "Preliminary Population Report—General Population Statistics and Trends," bound volume (Nashville, 1935). In 1930 Monroe County, which includes most of the lower valley, had a crude birth rate of 30.9, compared to 20.6 for the nation and 30.1 for the Southern Appalachian Region. In that same year, Monroe County's fertility ratio (the ratio of children aged 0–4 to women of childbearing age, 15–44) was 603.7, compared to 390.0 for the U.S. and 585.0 for the Southern Appalachian Region. Gordon F. DeJong, *Appalachian Fertility Decline: A Demographic and Sociological Analysis* (Lexington, Ky., 1968) Appendix A, Tables 1 and 2. For infant mortality data from fragmentary cemetery records, see TVA Division of Water Management, *Cemetery Relocations for the Tellico Project*.

24. On average farm size and the increase in improved farmland, see Tennessee State Planning Commission, "Preliminary Population Report." From 1900 to 1910, Blount County's persons per square mile increased from

16.8 to 36.4, while Monroe County's figure rose from 18.7 to 30.8. Ibid.

25. On tenancy in 1880, see U.S. Census, 1880. For 1900, see Tennessee State Planning Commission, "Preliminary Population Report." Interview with Charlotte Hughes, Mar. 16, 1983.

26. On railroad, see Sands, *History of Monroe County*, vol. I, pt. 2, p. 286.

27. On Babcock, see Sands, *History of Monroe County*, I, pt. 2, 480-89; Ray H. Jenkins, *The Terror of Tellico Plains: The Memoirs of Ray H. Jenkins* (Knoxville, 1978), 13-14. On the 1925 fire, see Sands, *History of Monroe County*, vol. I, pt. 2, p. 489.

28. On ALCOA, see Burns, *History of Blount County*, 232-33; Richard Buckner, "Calderwood, Tennessee: An ALCOA Company Town" (M.A. thesis, Univ. of Tennessee, 1981). The best work on ALCOA's Tennessee operations has been done by Maryville College historian Russell Parker.

29. On Stokely and POW camp, see Sands, *History of Monroe County*, vol. I, pt. 2, p. 520. On REA and radios, see interview with Louise Stamey, Mar. 16, 1983.

30. Burns, *History of Blount County*, 265-66; Sands, *History of Monroe County*, vol. I, pt. 2, p. 277-81, 283, 293; interview with James "Buster" Carver, Mar. 30, 1983.

31. Interview with James "Buster" Carver, Mar. 30, 1983; *U.S. Census: Labor Statistics*, 1920 and 1930. On the Madisonville Knitting Mill, see Sands, *History of Monroe County*, Vol. I, pt. 2, p. 177.

32. On wild animals, see interview with James "Buster" Carver, Mar. 30, 1983. On boar hunts, see advertisement for guides in *Madisonville Democrat*, Nov. 18, 1936, and Sands, *History of Monroe County*, Vol. I, pt. 2, p. 550 and 563.

33. TVA Industry Division, *Agricultural-Industrial Survey of Loudon County, Tennessee*, dir. W. R. Woolrich (Knoxville, 1935), 51-52, 180-82. On tenancy, see *U.S. Census: Labor Statistics*, 1940. Approximately one-fourth of Monroe County's tenants were related to their landlords, a very high percentage. For the rate of people on relief rolls in Jan. 1935, see Tennessee State Planning Commission, "Preliminary Population Report."

34. Tennessee State Planning Commission, "Preliminary Population Report;" *U.S. Census: Agriculture*, 1910 and 1940. TVA Industry Division, *Agricultural-Industrial Survey of Loudon County*, 2; TVA Industry Division, *Agricultural-Industrial Survey of Monroe County, Tennessee*, dir. W. R. Woolrich (Knoxville, 1934).

35. Land Purchase Records—Tellico, DPS.

36. For outmigration, see report prepared for TVA by M. I. "Mike" Foster, director of the Division of Navigation Development and Regional Studies, attached to his memorandum to H. N. Stroud, Jr., Aug. 12, 1971, OTAD. Copies of that report can be found in several other TVA files, including GMF. For the dependency ratio for Monroe County, see Tennessee Department of Employment Security, *Social and Economic Characteristics of the Tennessee Population* (Nashville, 1963). That study also deals with outmigration, but the figures do not appear to be as reliable as Foster's. For more on the aging population, see Citizens Advisory Council, "Loudon County," 22. This aging

was clearly recognized by the people themselves. See interview with Jean Ritchey, Mar. 31, 1983.

37. *U.S. Census: Labor Statistics*, 1970; interview with Ben and Carolyn Snider, Mar. 27, 1983.

38. For unemployment, see *U.S. Census: Labor Statistics*, 1970. For women in the work force, see ibid. and Citizens Advisory Council, "Loudon County," 45–46. On median family income in Loudon County (which by 1960 was higher than that of the state as a whole), see ibid., 48.

39. Land Purchase Records—Tellico, DPS. Interviews with Louis Moser Stamey, Leona Moser Cain, and Thomas Burel Moser, Mar. 16, 27, and 31, 1983. The Mosers were one of the first families in the lower valley, owned large holdings (which by the mid-twentieth century had dwindled through division), and once owned slaves. As Louise Stamey said, "The Mosers have been here forever—and I've been here half that time." On Moser slaves, see TVA Division of Water Management, *Cemetery Relocations for the Tellico Project*.

40. Interview with Louise Stamey, Mar. 31, 1983. See also TVA Division of Health, *Health and Medical Facilities in the Tellico Area* (Knoxville, 1971).

41. TVA Division of Health, *Health and Medical Facilities*. Interview with Charles Hall, Mar. 31, 1983. Interview with Ben and Carolyn Snider, Mar. 27, 1983.

42. "Hold Fast to the Right," arr. by Albert E. Brumley, 1938, for the Stamps-Baxter Music Co., in Albert E. Brumley, *Olde Time Camp Meetin' Songs* (Camdenton, Mo., 1971), 27. Lyrics in public domain.

43. J. B. F. Wright, "Precious Memories," in ibid., 59. Lyrics in public domain.

CHAPTER 4

1. The small community of Greenback (pop. 285 in 1960) is located about two miles off Hwy. 411 and ten miles, as the crow flies, from the proposed Tellico Dam site. The high school, which serves the Greenback community and the surrounding area, is quite small, usually graduating under 40 students per year.

2. For reports of the Greenback High School meeting of Sept. 22, 1964, see *Knoxville News-Sentinel*, *Knoxville Journal*, and *Chattanooga News-Free Press*, Sept. 23, 1964; *Maryville-Alcoa Times*, Sept. 24, 1964; *Bristol Herald Courier*, Sept. 25, 1964. For the sour editorial comment, see *Madisonville Democrat*, Sept. 24, 1964.

3. The statement was made by the editor of the *Knoxville Journal* to David Dickey. Interview with Dickey, March 25, 1983. Surprisingly, the *Journal* did oppose the Tellico Project.

4. Wagner was informed by Judge Kinnard and Joe Bagwell that approximately 90 percent of the people of Sweetwater and Madisonville were

pro-Tellico. See Wagner notation, Sept. 28, 1964, COF. On Tellico Plains, see interview with Charles Hall, Mar. 31, 1983.

5. On tenants, see interviews with Charlotte and Paul Hughes, Albert Davis, and Louise Stamey, Mar. 16, 1983.

6. TVA Land Purchase Records for Tellico, housed in Chattanooga, show which land parcels carried mortgages and which did not. The vast majority of the parcels were free of encumbrances.

7. The Sniders became vocal opponents of the Tellico Dam, believing that TVA had betrayed them by destroying their community rather than enhancing it. Interview with Ben and Carolyn Snider, Mar. 27, 1983.

8. Ibid.

9. The above are the authors' impressions, gleaned from interviews with numerous local opponents. Most of these interviews took place at the Fort Loudoun Visitors' Center in early 1983, with the cooperation of Joseph Distretti, state park historian for Fort Loudoun.

10. Interview with Ben and Carolyn Snider, Mar. 27, 1983. For TVA statements which might well have confused the local populace, see A. J. Gray to Files, May 28, 1963, and Loose to Files, Aug. 12, 1963, PPBF; Knoxville News-Sentinel, Nov. 22, 1963.

11. The meeting was recalled by Dickey in an interview on Mar. 25, 1983. Although Dickey could not remember the exact date of the meeting, he was certain that it took place prior to TVA's announcement.

12. Ibid. For their claim about tourism, see Knoxville News-Sentinel, Nov. 17, 1963.

13. For the trip to Arkansas, see interview with Dickey, Mar. 25, 1983. For Burch's claim concerning southeastern lakes and tourism, see Knoxville News-Sentinel, Nov. 17, 1983. David Dale Dickey, "The Little Tennessee River as an Economic Resource: A Study of Its Potential Best Use" (Univ. of Oklahoma, Norman, 1964), printed and distributed by the Tennessee Game and Fish Commission, copy in GMF. For its publication by the Game and Fish Commission, see A. R. Jones to L. J. Van Mol, Oct. 16, 1964, GMF.

14. Chance to Seigworth, Dec. 17, 1963, in folder marked "Tellico Project Opposition," COF.

15. On McDade, see ibid; interviews with Dickey, Mar. 25 and Apr. 6, 1983.

16. Telephone conversations with Dr. Richard Myers III and with Mrs. Richard Myers, Jr., Jan. 3, 1984. For Myers as a member of the opposition, see Knoxville News-Sentinel, Nov. 17, 1963. For Myers's opinions, see "Digging for History at the Tellico Blockhouse," Tennessee Conservationist 30 (Sept. 1964), 4–6.

17. On Bowaters, see Gilbert Stewart to Van Mol, Oct. 23, 1964, and Paul Evans to Van Mol, Oct. 26, 1964, GMF. On Mayfield, see transcription of telephone call between Mayfield and Wagner, Oct. 5, 1964, COF.

18. Tennessee Outdoor Record, Sept. 13, 1963; Knoxville News-Sentinel, Sept. 22, Oct. 13, Nov. 17, Nov. 22, 1963, and Jan. 9, 1964; Nashville Tennessean, Oct. 15, 1963.

19. For Dempster, Smith, and the car-towings, see Michael J. McDonald and Bruce Wheeler, *Knoxville, Tennessee: Continuity and Change in an Appalachian City* (Knoxville, 1983), 137. For Mrs. Lotspeich's conservatism see "The Conservatives of Knoxville," *Fortune Magazine* 46 (July 1952), 110–20.

20. Interview with Dickey, Mar. 25, 1983.

21. Ibid. Brewer did write an article which appeared in the *News-Sentinel* on Nov. 17, 1963, and which reported that opposition was mounting. But he took no public stand himself. TVA replied to that article on Nov. 29, 1963.

22. Interview with Walter Criley, Feb. 16, 1984.

23. Ibid.

24. Ibid.

25. Ibid.

26. For critique of Brewer's article, see Seigworth to Van Mol, Nov. 29, 1963, GMF. The critique was actually drafted by Charles J. Chance (chief, TVA Fish and Wildlife Branch) for Seigworth. On identifying opposition leaders, see Chance to Seigworth, Dec. 17, 1963, and Evans to Van Mol, Sept. 11, 1964, GMF. On sample resolution, see Kilbourne to John M. Carson, Jr. (a large property owner at Tellico who supported the project), May 28, 1964, OTAD. On influential local residents, see Carson to List, May 18, 1964, OTAD.

27. On OTAD's role, see Garn, "Tributary Area Development in Tennessee." On Carson see Carson to List, May 18, 1964, OTAD.

28. The Greenback meeting was well covered by the local press. See fn 2, this chapter. The psychotherapist, Dr. David Glenn, was on the faculty of Baylor University Medical School and was no friend of the Tellico Project. See Glenn to S. David Freeman, July 13, 1978, DWM.

29. TVA learned about the meeting the day after it was held, and details of the Rose Island gathering come from TVA reports. See Marvis Cunningham to Files, October 15, 1964, OTAD. Cunningham's informant was Mrs. James G. Carson of Vonore. On the purloined map see Kilbourne to Files, October 5, 1964, OTAD.

30. Cunningham to Files, October 15, 1964, OTAD. Mrs. Carson told Cunningham that Hicks "probably" was being paid for his opposition by ALCOA and Bowaters Paper Company. That charge was never proven, and was vehemently denied by both corporations. Incidentally, the name of the opposition group formed at the Rose Island meeting, Association for the Preservation of the Little Tennessee River, was not used that night, and some believe it was conceived somewhat later. See GMF, Nov. 16, 1964. However, the October 14, 1964, report in the *Maryville-Alcoa Times* used the title.

31. Cunningham to OTAD Files, Oct. 16, 1964, GMF.

32. Ibid.

33. At the formation of the Tri-Counties Development Association, as usual OTAD's Marvis Cunningham was TVA's "man on the ground." See Cunningham to Kilbourne, Nov. 12, 1964, OTAD, and Cunningham to OTAD,

Nov. 19, 1964, PPBF. For reports on the formation and purpose of the TCDA, see *Madisonville Citizen-Democrat*, Nov. 18, 1964; *Lenoir City News-Banner* and *Knoxville News-Sentinel*, Nov. 19, 1964. The TCDA met formally with TVA directors on Nov. 18, 1964.

34. Bill Hildreth to Wagner, Oct. 23, 1964, COF; Cunningham to Files, Nov. 24, 1964, GMF. On the TVA officeholders in the TCDA, see Wagner to Van Mol, Oct. 27, 1964; Kilbourne to Van Mol, Nov. 10, 1964; Van Mol to Kilbourne, Nov. 27, 1964; all GMF. For strategy and materials, see Cunningham to Files, Nov. 24, 1964, and TVA White Paper, Nov. 25, 1964, both PPBF.

35. The inference of personal benefit and the "booster mentality" can be drawn from the participants' respective occupations. Among the original TCDA officers was a banker, a dentist, an oil distributor, and the owner of a local telephone company. Of the local residents who in 1966 appeared in Washington to testify in favor of the Tellico Project before the Public Works Subcommittee of the House Appropriations Committee, four were mayors, two were county judges, one was a county trustee, two were newspaper editors, three were real estate brokers and developers, four were attorneys, four were bankers, and eleven identified themselves as businessmen. See "List of members of delegation appearing before the Public Works Subcommittee," May 2, 1966, in Frank Bird to Marvis Cunningham, May 5, 1966, GMF.

36. Interview with Charles Hall, Mar. 31, 1983.

37. For results of newspaper poll, see *Monroe Citizen*, Oct. 14, 1964. For summary of support and opposition letters, petitions, and resolutions received by TVA, see Ericson to Wagner, Oct. 19, 1964, OTAD.

38. Stewart to Van Mol, Oct. 23, 1964, and (attached) Evans to Van Mol, Oct. 26, 1964, GMF.

39. Bagwell to Van Mol, Nov. 1964, GMF. For state resolution, see "Excerpt from resolutions submitted to annual convention at the Tennessee Farm Bureau Federation, Nashville, Nov. 18, 1964, GMF. See also "News Release," Tennessee Outdoor Writers Association, Dec. 1964, in TVA Newspaper Clipping File, TL. A small eruption took place when the Tennessee Outdoor Writers Association quoted TVA Director Frank Smith as having said TVA "was glad to see the Farm Bureau go against the dam because it will help us in Washington. . . . You know the people up there don't care about the farmer." TOWA News Release, Dec. 1964, GMF. Smith vehemently denied having made the statement, calling the news release an "outright fabrication reflecting the desperation of selfish interests seeking to block a project which has great promise for the economic future of the area." TVA news release, Dec. 18, 1964, GMF. The authors find the initial statement uncharacteristic of Smith.

40. "Abstract of Roger Woodworth's Report, Dec. 29, 1964," attached to Evans to Van Mol, Jan. 4, 1964, GMF.

41. Williams to Kilbourne, Feb. 16 and 17, 1965, GMF.

42. Evans's note is appended to Monroe County Farm Bureau resolution, Nov. 5, 1964, GMF. The other statement (by Stewart) is in Stewart to L. B.

Nelson, manager of agricultural and chemical development, July 2, 1965, OACD.

43. For Knoxville's problems in the 1950s, see McDonald and Wheeler, *Knoxville, Tennessee*, Ch. 3.

44. Hammer and Associates, *The Economy of Metropolitan Knoxville* (Washington, Aug. 1962); see esp. 136-38, 203, 221, 231-32. See also McDonald and Wheeler, *Knoxville, Tennessee*, 101-103, on industrial sites. Local anti-Tellico forces used the Hammer Report to attack the Tellico Project. See Kilbourne to Van Mol, Nov. 10, 1964, GMF; Van Mol to Files, Nov. 30, 1964, PPBF. TVA claimed opponents had taken portions of the Hammer Report out of context. See Kilbourne to Van Mol, Nov. 10, 1964, GMF.

45. Interview with William Yandell, April 11, 1983. On special committee, see Wagner to Yandell, Oct. 22, 1964, COF. Yandell himself was torn on the issue. As a trout fisherman who had fished the "Little T," he did have "reservations" in 1964 about the project. At the same time, he was a close personal friend of TVA Director Frank Smith, though Yandell stated clearly that Smith never used their social friendship to influence Yandell on Tellico. Interview with Yandell, Apr. 11, 1983.

46. Dickey, "The Little Tennessee River as an Economic Resource." For Taylor's report, see Taylor to Van Mol, Nov. 13, 1964, GMF.

47. Taylor to Van Mol, Nov. 13, 1964, GMF. Van Mol to Directors agreed; see their margin comments on ibid. See also Van Mol to Taylor, Nov. 19, 1964, GMF.

48. The "fact sheet" and Dickey to Johnson, Mar. 24, 1965, GMF. For Wagner's reply to Dickey's letter to Johnson, see Wagner to Dickey, Apr. 14, 1965, GMF.

49. Interviews with Yandell, Apr. 11, 1983, and Dickey, Mar. 25, 1983. The material on Wagner's complaint to the chamber is from an Apr. 12, 1983, telephone conversation with Charles Heard, former executive vice-president of the Knoxville chamber; he was Dickey's immediate superior.

50. Jones to Wagner and Smith, Apr. 2, 1965, GMF. A notation at the bottom of this memo, by Van Mol on Apr. 9, 1945, reads, "PLE [Evans] discussed W/ARJ—Agreed 'official' protest will not be done." Jones to Jenkins, Apr. 15, 1965, and Jenkins to Jones, Apr. 19, 1965, GMF.

51. For Johnson's budget, see *Knoxville Journal*, Jan. 26, 1965.

52. On planning of Douglas's visit and the inner circle involved, see interview with Dickey, Mar. 25, 1983. A photograph in the Fort Loudoun Association Archives shows Douglas and his wife posing with a number of the principal organizers. For the uncomplimentary reference to Burch, see *Monroe Citizen*, Sept. 9, 1964.

53. Douglas's statement can be found in Wagner to Douglas, Apr. 7, 1965, GMF. For Douglas and warbonnet, see *Maryville-Alcoa Times*, Apr. 5, 1965. For more local coverage, see *Madisonville Citizen-Democrat*, Apr. 7, 1965; *Knoxville News-Sentinel*, Apr. 5 and 10, 1965; *Knoxville Journal*, Apr. 5 and May 11, 1965. Douglas did not present the Cherokees' petition to President Johnson but did give it to Secretary of the Interior Stewart Udall; *Knoxville*

Journal, May 11, 1965. In an article critical of Douglas's visit, the *Madisonville Citizen-Democrat,* Apr. 7, 1965, noted acidly that the media event was the first time a Cherokee Indian had visited the site in the last fifty years. That article was reprinted in the pro-TVA *Knoxville News-Sentinel,* Apr. 10, 1965.

54. For a sample of the national press coverage, see *St. Louis Post-Dispatch* and *Louisville Courier-Journal,* Apr. 5, 1965; *Washington Post,* Apr. 11, 1965; *Christian Science Monitor* Apr. 7, 1965. For the draft of Wagner's letter (never sent), see Wagner to Douglas, Apr. 7, 1965, GMF. The letter was drafted by Stewart and Evans and reviewed by at least four other individuals within TVA. For the decision to scrap the letter and meet with Douglas in person, see Wagner to Van Mol, Apr. 13, 1965, GMF. For Douglas's later statements, see William O. Douglas, *A Wilderness Bill of Rights* (Boston, 1965), 136; and Douglas to Smith, Nov. 27, 1965, GMF. For Douglas's announcement that he was writing an article on the Little Tennessee River for *National Geographic,* see *Knoxville Journal,* May 11, 1965.

55. Evans to Wagner, Oct. 21, 1965; Wagner to Grosvenor, Oct. 21, 1965; and Crossett to Wagner, Dec. 10, 1965; all, COF.

CHAPTER 5

1. The Venice story comes from Ellender's obituary in the *Washington Post,* July 28, 1972. For more on Ellender's background and on the Louisiana political maelstrom from which he emerged, see *Biographical Directory of the American Congress, 1774–1971* (Washington, 1971); T. Harry Williams, *Huey Long* (New York, 1969), 297, 449, 381, 397, 464, 856; Star Opotowsky, *The Longs of Louisiana* (New York, 1960), 64–65; V. O. Key, Jr., *Southern Politics in State and Nation* (Knoxville, new ed., 1984), 164; Jack Bass and Walter DeVries, *The Transformation of Southern Politics: Social Change and Political Consequences Since 1945* (New York, 1976), 158–85.

2. For the remark on Africans, see Ellender's obituary in the *New York Times,* July 28, 1972. Ellender's segregationist stance was certainly popular politically in Louisiana, a state which supported Strom Thurmond in 1948, George Wallace in 1968, and, as a reaction against Lyndon Johnson's racial liberalism, Barry Goldwater in 1964. See Bass and DeVries, *Transformation of Southern Politics,* 164–65.

3. In fact, TVA had reached the 1.9:1 benefit-cost ratio by adding benefits of $15 million for land sales, $30 million in recreation benefits, an additional $7.5 million in fishing and hunting benefits, and $2.1 million in water supply benefits to the "traditional" benefits of navigation, flood control, and power (together $27.4 million), for a total of $82.0 million in benefits and $42.5 million in costs. See Elliot to Van Mol, Feb. 17, 1965, WCPF.

4. General W. P. Leber to Col. Jesse Fishback, July 26, 1965, COF. For hearing, see *Hearings before the Subcommittee of the Committee on Appropriations on HR 9220,* U.S. Senate, 89th Cong., 1st Sess. (Washington, 1965), pt. 4, pp. 44–50.

5. In 1962 the Senate had clarified the process of computing a project's benefit-cost ratio in *Senate Document 97*. Some believe that TVA had a hand in drafting those clarifications.

6. Power, navigation, and flood control were always referred to as "traditional" benefits because they were specifically mentioned in TVA's 1933 charter.

7. *Who's Who in America, 1966–1967,* 34:1626.

8. Palo to Van Mol, Feb. 7 and Aug. 28, 1963, GMF.

9. On TVA's being forced to use these models, *Senate Document 97* (1962) defined "benefits" as "increases or gains not of associated or induced costs, in the value of goods and services which result from conditions with the project, as compared with conditions without the project." That sentence virtually forced TVA to use predictive models.

10. Palo to Van Mol, Feb. 7, 1963, GMF; interview with Taylor, Apr. 11, 1983; Ericson to Van Mol, Feb. 12, 1963, GMF.

11. Van Mol to Palo, May 7, 1963, GMF.

12. Ericson to Van Mol, Feb. 12 and Apr. 4, 1963, and Van Mol to Palo, May 7, 1963, all GMF. As it turned out, Ericson's estimate of power benefits (64 percent of the traditional benefits and 41 percent of the total benefits) was much too high. Palo's first detailed report (Aug. 28, 1963) put power benefits at $27.0 million, about $8.7 million less than Ericson's estimate. Palo's estimate would have dropped Ericson's total ratio to around 1.3:1. Indeed, even Palo's estimate of power benefits turned out to be excessive.

13. Van Mol to Palo, May 7, 1963, GMF.

14. Howes to Van Mol, Apr. 10, 1963, GMF. Howes was given permission to continue his work, but his report was not completed until July 1965. Even then, he lamely but honestly admitted that the models he had developed "were not acceptable if applied to shoreline development." See C. W. Nash to Van Mol, July 7, 1965.

15. Palo to Elliot, June 24, 1963, GMF. As it turned out, Palo's cost figures were much too low. Exclusive of condemnation (cost figures for those are not reported in TVA Land Purchase Records), TVA paid a total of $22,158,000 for all land acquired for the Tellico Project. The 16,500-acre figure was derived by measuring 1,000 feet back from the prospected reservoir's high water mark and then calculating the amount of acreage inside that strip.

16. Project Planning Branch, "Tellico—Economic Appraisal, June 28, 1963," WCPF.

17. *Tellico Project—Benefit-Cost Analysis,* Aug. 28, 1963, GMF. See also Palo to Van Mol, Aug. 28, 1963, GMF. The 16,500 acres were divided into several categories (better agricultural, moderately rolling pasture, steep woodland, prime industrial, basic industrial, etc.), and a per-acre value was assigned to each. This process undoubtedly accounts for the significant increase in land costs. The average per-acre price for the 16,500 acres was estimated to be $330.30, with the highest price being $1,250 per acre for a 823-acre plot near Vonore and the lowest $230.78 per acre for steep acreage.

18. Van Mol notes attached to *Tellico Project—Benefit-Cost Analysis,* after Sept. 5, 1963, meeting, GMF.

19. Van Mol to Wagner, Oct. 2, 1963, COF.

20. Ibid.

21. "AJW Questions re Tellico," July 7, 1963, GMF. Palo's August 28 report concluded that without recreation benefits the benefit-cost ratio for Tellico would drop to 0.97:1. Palo to Van Mol, Aug. 28, 1963, GMF.

22. David L. Pack (Resource Development Branch, Division of Land and Forest Reservoirs) to Jesse C. Mills (head, TVA Technical Library), signed by "Dick Austin for DLP," Sept. 19, 1963, TL.

23. Palo to Elliot, June 24, 1963, GMF.

24. Howes to Palo, Aug. 26, 1963, WCPF.

25. Riggs to Kilbourne, Aug. 2, 1963, WCPF.

26. On the joint study, see Seigworth to Van Mol, Oct. 23, 1963, GMF. See also TVA, "Discussion of the Phillips Report," in Tellico Project: Environmental Statement (1971), III, 3-9. Weeks, a Chattanooga attorney, was a key figure in that city's Trout Unlimited organization, which opposed the Tellico Project. Interview with Weeks, June 13, 1984. For Venable's statement, see interview with him, Aug. 23, 1983.

27. Interviews with Venable, Aug. 23, 1983, and June 20, 1985.

28. Interview with Kerry Schell, Mar. 29, 1983; "Discussion of the Phillips Report," in Tellico Project, III, 3-9.

29. The industries used were those which had located along the Tennessee River between 1953 and 1963. It was assumed there would be no navigation use during the first 10 years after the dam was closed, and $165 per acre per year for the next 35 years. The result was a final calculation of $78 per acre per year for 45 years. Unlike other benefit claims, the calculations were good enough if all 5,000 acres were occupied by water transportation-oriented industries within 10 years. For a full discussion of this problem, see Tellico Project, Environmental Statement, III, pt. 1, pp. 38-49 and pt. 3, pp. 13-21.

30. The list of ideal factors is from J. Porter Taylor's memo to Reed Elliot, Oct. 28, 1963, WCPF. According to some, the memo bears the stamp of Mike Foster, an economist whom Taylor had lured from the Office of Power to the Navigation Division. "He was hired . . . to focus on industrial development," recalled UT economist and former TVA employee Kerry Schell. Certainly Taylor admired Foster, calling him "one of the few people [at TVA] with walkin' around sense." Interviews with Schell, Mar. 29, 1983, and Taylor, Apr. 11, 1983. Ultimately Foster replaced Taylor as head of the Navigation Division.

31. Taylor to Elliot, Sept. 24 and Oct. 28, 1963, WCPF.

32. Howes to E. S. Brosell, Oct. 7, 1963, in PPBF.

33. Taylor to Elliot, Oct. 28, 1963, PPBF, and interview with Taylor, Apr. 11, 1983.

34. Interview with Kerry Schell, Mar. 29, 1983.

35. Evans to Van Mol, Jan. 4, 1965, GMF.

36. Evans to Van Mol, Jan. 4, 1965 (abstract of Woodworth's report of Dec.

29, 1964), GMF. Woodworth's report was revived ten years later as part of TVA's *Alternatives Report*.

37. Interview with H. A. Henderson, 13 June, OACD.

38. Gilbert ("Pete") Stewart, Jr., to L. B. Nelson, manager, Office of Agricultural and Chemical Development, July 2, 1965, and Nelson to Stewart, July 7, 1965, OACD.

39. "Contributions of Agriculture to Total Development," June 12, 1984, OACD. One of the stronger statements of agriculture's ability to surpass industry in terms of multiplier effects is the work done by Univ. of Georgia agricultural economists Bill R. Miller and Fred C. White. See "Dollar for Dollar: Additional Investment in Agriculture Can Bring Higher Returns Than Industry," *Georgia Agricultural Research Bulletin* 21 (1), Summer 1979. This piece contains the agricultural and manufacturing multiplier for each state excepting Alaska and Hawaii.

40. Interview with H. A. Henderson, June 12, 1984, OACD.

41. TVA, Office of Agriculture and Chemical Development, "Income and Employment for Four Agricultural Situations in the Tellico Project Area," OACD.

42. Ibid.; interview with H. A. Henderson, June 12, 1984, OACD.

43. Tims Ford, on the Elk River in West Tennessee, had been initiated as an OTAD project under Gen. Herbert Vogel. The project covered or affected seven counties, 2,249 square miles, and 96,000 people. See Wells, *TVA Tributary Development Program*, 6-7 and 95-98. Johnson's budget ultimately was $99.7 billion, with a predicted deficit of $5.3 billion.

44. For the defense of Tellico, see *Hearings before the Subcommittee*, Senate, 89th Cong., 1st Sess., pt. 4, p. 73.

45. *Hearings before the Subcommittee of the Committee on Appropriations*, U.S. House of Representatives, 89th Cong., 1st Sess. (Washington, 1965), pt. 3, pp. 22-30 and 147. The Melton Hill Dam was closed on May 1, 1963.

46. *Knoxville News-Sentinel*, Apr. 27 and June 8, 9, and 17, 1965.

47. *Hearings before the Subcommittee of the Committee on Appropriations*, U.S. House of Representatives, 89th Cong., 2nd Sess. (Washington, 1966), pt. 2, p. 761.

48. On flying to Washington, see *Knoxville News-Sentinel*, July 12, 1965. On appeals and pressures, see ibid., May 20, 1965; *Knoxville Journal*, May 21, 1965. The Chamber of Commerce did not formally endorse the Tellico Project, although the City Council did, on July 13, by a vote of four to three. See *Knoxville Journal*, July 14, 1965.

49. *Hearings before the Subcommittee*, Senate, 89th Cong., 1st Sess., pt. 4, pp. 44-50.

50. Ibid.

51. Ibid., pt. 4, p. 56. For Ellender's opposition to the project on the Senate floor, where he reiterated his belief that Tellico's actual benefit-cost ratio was probably 0.8:1, see *Congressional Record—Senate*, Aug. 23, 1965, 520567-8.

52. Rovere's comments on Jenkins are in Richard H. Rovere, *Senator Joe*

McCarthy (Cleveland, 1968), 208. Jenkins' testimony is from *Hearings before the Subcommittee*, Senate, 89th Cong., 1st Sess., pt. 4, pp. 87–92.

53.   *Congressional Record—Senate*, Aug. 25, 1965, pp. 20567–8; *Knoxville News-Sentinel*, July 16 and Aug. 24, 1965; *Nashville Tennessean*, July 12, 1965.

54.   *Public Works Appropriations Bill, 1966, Conference Report #1163, to Accompany HR 9220* (Washington, Oct. 13, 1965), 38; *Knoxville News-Sentinel*, Oct. 14, 1965; *Chattanooga News-Free Press*, Oct. 14, 1965; *Chattanooga Times*, Oct. 16 and 17, 1965. The appropriation for Tims Ford for fiscal 1966 was $5,570,000.

55.   *Hearings before the Subcommittee*, Senate, 89th Cong., 2nd Sess., pt. 4, p. 68. For the national Chamber of Commerce's attack, see *Knoxville News-Sentinel*, Aug. 24, 1965.

56.   Nash to Van Mol, July 7 and Aug. 13, 1965, and Ericson to Van Mol, Aug. 24, 1965, GMF.

57.   *Hearings before the Subcommittee*, House, 89th Cong., 2nd Sess., pt. 2, pp. 676–700.

58.   *Hearings before the Subcommittee*, Senate, 89th Cong., 2nd Sess., pt. 4, p. 43; *Knoxville News-Sentinel*, May 5, 1966.

59.   *Knoxville News-Sentinel*, May 12, 1966; Rechichar and Fitzgerald, *Consequences of Administrative Decisions*, 21; *Knoxville Journal*, Sept. 20, 1966. See also Smith to Ottinger, Sept. 22, 1966, and Ottinger to Smith, Oct. 6, 1966, GMF. During this conflict, Ottinger remarked that he had received "two hundred letters against [Tellico], none for." See Winnie Chandler memo to Van Mol, Sept. 20, 1966, GMF.

60.   *Hearings before the Subcommittee*, Senate, 89th Cong., 2nd Sess., pt. 4, pp. 43, 66–67; *Knoxville News-Sentinel*, May 5, 1966. The final public works appropriation bill (HR 17787) designated $3.2 million for Tellico and $8 million for Tims Ford, the latter down $1 million from the figure which had emerged from Evins's committee.

CHAPTER 6

1.   "Paul Evans to Retire as TVA Information Director," *TVA News*, Oct. 4, 1974; *Who's Who in America, 1966–67*. The following information was compiled from comments made by several of Evans's former coworkers, and from our own impressions of Evans from an interview in his home in Norris, Tennessee, on Apr. 7, 1983.

2.   Evans to Seeber, Apr. 12, 1971, GMF.

3.   Seeber to Directors, Apr. 13, 1971, and Seeber to Derryberry, Apr. 13, 1971, GMF.

4.   For Evans's "neo-environmentalist" comment and his recognition of the opposition as skillful and well organized, see interview with Evans, Apr. 7, 1983.

5.   Simultaneously TVA was facing similar opposition in western North

Carolina. There, a group of upper- and upper-middle-class retired citizens was able to block the agency's Mills River Project. "I thought we weren't going to get out of there in one piece," remembered OTAD's Richard Kilbourne, in an interview, Feb. 12, 1982. Timberlake, to be jointly planned and developed by TVA and a private developer and to be located on the shore of the reservoir, will be discussed in Ch. 7.

6. Interviews with Wade K. Swafford, Mar. 30, 1983; Paul and Charlotte Hughes, Alfred Davis, and Louise Stamey, Mar. 16, 1983; and Jean Ritchey and Reece Nicholson, Mar. 31, 1983.

7. Elliot to Rozek, May 1, 1967, WCPF.

8. Rozek to Van Mol, May 4, 1967, WCPF. For board's decision, see Van Mol to Todd, May 25, 1967, WCPF.

9. For recommendations to expand the taking line, see, for example, Edward Lesesne to Files, Nov. 15, 1967; David H. Talley (a TVA land appraiser) to Paul F. Meredith, Dec. 6, 1967; Foster to Elliot, Oct. 11, 1968; and Parrott to Perry, Dec. 13, 1968; all WCPF.

10. Interview with Davis, Mar. 16, 1983.

11. The Project Planning Branch believed that 825 tracts would be acquired. The branch then "hastily assumed that families would be living on $2/3$ of these—or 550 families. This figure apparently was rounded upward to 600." Mattern to Elliot, July 21, 1967, WCPF.

12. On rising costs and project slowdown, see *Knoxville News-Sentinel,* June 21, 1967. On the recommendation to shrink the taking line, see Parrott to Perry, Dec. 13, 1968, WCPF. On Foster's simultaneous moves to increase land purchases, see Taylor to Elliot, Dec. 17, 1968, and Foster to Elliot, Dec. 18, 1968, WCPF.

13. Wetherholt to Files, Sept. 16, 1968; Foster to Heads of Offices and Divisions, Jan. 17, 1969; Palo to Foster, Jan. 16, 1969; Ericson to Foster, Jan. 21, 1969, all in Benefit-Cost Material—Tellico, NDRSF.

14. For appropriations, see *Chattanooga Times,* Nov. 16, 1968, and *Johnson City Press-Chronicle,* Jan. 21, 1969. On monthly payroll and average number of people working on Tellico, see *Knoxville Journal,* Mar. 7, 1968. For announcement of construction slowdown or stoppage, see *Chattanooga Times,* Nov. 16, 1968. For 1971 completion date, see *Knoxville News-Sentinel,* June 21, 1967. For later date, see *Johnson City Press-Chronicle,* Jan. 21, 1969. For percentage of land acquired, see *Chattanooga Times,* Feb. 20, 1969. For a brief summary of Tellico land purchases, with accompanying tables, see Appendix.

15. Minutes of Tellico Area Planning Council, Feb. 18, 1969; TAPC to All Members, Mar. 25, 1969; Sproul to Seeber, Mar. 26, 1969; all GMF. For original appropriation, see *Johnson City Press-Chronicle,* Jan. 21, 1969.

16. For the first whiff of the opposition's new tactics, see *Knoxville News-Sentinel,* June 8, 1967. For letter to Duncan, see Giles to Duncan, Sept. 23, 1968, a typed copy of which is in WCPF. As Duncan was a strong supporter of the Tellico Project, Giles's letter had little effect. See also Mary Anne Hensley to Congressman Bill Brock, Apr. 9, 1968, WCPF.

17. On Akard's 1964 behavior, see Kilbourne to Files, Oct. 5, 1964, and Cunningham to Files, Oct. 15, 1964, OTAD. For more details, see Ch. 4. For Elmore's memo, see Elmore to Barron, Feb. 21, 1968, OTAD. For the freezing out of Akard, see Lowry to Stern, Apr. 30, 1968, and Barron to Akard, May 1, 1968, OTAD.

18. Palo to Van Mol, May 31, 1962, and Palo to Van Mol, Feb. 7, 1963, GMF. On memo concerning Brock's inquiry, see Office of the General Manager to Frank E. Smith, June 12, 1969, WCPF. For UT economist's view, see interview with Dr. Keith Phillips, June 21, 1983.

19. *Nashville Tennessean*, Apr. 14 and 17, 1969; *Knoxville Journal*, Apr. 15, 1969.

20. For Douglas's "worst offender" statement, see Douglas, *A Wilderness Bill of Rights*, 136. On his visit, see *Nashville Tennessean*, Apr. 14 and 17, 1969, and *Knoxville Journal*, Apr. 15, 1969.

21. William O. Douglas, "This Valley Waits To Die," *True Magazine*, 50 (May 1969), 40–43, 91, 95–98. See Hall to Editor, n.d. [1969], and McBride to Hall, Apr. 28, 1969, COF. For one example of TVA's counterattack on Douglas, see *Memphis Press-Scimitar*, Aug. 13, 1969.

22. For the popular stir over the Douglas article, see correspondence in COF, Aug. to Oct., 1969, esp. Folder No. 230. For 41 organizations, see "Organizations Actively Opposed to the Tellico Dam Project," July 7, 1970, GMF.

23. On land costs see L. E. Clark, Jr. (assistant supervisor of appraisals), to J. R. Perry (chief, Land Branch), May 2, 1969, WCPF. For local landowners, see Brock to Wagner, May 26, 1969; Wagner to Brock, June 9, 1969; Smith to Van Mol, June 10, 1969; Office of the General Manager to Smith, June 12, 1969; all WCPF.

24. For Foster's appointment to coordinate Tellico, see minutes, Tellico Area Planning Council, Feb. 18, 1969, GMF. For details on Foster, see his obituary by Carson Brewer in *Knoxville News-Sentinel*, Oct. 9, 1979. Foster did not live to see the dam closed.

25. The Eastern Band of the Cherokees were the descendents of those people who either escaped from or were overlooked in the Cherokee removal of 1838, the famous "Trail of Tears." The Cherokee Nation, centered in Oklahoma, had become wealthy through the discovery of oil, whereas the Eastern Band remained on a reservation in North Carolina, created for them in 1876, and lived principally from tourist trade.

26. For local politicians, see Conference Schedule, May 1, 1970, GMF. On Cherokees, see Walter S. Jackson (principal chief) to Wagner, Jan. 14, 1970; Wagner to Jackson, Jan. 23, 1970; Evans to Foster, Jan. 27, 1970; all PPBF.

27. Ottinger to Train, July 7, 1970; Train to Van Mol, July 14, 1970; Dingell to Wagner, July 13, 1970; all GMF. For replies and reactions, see Wagner to Dingell, July 28, 1970; Wagner to Ottinger, Aug. 4, 1976; Wagner to Train, Aug. 4, 1970; Franson to Ottinger, Aug. 8, 1970; all GMF.

28. For anti-Tellico letters spurred by Douglas's article, see NDRSF, mid-1969 to mid-1970. For TVA attempts to discredit Douglas, see *Memphis Press-*

Scimitar, Aug. 13, 1969, and Frank Smith (TVA director) to Rep. John Dingell, May 22, 1969, NDRSF.

29. George Laycock, *The Diligent Destroyers* (Garden City, N.Y., 1970), 195–96. For TVA's reaction, see Frank Smith to Laycock, Mar. 4, 1970, in NDRSF. On Ralph de Toledano's article and reaction, see letters in NDRSF, mid-1969 to mid-1970.

30. For copies of Berry's resolutions, see GMF, Apr. 1971. For non-inclusion of the Little Tennessee River in the 1969 Scenic River Act, see form letter from A. Heaton Underhill, Aug. 1971, OTAD. For copies of Hart's letters, see Hart to Berry, May 6, 1971; Hart to Sen. Brown Ayres, May 6, 1971; Hart to Knox County Delegation, May 6, 1971; all OTAD. In May 1971 a Tennessee House Resolution proposed to limit TVA's taking line at Tellico. It did not pass. See Marquis to Foster, May 21, 1971, GMF. For Tellico area reaction to Berry resolutions, see *Loudon County Herald,* May 13, 1971.

31. *Knoxville News-Sentinel,* Apr. 25 and May 5, 1971; *Lenoir City News-Banner,* May 6, 1971.

32. For opposition groups, see "Organizations Actively Opposed to the Tellico Dam Project," July 7, 1970, GMF. For the Frank Smith incident, see Smith to Seeber, June 23, 1971, COF. For difficulties encountered by TVA people, see interview with Louis Gwin (former employee in TVA's Information Office), July 14, 1983.

33. Evans to Seeber, Apr. 12, 1971; Seeber to Directors, Apr. 13, 1971; Seeber to Derryberry, Apr. 13, 1971; all GMF.

34. For early work on the Environmental Impact Statement, see Derryberry to Seeber, June 17, 1971, GMF. For completion see Seeber to Timothy Atkinson (Council on Environmental Quality), June 18, 1971, GMF; TVA news release, June 18, 1971, COF. For TVA's claim of voluntarism, see Seeber to John L. Franson (National Audubon Society), June 30, 1971, COF. For night and overtime work, see Seeber to Robert Marquis and Ashford Todd, June 18, 1971, GMF. For entire TVA position, see *Knoxville News-Sentinel,* June 18, 1971.

35. *Knoxville News-Sentinel,* June 18, 1971; *Chattanooga Times,* June 26, 1971; Wagner to H. R. Payne (president, Smoky Mountain Hiking Club), June 24, 1971, PPBF. For internal memo on benefit-cost ratio, see John L. Furgurson to Corydon Bell, Jan. 19, 1971, PPBF. On public relations efforts, see James Gober to A. J. Gray, July 2, 1971, and "Public Relations on Tellico Project," Sept. 1, 1971, NDRSF, General Correspondence—Tellico. On increase in benefit-cost ratio, see "Tellico Project—Economic Analysis—Capitalized Basis," May 10, 1971, in NDRSF, Benefit-Cost Statement, Tellico.

36. On original lawsuit, see *Knoxville Journal, Chattanooga News-Free Press,* and *Knoxville News-Sentinel,* Aug. 10, 1971; *Chattanooga Times,* Aug. 12, 1971. For some reactions to the Environmental Impact Statement, see Thomas J. Armstrong (HUD) to F. E. Gartrell, July 26, 1971, and George Marienthat (EPA) to Gartrell, Nov. 12, 1971, both GMF. On Prichard, see Prichard to Wagner, Aug. 27, 1971; COF. For leaks on Phillips, see *Chattanooga News-Free Press,* Aug. 1, 1971, and esp. margin comments on article from Foster to Gray, NDRSF.

37. Much of the above and following material is gleaned from an interview with Phillips, June 21, 1983.

38. Ibid.

39. Ibid.

40. "The Tennessee Valley Authority's Tellico Project: A Reappraisal" [commonly known as the Phillips Report] can be found *in toto* in TVA, *Environmental Statement, Tellico Project, Volume III* (Knoxville, 1972), III-I-3 to 9.

41. Ibid., III-I-17, 25-26, 35-36.

42. For TVA's version of how the Phillips Report was obtained by the agency, see ibid., III, Introduction. For mailings of the report and Phillips's appearances before opposition groups, see interview with Phillips, June 21, 1983; *Knoxville Journal* article, Aug. 7, 1971.

43. For McGloughlin's investigation and Foster's comments to Gray, see margin of xeroxed copy of story in *Chattanooga News-Free Press*, Aug. 1, 1971, in NDRSF. For results of investigation and the Madisonville editor's agreement to publish the attack on Phillips, see Gober to Gray, Oct. 19, 1971, NDRSF. A college classroom project not dissimilar to Phillips's was carried out at Tennessee Wesleyan College by biology professor Dr. John Woods. The conclusions, released in mid-Sept. 1971, were not far from those reached by Phillips's students. See *Chattanooga News-Free Press*, Sept. 17, 1971; *Knoxville Journal*, Sept. 20, 1971.

44. See TVA, *Environmental Statement, Tellico Project, Volume III*, esp. III-3-Introduction, 1, 2, 4, 7, 24. The Canal Zone, of course, was leased and not purchased.

45. The John Birch Society was an extreme anticommunist pressure group that garnered a good deal of attention and influence in the 1950s, fitting in well with the general spirit of McCarthyism. The organization lost considerable prestige and power when it attacked General George C. Marshall and President Dwight Eisenhower.

46. Smith to White, Aug. 9, 1971, COF. Evans to Seeber, Dec. 16, 1971, PPBF. Prichard to Wagner, Aug. 27, 1971, COF. Seeber to Jenkins, Sept. 2, 1971, GMF. Taylor to Foster, Oct. 15, 1971, OTAD. Venable's editor, Ralph Millett, was listed by Reed Elliot as a possible pro-TVA witness. See Elliot to Foster, Oct. 12, 1971, NDRSF. Venable's denial was in interview with Venable, Aug. 23, 1983, and his remarks about being prevented from writing about Tellico are from a June 20, 1985, interview.

47. Wagner to Rep. William L. Springer, Feb. 8, 1971, PPBF. On telephone poll, see Charles Hall to President Richard Nixon, Oct. 20, 1971, OTAD. For "New York groups," see *Maryville-Alcoa Times*, Aug. 26, 1971.

48. Frank Smith to John B. Connally, Jr. (Secretary of the Treasury), Sept. 8, 1971; Marquis to Seeber, Sept. 3, 1971; Johnnie M. Walters (IRS) to Wagner, Sept. 27, 1971; Daniel H. Hollums (IRS) to Wagner, Oct. 15, 1971; all GMF.

49. Dunn to Wagner, Dec. 7, 1971, COF; Wagner to Dunn, Dec. 17, 1971, OTAD; *Knoxville Journal* Dec. 15 and 16, 1971; *Chattanooga News-Free*

Press, Dec. 15, 1971; *Knoxville News-Sentinel*, Dec. 15 and 19, 1971; *Chattanooga Times*, Dec. 18, 1971. For charges of political motivation, see *Maryville-Alcoa Times*, Dec. 15, 1971; *Monroe County Democrat*, Dec. 22, 1971. Wagner waited so long to reply to Dunn because he had the flu and was away from the office for several days. The letter from Wagner was drafted by Claussen. See Seeber to Foster, Dec. 6, 1971, NDRSF. For letter-writing campaign, see M. C. Hargett (Tenn. Valley Trades and Labor Council) to Dunn, Dec. 21, 1971; Fred H. Harris (Tenn. Industrial Development Council) to Dunn, Dec. 23, 1971; Dr. William Joubert (Bowater) to Dunn, Jan. 5, 1972; Greater Knoxville Chamber of Commerce to Dunn, Jan. 25, 1972; all NDRSF. There is evidence that Dunn had seen the Phillips Report. See *Nashville Tennessean*, Oct. 17, 1971.

50. *Chattanooga News-Free Press*, Dec. 8, 1971; *Nashville Tennessean*, Dec. 12, 1971. Privately Seeber labeled the report "absurd." See Seeber to Directors, Dec. 2, 1971, GMF.

51. *Knoxville Journal*, Dec. 22, 1971; *Knoxville News-Sentinel*, Dec. 31, 1971.

52. Rechichar and Fitzgerald, *Consequences of Administrative Decision*, 30–31; Baird to Wagner, Jan. 15, 1972; COF.

53. *Knoxville Journal*, Jan. 8, 1972; *Chattanooga Times*, Jan. 8, 1972; *Knoxville News-Sentinel*, Jan. 8, 1972.

54. *Environmental Defense Fund v. TVA*, 339 F. Supp. 806, 1971; *Environmental Defense Fund v. TVA*, 468 F. 2nd 1164, 1972. *Knoxville News-Sentinel* and *Chattanooga News-Free Press*, Mar. 3, 1972.

55. *Chattanooga Daily Herald* and *Knoxville News-Sentinel*, Jan. 11, 1972. On *Sports Afield*, see *News-Sentinel*, Jan. 9, 1972.

56. McBride to J. Wiley Bowers, Jan. 13, 1972, GMF. For land acquisition, see Tellico Land Purchase Records, Chattanooga. On progress in acquiring tracts before the lawsuit, see Ashford Todd to Seeber, Jan. 17, 1972; GMF. Legal help on the appeal was provided by the Audubon Society. See *Knoxville Journal*, Mar. 15, 1972.

57. For resurfacing of Phillips, see *Huntsville Times*, May 10, 1972. For TVA's response to the Phillips Report, see *Knoxville News-Sentinel*, Feb. 11, 1972. On Elrod and TVA board, see *Nashville Tennessean*, Mar. 19, 1972; *Knoxville Journal*, Feb. 29, 1972. Elrod's chances were almost nil, and Smith's spot was filled by Rogersville Republican and State Conservation Director William Jenkins. For other opposition speakers, see *Athens Post-Athenian*, Feb. 7, 1972; *Chattanooga News-Free Press*, Apr. 9, 1972. On Etnier see *Daily Beacon* (the student newspaper for the Univ. of Tennessee, Knoxville campus), Jan. 21, 1972; *Maryville-Alcoa Times*, June 9, 1972.

58. On Dunn, see *Chattanooga Daily Herald*, Jan. 11, 1972; *Huntsville Times*, Jan. 14, 1972; *Knoxville News-Sentinel*, Jan. 11 and 14, 1972; *Athens (Tenn.) Post-Athenian*, Feb. 1, 1972. On Berry, see above. On Ashe, see *Knoxville News-Sentinel*, Mar. 10, 1972. On Brock, see *Knoxville Journal*, Mar. 16, 1972; *Knoxville News-Sentinel*, Mar. 16, 1972; *Nashville Tennessean*, Mar. 19, 1972.

59. *Knoxville Journal,* Feb. 19, 1972.

60. Peter Mattheissen, "This River Is Waiting To Die," *Sports Afield* 167 (Mar. 1972), 12–18.

61. Roland Elliott to W. L. Russell (president of the Tennessee Citizens for Wilderness Planning and a longtime foe of the project), July 21, 1972, together with copies of all the petitions, GMF.

62. For the results of Baker's poll, see *Knoxville Journal,* Mar. 31, 1972. For TVA's poll, see Wagner to Seeber, Mar. 31, 1972, GMF. For Roll's results, see Roll to John Doig (of Boeing), Mar. 6, 1972, GMF.

63. The term "Overhill Cherokees" refers to the group that probably crossed the Appalachian Mountains in the early eighteenth century and settled along the Great Valley of East Tennessee. The date of the group's arrival would become a major point of contention in an argument among some of the Cherokees, TVA, and the University of Tennessee archaeologists, concerning which of the unearthed archaeological relics were Cherokee and which were not. For popular Indian depiction, see Dee Brown, *Bury My Heart at Wounded Knee: An Indian History of the American West* (New York, 1970), and Thomas Berger, *Little Big Man* (New York, 1964). The film of the latter, which glorified Indian life, starred Dustin Hoffman and Faye Dunaway.

64. In 1838, in the famous "Trail of Tears," most of the Cherokees were moved from parts of Georgia and North Carolina to Oklahoma. That group officially designated itself and was recognized by whites as the Cherokee Nation. But numerous Cherokees hid in the mountains to avoid forcible removal. As a compromise, they were later designated the Eastern Band of the Cherokees, settled on a reservation in western North Carolina, and allowed to set up a government structure independent of that of the more numerous and ultimately wealthier Oklahoma group. For excellent background on this and other aspects of Cherokee history important to this story, see John R. Finger, *The Eastern Band of Cherokees, 1819–1900* (Knoxville, 1984).

65. For Oklahoma Cherokee objections, see *Knoxville News-Sentinel,* Feb. 1, 1972. For TVA person's margin comments, see xeroxed copy of this article, NDRSF. For visit, see Corydon Bell to Reed Elliot, Mar. 16, 1972, PPBF. For negotiations over payment, see transcription of telephone call, Earl Boyd Pierce (General Counsel of the Oklahoma Cherokees) to Seeber, Jan. 26, 1972, and Cherokee Nation TVA Committee to TVA, Apr. 14, 1972, both COF. For TVA recommendation to the Cherokee Nation, see TVA Subcommittee to Cherokee Nation TVA Committee, Apr. 14, 1972, NDRSF. See also *Tulsa Daily World,* Apr. 29, 1972, and *Tulsa Tribune,* Apr. 19, 1972. For report of visit, see *Daily Beacon,* Apr. 15, 1972. For non-opposition by the Cherokee Nation, see *Knoxville News-Sentinel,* May 3, 1972.

66. On Hagerstrand, see Earl Boyd Pierce to Foster, July 7, 1972, GMF; Gober to Gray, Sept. 15, 1972, and General Manager to Foster, May 26, 1972, NDRSF; interview with Dr. Alfred Guthe (Univ. of Tennessee archaeologist), Apr. 1, 1983.

67. TVA kept well informed on this factional rift. See Corydon Bell to Glenn O'Neal, Nov. 1, 1972, PPBF; Gober to Gray, Sept. 15, 1972, NDRSF; Claussen to Pierce, Nov. [?], 1972, and Hagerstrand to Gober, Dec. 19, 1972, PPBF.

68. Gober to Gray, Sept. 15, 1972, NDRSF; Mark Harrison (of the Assn. for the Preservation of the Little Tennessee River) to Powell, Sept. 26, 1972, GMF; John P. Hart (vice president of WBIR Radio and TV in Knoxville) to Wagner, Oct. 13, 1972, COF; Claussen to Pierce, Nov. [?], 1972, and Hagerstrand to Gober, Dec. 19, 1972, both PPBF.

69. Interview with Guthe, Apr. 1, 1983.

70. Ibid. See also Beverly Burbage, "The Archaeological Aspects of the Tellico Reservoir," TVA internal report, in PPBF. After the passage of NEPA in 1969, TVA was *required* to conduct cultural impact studies. Also Executive Order 11593 mandated that such surveys be conducted. In fairness, it should be noted that TVA was genuinely concerned *before* the passage of NEPA.

71. Interviews with Guthe, Apr. 1, 1983, and Dr. Jefferson Chapman, Apr. 4, 1983. For samples of Greene's reports to Bell, see Bell to Seeber, Jan. 23, 1973, and Bell to Files, Feb. 5, 1973, both in Archaeological Explorations Folder #3, WCPF; Bell to Files, Nov. 1, 1972, PPBF.

72. Interview with Guthe, Apr. 1, 1983. On TVA's nonprosecution of "pot hunters," it should be pointed out that the 1906 Federal Antiquities Act made such activities illegal but carried no penalties. TVA's Office of General Counsel did write stern letters to those who were caught, but that appears to have done little good. After all, an ancient arrowhead in good condition could be worth hundreds of dollars. See Beauchamp Brogan to Orvie Rice, Apr. 11, 1973, GMF.

73. On Dunn, see *Knoxville News-Sentinel*, Aug. 12, 1972; Cookeville *Herald-Citizen*, Oct. 2, 1972. On the Eastern Band's formal opposition, see *Knoxville Journal*, Aug. 29, 1972. On the resolutions passed by the National Congress of American Indians, see Charles E. Trimble (executive director) to Wagner, Dec. 19, 1972, COF. On CBS News, see Evans to John Synder (CBS), Oct. 25, 1972, COF.

74. Harrison to Powell, Sept. 26, 1972, GMF.

75. For Powell biography, see his obituary in *The Cherokee One Feather* (a newspaper published in Cherokee, North Carolina, that leaned toward the anti-Powell faction), Apr. 11, 1973; a copy is in NDRSF.

76. On Powell's request for a meeting with Wagner, see Powell to Wagner, Sept. 29, 1972 (three days after Harrison wrote his letter to Powell), GMF. For the mollification of Powell and a majority of the tribal council, see Bell to Files, Nov. 1, 1972, PPBF.

77. For Littlejohn's "official" biography, see cover story on him in magazine section, *Memphis Commercial Appeal*, Dec. 10, 1972. See also *Tennessee Traveler*, Oct. 1972.

78. Statement of Officer John Jones, Apr. 4, 1973, GMF.

79. *Knoxville Journal*, Aug. 30, 1972; transcript of telephone call from Morgan to Wagner, Aug. 30, 1972, COF; *Knoxville News-Sentinel*, Aug. 31,

1972; Morgan to Littlejohn, Apr. 5, 1973, Archaeological Explorations Folder #3, WCPF. Morgan later defended the appointment of William Jenkins to the TVA board. See Morgan to Alice W. Milton, Aug. 7, 1973, COF. The opposition had accused Jenkins of being connected to strip-mining interests.

80. For Greene's report on Littlejohn's waning influence among the Cherokees, see Bell to Seeber, Jan. 23, 1973, Archaeological Explorations Folder #3, WCPF. For the activities of Littlejohn and Mashburn, see interview with Guthe, Apr. 1, 1973; and transcript of "Open Line" radio show, WETE Knoxville, Feb. 28, 1973, in Archaeological Explorations Folder #3, WCPF. For Guthe's defense, see transcript of "Open Line" for Mar. 7, 1973, in ibid. For the "most militant" statement, see *Memphis Commercial Appeal*, Dec. 10, 1972.

81. *Washington Post*, Feb. 18, 1973; *Playboy* 20 (Mar. 1973), 48, 50; Brown to Bell, Feb. [?], 1973, in Archaeological Explorations Folder #3, WCPF.

82. For rallies and petition see *Johnson City (Tenn.) Press-Chronicle*, Feb. 3, 1972; *Knoxville News-Sentinel*, Feb. 6, 1972; *Monroe County Democrat*, Feb. 9, 1972; *Loudon County Herald*, Feb. 10, 1972. On attempts to discredit Phillips Report, see TVA *Environmental Statement, Tellico Project* (Knoxville, Feb. 1972), Vol. 3; *Knoxville News-Sentinel*, Feb. 11, 1972; TVA news release, Jan. 24, 1972, GMF. For outside archaeologists, see interview with Guthe, Apr. 1, 1983, and the very interesting Robert Stephenson (one of three consultants TVA brought in) to Guthe, Jan. 24, 1973, GMF. On Wagner's "reasonable balance," see *Knoxville News-Sentinel*, Nov. 30, 1971, and *Chattanooga Times*, Feb. 12, 1972. For the unfortunate incident in Chattanooga, see *Knoxville Journal*, Sept. 13, 1972; Seeber to Waller, Sept. 15, 1972, and John Tansil to Wagner, Sept. 15, 1972, both in Public Interest Folder 4, PPBF.

83. For a sample of pro-dam letters to Nixon, see Yellow Creek Watershed Authority to Nixon, Mar. 31, 1972, and similar letters from late March to mid-April, GMF. For McBride's plea, see McBride to Richard H. Hughes (chairman of the board, Beverage Products Corp., Tulsa), Apr. 4, 1972, GMF.

84. *Knoxville Journal*, July 27, 1972. See also Seeber to Dr. Archie Dykes (chancellor of UT, Knoxville), May 28, 1972, GMF.

85. For the *EDF v. Corps of Engineers* decision, see Seeber to Wagner, June 2, 1972, COF. For belief that this decision would be "helpful" to Tellico, see ibid. and *Maryville-Alcoa Times*, July 3, 1972. On flooding in Chattanooga (TVA claimed Tellico would have reduced the damage by "more than half"), see TVA news release, Mar. 21, 1973, and Wagner to Brock, Feb. 25, 1974, COF.

86. *Knoxville News-Sentinel*, June 18, 1978.

CHAPTER 7

1. *U.S. News and World Report*, July 6, 1970.

2. Ibid. Nixon specifically opposed "large-scale government purchases of land as a means of controlling development."

3. Blowers to Gray, July 20, 1970, NDRSF.

4. For a discussion of Norris, see McDonald and Muldowny, *TVA and the Dispossessed*, 217–35, esp. 219. For the Morgan-Roosevelt agreement, see Morgan, speech, Aug. 31, 1933, in *Speeches and Remarks by Arthur E. Morgan*, TL.

5. McDonald and Muldowny, *TVA and the Dispossessed*, 234–35.

6. For the model city as "Wagner's baby," see interview with Gray, Mar. 9, 1982.

7. Gray to Durisch, Jan. 26, 1965; Durisch to Van Mol, Jan. 26, 1965; Van Mol to Durisch, Jan. 29, 1965; Durisch to Van Mol, Feb. 1, 1965; all GMF.

8. Notes on general manager's staff meeting, Mar. 15, 1965, GMF. See also interview with Gray, Mar. 9, 1982.

9. For New Town movement and some companies involved, see Gray to Van Mol, Feb. 9, 1966, NDRSF. For General Electric's plans and criticisms of them (principally by Sen. Abraham Ribicoff, Conn.), see *Washington Post*, Dec. 8, 1966.

10. Stern to Van Mol, Nov. 29, 1966, GMF; Smith to McBride and Wagner, Dec. 9, 1966, COF; Reiger (of GE) to Wagner, Jan. 31, 1967, GMF; G. T. Bogard to Wagner, Feb. 10, 1967, COF.

11. Smith to Van Mol, Sept. 7, 1966, COF; McBride to Van Mol, Nov. 13, 1966, and Smith to Van Mol, Oct. 12, 1966, both NDRSF; Albert M. Cole (president of Reynolds) to Smith, Mar. 6, 1967, and Leslie Larsen (of David Rockefeller Associates) to Smith, Dec. 20, 1966, both COF; Stern to Wessenauer, June 28, 1967, and Van Mol's notes on Wagner luncheon, Nov. 1, 1967, both GMF.

12. Rozek to Van Mol, Jan. 11, 1967, GMF. Rozek had been appointed by Van Mol as a day-to-day coordinator of the Tellico Project, "an important multipurpose element in TVA's resource development." Van Mol to Heads of Offices and Divisions, Nov. 2, 1966, GMF.

13. J. R. Perry (chief, Land Branch) to Stern, Jan. 4, 1967, GMF; "Proposed Land Development in the Tellico Project Area," background paper for staff discussion, Jan. 26, 1967, and attached note from Van Mol on board's awareness of increased land purchases and costs, Feb. 6, 1967, GMF.

14. Notes on Meeting with Carl Feiss, Apr. 22, 1967, NDRSF. The HUD legislation in question was Title IV of the Federal Housing Act.

15. Notes on Meeting with Feiss, Apr. 22, 1967, NDRSF.

16. The problems of industry and housing were recognized by Frank Smith. See Smith to Van Mol, May 4, 1967, and Van Mol to Smith, May 18, 1967, GMF. Van Mol opposed any promotional activities for Tellico New Town "until the Tellico lands are substantially in TVA ownership." On some TVA employees' desire that the new community be a TVA retirement enclave, see interview with Paul Evans, Apr. 7, 1983.

17. Gladstone and Associates, "Preliminary Design Program," Nov. 1, 1967, NDRSF. Lee Kribbs of the State Planning Office agreed with Gladstone's belief that Tellico New Town would crack the Knoxville housing market. See Kribbs to Task Force, Dec. 1, 1967, NDRSF. For Gladstone's retention, see

Stern to Beatrice Haskin (chief, Employment Branch), Dec. 26, 1976, NDRSF.

18. Carl Feiss, "The Tellico Community Program Memorandum," attached to Feiss to Gober, Jan. 17, 1968, and Gober to Feiss, Jan. 25, 1968, NDRSF.

19. Stern to Withers Adkins (TVA chief architect), Apr. 11, 1968, NDRSF.

20. "New Town Discussions with FHA and HUD in Washington," in Gober to Stern, Jan. 26, 1968; Gober to Thomas Armstrong (HUD), Mar. 8, 1968; and Gober to Feiss, Feb. 15, 1968; all NDRSF.

21. "Tellico Community Study, Community Fiscal Viability Analysis," draft, Apr. 23, 1968, NDRSF. The $1 million surplus projected for 1995 included tax revenues from industrial lands within the New Town limits. Without those, the surplus projection would have shrunk to $259,000.

22. TVA Task Force, "Tellico New Community: A Proposal," May 1968, in Division of Water Control Correspondence Files; Gober to Feiss, May 17, 1968, and Gober to Gladstone, May 17, 1968, both NDRSF. For Van Mol's comments, see informal memo by Richard E. Brown, June 26, 1968, NDRSF.

23. Brown's memo, June 26, 1968, NDRSF.

24. Taylor to Stern, July 15, 1968, NDRSF.

25. Gray to Foster, Nov. 3, 1968, and Foster to Van Mol, Nov. 7, 1968, NDRSF.

26. Mayer to Gray, Dec. 6, 1968, and Wagner to Mayer, Jan. 13, 1969, NDRSF. Albert Mayer was author of *The Urgent Future* (New York, 1967).

27. Gober to Stern, Jan. 20, 1969, NDRSF.

28. For one opinion, see Harold Fleming (of the Potomac Institute) to Gober, May 15, 1969, and Gober to Fleming, June 4, 1969, NDRSF. On the Chapel Hill sociologists, see telephone conversation with Gray, June 27, 1984.

29. Gober to Foster, Jan. 27, 1969, NDRSF.

30. Sproul to Foster, Jan. 25, 1971, NDRSF.

31. For Gober's speech, see *Knoxville Journal* Feb. 19, 1969. For Van Mol's comment on Lenoir City, see notation attached to Evans to Van Mol, Feb. 18, 1969, GMF. On TVA and services, see "Rationale for TVA Expenditure," draft, Feb. 11, 1970, NDRSF. Of course TVA was still counting on HUD for considerable assistance with the water and sewage system.

32. Foster to Van Mol, June 4, 1969, NDRSF.

33. On Gladstone, see Gladstone to Gober, June 5, 1969; Gober to Foster, June 6, 1969; Foster to Haskin, July 12, 1969; all NDRSF. For the doubling of estimates, see Gober to Those Listed, Aug. 19, 1969; Charles Blowers (TVA) to Chris Weeks (Gladstone), Sept. 19, 1969; Gladstone to Gober, Oct. 8, 1969; all NDRSF.

34. On selection of Timberlake, see this chapter, Note 2. On planning, see Timberlake Task Force Meeting, Minutes, Sept. 15, 1969, NDRSF. On shoreline improvement, see Mattern to Gober, Oct. 16, 1969, NDRSF.

35. At the turn of the century the term "dope" was a common synonym for a cola drink. It is still used frequently in East Tennessee. The term probably reflected the fact that some soft drink companies put small amounts of cocaine in their mixtures.

36. Timberlake Task Force Meeting, Minutes, Sept. 15, 1969, NDRSF. Gladstone and Associates estimated the cost of "educational facilities" to be $15 million. The figure seems low. See Weeks to Gober, Nov. 4, 1969, NDRSF.

37. Weeks to Gober, Nov. 4, 1969, NDRSF; Ericson to Seeber, May 6, 1970, GMF. Lynn Seeber had replaced Van Mol as general manager; he was an ardent supporter of both the Tellico Project and Timberlake.

38. See report and margins of Cody and Byrnes (Best and Associates) to Gober, Sept. 26, 1969, and Adkins to Gober, Nov. 24, 1969, NDRSF. Adkins did admit that Cody's and Byrnes's critique of Timberlake transportation planning was well taken, in that the subject "had not received sufficient consideration" and needed "further study."

39. Blowers to Brown, Nov. 18, 1969, NDRSF. Feiss had made the statement about "a string of TVAs" at a convention of the American Institute of Planners. See Gober to Feiss, Nov. 29, 1967; Earle Draper to Feiss, Nov. 30, 1967; Feiss to Draper, Dec. 8, 1967; all NDRSF.

40. The $145 million figure is from TVA Task Force, "Timberlake: A Preliminary Planning Report, December 1969," NDRSF. For the consideration of an additional 9,300 acres, see Perry to Ashford Todd, Jr., Jan. 30, 1970, NDRSF. For Kimmons's remarks, see Kimmons to Foster, Feb. 9, 1970, NDRSF.

41. Nixon's negative view of the "model cities" movement was expressed in "Nixon's Plans for Future of U.S. Cities and Small Towns," *U.S. News and World Report*, July 6, 1970. Knoxville itself was about to pick up on the "gentrification" movement. In the Knoxville Metropolitan Planning Commission Files, see Mayor's Downtown Task Force, *A Prospectus for Central Knoxville* (Knoxville, 1972); Regional Urban Development Assistance Team, *After Expo* (Knoxville, Mar. 1977); UT Graduate School of Planning Research Center, "A Housing Market Study for Lower Second Creek Redevelopment Plan," in Metropolitan Planning Commission, *Knoxville's Center City: Data and Technical Information Report* (Knoxville, Aug. 1981). See also McDonald and Wheeler, *Knoxville*, 157, 163-71, and A. J. Gray, "Lower Second Creek Redevelopment," lecture presented at the University of Tennessee, Knoxville, Apr. 1984. Retired from TVA, Gray was a consultant to the City of Knoxville on redevelopment of the 1982 World's Fair site.

42. For the Wagner-Ellender exchange, see *Hearings on HR 18127*, Subcommittee of the Committee on Appropriations, 91st Cong. 2nd Sess., 596-633.

43. Ibid.

44. Ibid.

45. Gober to Files, July 16, 1970, NDRSF. The three representatives from Boeing were Russell Knapp, Jesse Henry, and Ed Keen. For Boeing's statement, see Henry to Gober, July 22, 1970, NDRSF. The corporation had the goal of 25 percent of its business in nonflight industries by 1975 and was becoming involved in solid and liquid waste management, computer assisted garbage collection, mass transit systems, and hydrofoil development. *Knoxville News-Sentinel*, Jan. 9, 1972.

46. Gerson to Gober, Aug. 21, 1970; Gober to Diggs, Aug. 28, 1970; Foster

to Seeber, Sept. 1, 1970; Seeber to Foster, Sept. 8, 1970; all NDRSF. For refusals, see ibid. A sample of other corporations contacted includes Hallmark Cards, Friendswood Development Co., International Paper, Atlantic-Richfield, Goodyear Tire and Rubber, W. R. Grace, Phillips Oil, Kaiser, Metro-Goldwyn-Mayer, and Radio Corp. of America; ibid. For a typical letter of solicitation, see Seeber to John D. Harper (ALCOA), June 15, 1973, GMF.

47. Gober to Williams, Dec. 30, 1970, and Gober's "Notes on Conversation with Red Williams, Dec. 15, 1970," NDRSF. For Van Mol's phrase, see Van Mol memorandum, June 26, 1968, NDRSF.

48. Gober to Gray, Feb. 23, 1971, NDRSF. A 15 percent return would have been reasonable, seeing that some development corporations were looking for returns of 66 to 71 percent. See Gray to Van Mol, Feb. 9, 1966, NDRSF. For the 23,000-acre figure, see Blowers to Rogers, Nov. 30, 1970, NDRSF.

49. Foster to Gray, Mar. 30, 1971; Gober to Gray, Mar. 16, 1971; "News Release," Apr. 1, 1971; O. C. "Ollie" Boileau (Boeing) to Seeber, May 21, 1971; Gray to Foster, June 17, 1971; Foster to Seeber, July 6, 1971; Shirley Weiss (Chapel Hill Center for Urban and Regional Studies) to Gober, Aug. 2, 1971; all NDRSF.

50. Gladstone to Gober, Aug. 1971, and Gober to Gladstone, Aug. 17, 1971, NDRSF.

51. Gober to Staff, Oct. 27, 1971, and Gober to Gladstone, Nov. 26, 1971, NDRSF.

52. Gober to Gladstone, Nov. 26, 1971, NDRSF.

53. For a report of the Jan. 12 meeting, see Kercher to Files, Jan. 21, 1972, NDRSF.

54. Fantus Corporation, "Appraisal of Industrial Potential, April 1972," copy in NDRSF.

55. Kercher to Files, Jan. 21, 1972, NDRSF.

56. J. E. Parazynski to F. D. Williams, Apr. 14, 1972, NDRSF.

57. Real Estate Research Corporation, "Industrial Potential of Timberlake New Community, April 1972," NDRSF.

58. E. S. Keen (Boeing) to John Wellborn (State Office of Urban and Federal Affairs [OUFA]), May 1, 1972; Criley, "Review of the Tellico Recreation Report for the Boeing Company, May 22, 1972"; Keen to Richard Henderlight (State OUFA), June 28, 1972, with draft of "memorandum of understanding" attached; all NDRSF.

59. Lesesne to Kercher, "Comments on Report and Preservation of Martin Hagerstrand on Opportunities for Indians in the Timberlake Project," NDRSF.

60. For Gober's earlier comment, see Gober to Weiss, Jan. 24, 1972, NDRSF. For proposal, see Boeing Corporation, A Proposal for Timberlake, Feb. 15, 1973, and Robert F. Worthington to Williams, Jan. 29, 1973, both NDRSF.

61. Ibid. See also "Contract Timberlake"; "A Concept for a Timberlake-Tellico Environmental Program," Apr. 1973; and "Energy Conservation"; in Topical Files, NDRSF. See also "Timberlake Environmental Program," preliminary draft, Apr. 10, 1974; John H. Gibbons (Federal Energy Office)

to Colaw, June 12, 1974; Doig to Williams, May 23, 1974; all NDRSF.

62. Stephenson to Gober, Feb. 27, 1973, and Nichols to Gober, Mar. 26, 1973, NDRSF.

63. Wiley to Colaw, July 9, 1973, and Colaw to Gober, July 18, 1973, NDRSF.

64. David Springate (consultant hired by TVA) to Colaw, July 3, 1973, NDRSF.

65. Boileau to Seeber, June 8, 1973, and Seeber to Boileau, June 15, 1973, NDRSF.

66. Foster to Seeber, Sept. 6, 1973, NDRSF.

67. Interview with Niles Schoening, June 12, 1984. On Lambert, see Gober to Foster, July 7, 1967, NDRSF. The only tactic available to people about to be annexed was to incorporate, as the town of Farragut did when it was threatened with annexation by Knoxville in the early 1980s.

68. Interview with Schoening, June 12, 1984. TVA's initial draft was done by C. M. Stephenson who, according to Schoening, was "riding point" on this.

69. Wiley to Colaw, Apr. 30, 1974, and July 5, 1974; TVA Regional Planning Staff, "A Summary of Possible Sources of Federal Aid," Aug. 1974, all NDRSF.

70. Rasnic to Files, Feb. 11, 1974, NDRSF.

71. Wiley to Files, June 12, 1974, and Wiley to Gray, July 28, 1974, NDRSF. On TVA's and Boeing's desire to control the whole planning process and still receive the protection of state government, see interview with Schoening, June 12, 1984.

72. Trevino (HUD) to Williams (Boeing), Feb. 26, 1974, and Colaw to Foster, Mar. 20, 1974, NDRSF.

73. Colaw to Files, Sept. 3, 1974, NDRSF.

74. G. L. Keister (Boeing) to L. D. Alford and J. E. Andrew (TVA), Oct. 7, 1974, NDRSF.

75. On the failure of the "model cities" movement, see the *Washington Post*, Nov. 15-16, 1974.

76. Boileau to Seeber, Mar. 3, 1974, NDRSF; *Knoxville Journal*, Mar. 6, 1975.

77. *Knoxville Journal*, Mar. 6, 1975; Wagner to North Callahan, Apr. 2, 1975, NDRSF.

CHAPTER 8

1. S. David Freeman was appointed in July 1977 and confirmed the next month. He replaced Don McBride, who had retired in 1975. The TVA directors served nine-year terms, and the two-year gap between McBride's retirement and Freeman's confirmation meant that the latter's term expired in 1984. For Freeman's style and activities, see interviews with Louis Gwin

(assistant director of the TVA Office of Information), July 14, 1983, and S. David Freeman, February 25, 1983. For the incident in Wagner's office, see interview with Freeman, ibid.

2. The Office of General Counsel (which TVA people refer to simply as "Law") is the only TVA body which reports directly to the board and not to the general manager.

3. For the statement's critics and TVA's defense, see TVA, Office of Health and Environmental Science, *Environmental Statement — Tellico Project, Feb. 10, 1972* (Chattanooga, 1972), vol. I.

4. Prichard to Brock, Aug. 27, 1971, GMF. For TVA's reaction, see McBride to Seeber, Mar. 16, 1972, GMF. For Bell's remarks, see Bell to Files, Aug. 10, 1973, PPBF. Technically the nomination was submitted by Michael J. Smith, executive director of the Tennessee Historical Commission. See ibid. and Bell to Seeber, Sept. 7, 1973, PPBF.

5. According to Bell, Lawrence Aften (NPS executive order consultant) and Jerry Rogers (NPS chief of registrations) "were agreeable to 'shelving' the nominations until after the trial." Less clear was the role of Robert Utley (NPS director of archaeology and historic preservation), a well-known historian of American Indians. See Bell to Seeber, Sept. 7, 1973, WCPF. Executive Order 11593 was for "Protection and Enhancement of the Cultural Environment." Section 2, Part B, was the important passage. Public Law 89–665 was "An Act to Establish a Program for the Preservation of Additional Historic Properties Throughout the Nation." For TVA's intention not to comply, see Bell to Files, Oct. 1, 1973, Archaeological Explorations Folder #4, WCPF. For claims of full compliance, see Paul Evans to *Mountain Life and Work*, Nov. 13, 1973, in ibid. The Advisory Council on Historic Preservation did not agree. See Tapman to Taylor, Sept. 13, 1973, WCPF. For the court's ruling of compliance, see Porter to Kenneth Tapman (NPS compliance officer), Nov. 23, 1973, Archaeological Explorations Folder #4, WCPF.

6. For Cherokee complaints, see Guthe to Bell, July 9, 1973, Archaeological Explorations Folder #4, WCPF. For TVA's defense, see Bell to Files, Jan. 8, 1973, PPBF; John L. Cotter, "Observations on the Status and Potential of Archaeology and Interpretation at the Tellico, Chota, and Related Sites" (NPS report, June 8, 1973); Bell to J. O. Brew, John M. Corbett, and Robert L. Stephenson (Board of Archaeological Consultants), Sept. 5 and 14, 1973, and Bell to Notes, Sept. 18, 1973, both in PPBF; "Archaeological Environmental Impact Statements," Oct. 31, 1973, Archaeological Exploration Folder #4, WCPF.

7. Moser to Duncan, July 12, 1973, and McBride to Duncan, Aug. 1, 1973, GMF. Moser's brother-in-law was Lon Stamey, now deceased, who received $16,400 for his land. Stamey's widow (Moser's sister) continues to maintain the accuracy of her brother's charges. See interview with Louise Stamey, Mar. 31, 1983. The rumors noted above continue to this day, but no additional evidence has been found to substantiate them.

8. Alexander to Wagner, Mar. 7, 1972, and throughout GMF. For the sale to TVA, see material for tract #TELR 2501, Land Branch Records (Tellico), Chattanooga. In Alexander's first bid for the governorship, he was defeated

by Democrat Ray Blanton. Alexander tried again four years later and, with a well-run campaign and with the Blanton administration awash in scandal, he was successful.

9. Evans to Seeber, June 15, 1973 in Project General I, WCPF.

10. *New York Times*, Oct. 26, 1973. See also Porter to Tyree, Nov. 23, 1973, Archaeological Explorations Folder #4, WCPF.

11. Interview with Louis Gwin, July 14, 1983, and Sam Venable, Aug. 23, 1983.

12. Rechichar and Fitzgerald, *Consequences of Administrative Decision*, 40-41.

13. For two examples of opponents' dilemma, see Sara Grigsby Cook, Chuck Cook, and Doris Gove, "What They Didn't Tell You About the Snail Darter & the Dam," *National Parks and Conservation Magazine* 51 (May 1977), 10-13; and Zygmunt J. B. Plater, "Reflected in a River: Agency Accountability and the TVA Tellico Dam Case," *Tennessee Law Review* 49 (1982), 747-87. For Sept. 1974 cost figure, see Kimmons to Seeber, Sept. 6, 1974, Construction Specifications—Tellico, NDRSF.

14. Much of the material on Plater comes from an interview with him and from interviews with Profs. Forrest Lacy and Fred Thomforde of the UT College of Law. Former students, coworkers, and friends have filled in the gaps. See interview with Peter Alliman, Aug. 18, 1983.

15. For a good sense of the era, see Bob Dylan, "The Times They Are A'Changin'." Columbia CL 2105.

16. See Note 14.

17. See Note 14. For Plater's early consulting work for TVA and his own personal involvement, see Plater, "Reflected in a River," 748. For Plater's philosophy, see ibid., 747n., 749n., 752n., esp. 757n., 758n.

18. For Etnier's "correct distance," see interview with Peter Alliman, Aug. 18, 1983. On the snail darter's classification, see *Knoxville News-Sentinel*, Oct. 27, 1974; *Daily Beacon*, Nov. 15, 1974; *Chattanooga Times*, Dec. 4, 1974. For Plater's immediate excitement, see interview with Prof. Fred Thomforde, June 24, 1983. For Plater's tongue-in-cheek statement, see Plater, "Reflected In a River," 771n.

19. Interview with Alliman, Aug. 18, 1983. Alliman's future father-in-law owned considerable holdings bordering TVA's taking line, some of which he developed into the Bass Harbor subdivision. To the credit of both men, Alliman and Lee remained on the best of terms.

20. For TVA plans for Fort Loudoun, see Jack Rountree to Files, Jan. 21, 1975, PPBF.

21. For opponents' opinion of the courts and eminent domain, see interview with Alliman, Aug. 18, 1983. For TVA ultimatum to the Fort Loudoun Association, see Jack Rountree to Files, Jan. 21, 1975, PPBF. For efforts to ease Alice Milton out, see Report of Phone Conversation between "PG" and Mrs. W. W. Stanley (a past president of the association), Feb. 24, 1976, COF. For the "last weapon" theory, see Plater, "Reflected In a River," 769-70.

22. Ibid., 772.

23. *Knoxville News-Sentinel*, Oct. 27 and Dec. 4, 1974; *Chattanooga Times*, Dec. 4, 1974; *Daily Beacon*, Nov. 15, 1974; E. Bruce Foster to Seeber, Dec. 5, 1974, COF. For attempts to reach Blanton, see *Knoxville News-Sentinel*, Feb. 27, 1975. For Blanton's earlier comment about Tellico, see ibid., Feb. 19, 1972.

24. For Seeber's refusal to stop cutting see Seeber to Foster, Dec. 12, 1974, COF. Wayne Starnes contested the claim that siltation would not hurt the snail darter. See *Knoxville News-Sentinel*, Mar. 14, 1975. On work speedup, see Plater, "Reflected In a River," 768. For search for other snail darters, see *Nashville Tennessean*, Mar. 14, 1975. For transplanting, see *Knoxville News-Sentinel*, June 3, 1975. On Raney, see his signed affidavit dated Dec. 14, 1974, in Public Interest Files, WCPF; *Nashville Tennessean*, Mar. 14, 1975; Raney to Director, U.S. Fish and Wildlife Service, Feb. 16, 1976, GMF. For Seeber on Etnier, see *Knoxville News-Sentinel*, Aug. 20, 1975. What Seeber said was that Etnier had always been against the dam and that "it appears the whole matter is little more than an effort by a group of professionals fully aware of scientifically accepted standards and methods who have chosen to ignore them in order to attempt once again to stop a project which has already been approved." For use of term "so-called snail darter," see Seeber to Carol Stephens, June 13, 1975, and Seeber to Paul H. Clark, June 30, 1975, GMF. For Raney's salary as consultant ($4,837), see John Van Mol to Christopher Bergesen, Dec. 3, 1976, GMF. For total cost (to Dec. 1976) of snail darter transplant, see ibid.

25. Evins to Blanton, Mar. 6, 1975, NDRSF. For a similar but softer approach, see Wagner to Blanton, Mar. 13, 1975, NDRSF. For Criley's recollections, see interview with Criley, Feb. 16, 1984; Rechichar and Fitzgerald, *Consequences of Administrative Decision*, 43–44. For Duncan's blast, see Duncan to Reed, Mar. 12, 1975, NDRSF. Reed was a former schoolmate of Joseph Congleton, head of Knoxville chapter of Trout Unlimited and longtime Tellico foe. For Rabel's story, see Sept. 30, 1975, transcript of CBS Morning News, COF. TVA's John Van Mol believed the program to have been a "balanced account." See Van Mol to Seeber, Sept. 30, 1975, COF.

26. This entire account is from an interview with Paul Evans, Apr. 7, 1983.

27. Ibid.

28. The prospectus, research notebooks, and drafts were donated to and now reside in the McClung Historical Collection, Lawson-McGhee Public Library, Knoxville. There are restrictions on their use.

29. The draft of the chapter in question can be found at the McClung Historical Collection. Reproduction, however, requires the Brewers' written permission. The debate among the officers of the East Tennessee Historical Society was recounted to us by Jesse Mills, TVA technical librarian and ETHS member, and substantiated in a telephone conversation with Carson Brewer. Incidentally, all indications are that Mills acted with complete correctness despite his dual loyalties, in what must have been an uncomfortable situation.

30. The tenure system is a fixed feature of American higher education.

After a probationary period (usually five years), a faculty member is evaluated on the bases of teaching, scholarly research and publications, committee work, and public service. Once a faculty member achieves tenure at an institution, he or she cannot be dismissed except for financial reasons or for gross misconduct. Precise criteria for granting tenure vary from institution to institution, but in the mid-1970s requirements generally became more rigorous, especially with regard to research and publication. Faculty members who do not receive tenure must leave the institution, though they are often given an additional year to find other employment. At the UT Law School, faculty members were considered for tenure at the beginning of their third years.

31. The particulars of the meeting were related to the authors in interviews with various tenured members of the Law School faculty. The vote, however, was provided by Peter Alliman, at the time a law student, in an interview on Aug. 18, 1983. Alliman may have gotten that information from Plater himself, although "leaks" from tenure meetings are not at all uncommon. We must add that none of the faculty members we interviewed named names or spoke in anything but generalities.

32. One former student of Plater's told us that he began one class by announcing "I'll give an A to anybody in this class who can find a legal way to stop the Tellico Dam." However, one senior law school faculty member told us that he considered that statement a "legitimate teaching technique in that it stimulated thought and discussion."

33. On Alliman's role, see interview with him, Aug. 18, 1983.

34. For TVA's opposition to critical habitat status, see *Knoxville News-Sentinel*, Dec. 23, 1975, and Feb. 6, 1976. For around-the-clock work, see *News-Sentinel*, Aug. 3, 1976, and Wagner's testimony before Culver subcommittee, July 1977, GMF. For Tennessee Endangered Species Committee claims, see Cook et al., "What They Didn't Tell You About the Snail Darter and the Dam," 11. For "volunteers," see interview with Louis Gwin, July 14, 1983, and Louis Garono to Dr. Louis Nelson, Sept. 13, 1976, NDRSF. On consultation with Fish and Wildlife Service, see Richard Fitz to Files, Sept. 15, 1976, NDRSF, and Kenneth Black to Seeber, May 27, 1976, GMF. For costs of transplants and the incubation scheme, see John Van Mol to Christopher Bergesen, Dec. 3, 1976, GMF. For project activities yet to be completed, see Jerry W. Wilson to Files, July 9, 1976, NDRSF; Conro Olive to J. Porter Taylor, May 13, 1976 in Family Relocation Files, DRPF.

35. *Chattanooga News-Free Press*, Feb. 26, 1976; *Nashville Tennessean*, Feb. 26, 1976; *Daily Beacon*, Feb. 26, 1976; *Knoxville News-Sentinel*, May 25, 1976; *Chattanooga Times*, May 26, 1976. The intent to appeal obliged Judge Taylor to allow TVA to continue construction but prohibited closing the dam. *Knoxville News-Sentinel*, Aug. 3, 1976. For Plater's "confidential source," see Plater, "Reflected In a River," 768.

36. For Celebrezze's opening remarks, see *Knoxville News-Sentinel*, Oct. 15, 1976. The trial was well covered in the local press.

37. *Chattanooga News-Free Press*, Jan. 31, 1977; *Newsweek*, Feb. 21, 1977.

38. *Washington Star*, Feb. 7, 1977; *Birmingham News*, Feb. 14, 1977; *Nashville Banner*, Feb. 1, 1977; *Memphis Commercial Appeal*, Feb. 2, 1977.

39. *Knoxville Journal*, Feb. 2, 1977.

40. *Knoxville News-Sentinel*, Feb. 4, 1977. Taylor issued the permanent injunction on Feb. 24, 1977. *New York Times* and *Washington Post*, Feb. 25, 1977.

41. See Seeber to Lynn A. Greenwalt (director, U.S. Fish and Wildlife Service), Feb. 28, 1977, and the attached "Petition to Delist the Little Tennessee River . . . as Critical Habitat for the Snail Darter," GMF.

42. For TVA pushing for exemption, see *Knoxville News-Sentinel*, Jan. 31, 1977; *Newsweek*, Feb. 21, 1977. For Baker's statements, see *Johnson City (Tenn.) Press-Advocate*, Feb. 3, 1977; *Oak Ridger*, Feb. 10, 1977; *Knoxville News-Sentinel*, Feb. 11, 1977.

43. Tennessee Endangered Species Committee Newsletter, Mar. 14, 1977, in TESC Files. For Quillen's reply to Burgner and the margin comment, see Quillen to Burgner, Feb. 16, 1977, in TESC Files.

44. Evins to Lance, Mar. 29, 1977, NDRSF; Johnson to Carter, Mar. 21, 1977, and Brogan to Carter, Apr. 5, 1977, GMF. No written record of TVA's request to Brogan has been found, because it was likely made by telephone. Moreover, according to then-head of the History Department LeRoy P. Graf, Brogan never inquired of him whether Johnson was speaking "officially" or not. For lobbying by TVA allies, see news release of the National Rural Electric Cooperative Association (representing 1,000 rural electric systems), Mar. 10, 1977, GMF.

45. U.S. General Accounting Office, *Draft—The Tellico Dam Project—A Brief Assessment of Costs, Alternatives and Benefits* (July 1977), esp. Ch. 4. For TVA's response, see "Comments on Revised GAO Report, Tellico Project," WCPF. The final report was issued on Oct. 14, 1977. For essence of GAO's recommendations, see Comptroller General Elmer B. Staats to Wagner, Oct. 14, 1977, WCPF. For a further interpretation, see Duncan and Lloyd to Wagner, Oct. 20, 1977, WCPF.

46. For Seeber's ordering of a new benefit-cost study, see *Knoxville Journal*, Sept. 22, 1977. For the new method of computation and the 7:1 figure, see *Hearings of the Senate Subcommittee on Resource Protection*, a copy of which is in GMF.

47. Records of the Culver subcommittee hearings are in GMF. For Guthe's letter, see Guthe to Culver, July 20, 1977, GMF. Guthe had been removed as project director and replaced by Chapman, ostensibly for not completing his archaeological reports on time and for not being sufficiently watchful of his budget. Intra-university politics may also have been involved. There was never a question concerning the archaeological competence of either Guthe or Chapman. Chapman refused to comment on the reasons for the switch of project directors. See Carl Thomas (UT dean for research) to Guthe, Oct. 3, 1977, Archaeological Exploration Folder #6, WCPF.

48. Interview with Alliman, Aug. 18, 1983. For report of the Holiday Inn meeting, see *Waynesville (N.C.) Mountaineer*, June 13, 1977.

49. Van Mol to Seeber, June 21, 1977; Seeber to Lesesne, June 24, 1977; Lesesne to Seeber, June 28, 1977; J. Bennett Graham to Files, July 22, 1977; all in Archaeological Explorations Folder #6, WCPF.

50. For opposition claims that TVA spent $900,000 on lobbying, see *Daily Beacon*, Apr. 26, 1977. For information on the number of tracts purchased and condemned, see Rosek to Seeber, Nov. 7, 1977, GMF. Some of those whose property had been condemned had been allowed to stay on their land, a decision which raised problems later.

51. Interview with Freeman, Feb. 25, 1983. For results of the Ford Foundation study, see [Freeman], *Electric Power and the Environment* (Washington, Aug. 1970), esp. Ch. 7. For TVA's disagreement, see ibid., 46n.

52. Interview with Freeman, Feb. 25, 1983.

53. The above is the result of conversations with many present and former TVA workers at all ranks, most of whom talked with us "off the record." Louis Gwin characterized the respective styles of the two men in this way: "If you were a boy and wanted to borrow the family car, Mr. Wagner—a grandfatherly type—would say, 'Son, I'm sorry, but you can't borrow the car,' whereas Dave could make you feel deficient for asking." Interview with Gwin, July 14, 1983.

54. Freeman quotes from interview with him, Feb. 25, 1983. He was recalling statements he'd made earlier. The director kept a photograph of himself with Carter on his office wall.

55. Interview with Alliman, Aug. 18, 1983.

56. For Freeman's views of TVA's mission and his private inclinations on Tellico, see interview with Freeman, Feb. 25, 1977. For one example of the use of Freeman's noncommittal statement, see Wagner to Rep. Duncan and Lloyd, Nov. 10, 1977, Projects General File, WCPF.

57. Freeman to Seeber, Oct. 27, 1977, Projects General File, WCPF.

58. Rechichar and Fitzgerald, *Consequences of Administrative Decision*, 51.

59. Ibid., 52. For filing of brief, see *Knoxville News-Sentinel*, Jan. 25, 1978.

60. On Carter's tactic, see *Atlanta Constitution*, Mar. 26, 1978; *Knoxville Journal* and *Knoxville News-Sentinel*, Mar. 28, 1978. For denial of pressure, see *Lenoir City News-Banner*, Mar. 30, 1978. For a public airing of the controversy, see transcript of "The MacNeil-Lehrer Report," PBS broadcast on Apr. 13, 1978, NDRSF. For Wagner's refusal, see Wagner to Andrus, Mar. 31, 1978, COF. The letter was made public in early April. See *Chattanooga News-Free Press*, *Oak Ridger*, *Knoxville News-Sentinel*, *Knoxville Journal*, and *Maryville-Alcoa Times*, Apr. 6, 1978.

61. Freeman to Andrus, Apr. 6, 1978, GMF; *Nashville Tennessean*, *Washington Post*, *Chattanooga News-Free Press*, and *Knoxville News-Sentinel*, Apr. 7, 1978.

62. Foster's marginalia is on Freeman to Frank W. Goddard, June 1, 1978, NDRSF. For an agriculturist's point of view, see Roger C. Woodworth to John L. Furgurson, June 9, 1978, Administrative Files, OACD. For Harned's letter, see Harned to Freeman, Sept. 29, 1978, COF.

63. For Duncan's letter and the poll results, see Duncan to Freeman, June 19, 1978, NDRSF. The poll was conducted in April 1978. For other critical letters, see Pennington to Freeman, June 5, 1978, in Public Interest File, WCPF; Kennedy to Freeman, June 26, 1978, and Wilson to Freeman, July 22, 1978, NDRSF; *Nashville Banner* Apr. 13, 1978; *Knoxville News-Sentinel, Nashville Tennessean,* and *Chattanooga News-Free Press,* Apr. 14, 1978. On Madisonville meeting, see *Knoxville Journal* and *Nashville Tennessean,* June 21, 1978; as well as Sarah Cardin to Freeman, June 21, 1978; Francis B. Schoidt to Freeman, June 23, 1978; and Freeman to Harwell Proffitt, July 6, 1978; all NDRSF.

64. *Knoxville Journal,* Apr. 11, 1978; Vines to Freeman, Apr. 11, 1978, and Plant to Freeman, Apr. 18, 1978, GMF. For statement, see *Knoxville News-Sentinel,* June 24, 1978.

65. It is now quite fashionable to criticize TVA and the New Deal which gave it birth. See Erwin Hargrove and Paul Conkin, eds., *TVA;* William U. Chandler, *The Myth of TVA: Conservation and Development in the Tennessee Valley, 1933–1983,* (Cambridge, Mass., 1984).

66. For Jenkins's resignation, see *Knoxville Journal,* May 6, 1978; *Nashville Banner* May 8, 1978. For meetings with the Interior Department, see *Knoxville News-Sentinel,* May 7, 1978; *Knoxville Journal,* May 11, 1978. For Wagner's refusal to compromise, see *Chattanooga News-Free Press,* Apr. 26, 1978; *Nashville Tennessean,* Apr. 27, 1978. Wagner to Rep. Tom Bevill (chairman, House Subcommittee on Public Works), May 18, 1978, GMF. For his new tack, see ibid. An earlier "alternatives report" had been prepared in January 1972, but the purpose of that effort had been to "prove" the original concept was superior. See "Tellico Project—Economic Analysis," Jan. 12, 1972, PPBF.

67. Rechichar and Fitzgerald, *Consequences of Administrative Decision,* 54–55.

68. *TVA v. Hill,* 437 U.S. 153, 1978. For reports on the decision see *Nashville Banner, St. Louis Post-Dispatch,* and *Knoxville News-Sentinel,* June 15, 1978; *Newsweek,* June 26, 1978.

69. *Chattanooga Times, Knoxville Journal,* and *Oak Ridger,* Oct. 16, 1978.

70. For one example of Freeman's statements, see "Statement by S. David Freeman . . . Before the Subcommittee on Fisheries, Wildlife Conservation, and the Environment . . . ," June 23, 1978, a copy of which is in WCPF; *Knoxville News-Sentinel,* June 24, 1978. For circulation of "Alternatives Report" draft, see Jane Parker to Colaw, July 13, 1978. For establishment of the task force to write the draft, see Gardner Grant to Freeman, June 26, 1978. For Baker's complete support of the Tellico Project, see Baker to Robert Pennington, May 11, 1978, Public Interest Folder, WCPF. In the "dry dam" alternative, theoretically the dam could be closed if a threat of flooding downriver necessitated it.

71. Baker to Freeman, June 29, 1978, COF; *Knoxville News-Sentinel,* June 30, 1978.

72. Interview with Freeman, Feb. 25, 1983.

73. Freeman to Baker, July 7, 1978, COF. See related memos, suggestions, and marginalia in COF.

74. For ABC News report, see transcript for July 19, 1978, COF. For distribution of "Alternatives Report" draft, see Claussen to Parker, Aug. 18, 1978, Alternatives Folder #1, Division of Water Resources. For number of comments, see Freeman to Duncan, Sept. 4, 1979, COF. For some reactions of foes, see Plater to Freeman, Sept. 26, 1978, COF, and Congleton to Freeman, Aug. 31, 1978, in Public Interest Folder 6, WCPF. For Plater's second thoughts, see Plater, "Reflected in a River," 780–82. For final report, see TVA, *Alternatives for Completing the Tellico Project* (Knoxville, Dec. 1978).

75. Mattern to Freeman and Elliot to Freeman, Sept. 25, 1978, Administrative Files, OACD; Vogel to Herbst and Freeman, Aug. 30, 1978, COF.

76. "An Outline for Citizen Participation in the Tellico Project," July 27, 1978, OTAD. See also Philip Hyatt to Claussen, Nov. 15, 1978, OTAD, and *Lenoir City News*, Nov. 8, 1978.

77. Lee to Freeman, Sept. 13, 1978, and Heiskell to Freeman, Sept. 19, 1978, COF; Wilson to Freeman, Sept. 22, 1978, Administrative Files, OACD.

78. Transcript of meeting, Jan. 23, 1979, GMF. For Andrus's statement, see *Washington Post* and *Wall Street Journal*, Jan. 24, 1979.

79. *Washington Post*, Jan. 24, 1979. Baker introduced such legislation on Jan. 30. See *Knoxville Journal*, Jan. 30, 1979. John Duncan introduced an identical bill in the House. Baker's bill to exempt failed in committee on May 9. *Nashville Tennessean*, May 10, 1979.

80. Stafford to Freeman, June 30, 1978, COF; *Loudon County Herald*, Feb. 24, 1979; Freeman to Parker, Dec. 26, 1978, GMF; Will A. Hildreth (president, First National Bank of Loudon County) to Freeman, Feb. 22, 1979, COF.

81. *Knoxville News-Sentinel*, June 19, 1979.

82. *Kingsport Times*, June 26, 1979. On Herod's choreography of the tactic, see interview with Alliman, Aug. 18, 1983.

83. *Knoxville Journal*, July 18, 1979; *Knoxville News-Sentinel*, Sept. 11, 1979.

84. For Carter's behind-the-scenes moves, see *Knoxville News-Sentinel*, July 12, 1979.

85. On Carter's motivation for signing, see interview with Freeman, Feb. 25, 1983. On the telephone call to Plater, see interview with Alliman, Aug. 18, 1983. On Plater's bitter recounting of most of the details, see his "Those Who Care About Laws or Sausages Shouldn't Watch Them Being Made," *Los Angeles Times*, Sept. 2, 1979. For more on Carter, see *Chattanooga News-Free Press*, Sept. 26, 1979.

CONCLUSION

1. *Los Angeles Times*, Sept. 20, 1979; *Boston Globe*, Sept. 14, 1979; *Chicago Tribune*, Sept. 17, 1979; *New York Times*, Sept. 14, 1979; *Philadelphia Inquirer*, Sept. 20, 1979; *Christian Science Monitor*, Sept. 12, 1979; Philip

Ryan to Hamilton Fish, Nov. 13, 1979, COF. There are literally hundreds of similar letters (some addressed to members of Congress, some to Carter, some to Freeman) in COF.

2. For "campout," see CBS Evening News, Oct. 20, 1979, in Vanderbilt University Television News Archive.

3. Interview with Freeman, Feb. 25, 1983.

4. Conversation with Clarence Gibbons of CBS camera crew, July 27, 1983.

5. CBS Evening News with Walter Cronkite, Nov. 13, 1979, in Vanderbilt University Television News Archive. TVA was angered by Mrs. McCall, feeling that the agency had done its best to satisfy her. She had received $86,068.30 for her 91-acre farm, had been driven by TVA staffers over 1,000 miles in search of replacement property, and had been put up for two weeks at the Hyatt-Regency Hotel in Knoxville (where she was allowed to charge everything, including her hairdresser's fees). TVA felt that the CBS interview was poor repayment for their efforts.

6. Interview with Hall, Mar. 31, 1983.

7. Interview with Charlotte Hughes, Mar. 16, 1983.

8. Interviews with Freeman, Feb. 25, 1983, and with Gwin, July 14, 1983. Other people have characterized Tellico as "traumatic" (Rechichar and Fitzgerald, Consequences of Administrative Decision, 70), "deranging" (TVA's Jesse Mills, quoted in ibid., 70) and an "opportunity lost" (Plater, "Reflected In a River," 787).

9. See Chandler, The Myth of TVA.

10. Lance E. Davis, "And It Will Never Be Literature—The New Economic History: A Critique," in Ralph L. Andreano, ed., The New Economic History: Recent Papers in Methodology (New York, 1970), 74–75.

11. For board resolution of Aug. 25 and the contract (TV 60000A), see DRPF. The agreement contained several restrictions so that TVA could maintain some control over development. As land was sold, the money would be passed from TRDA to TVA to the U.S. Treasury. TRDA had a nine-member board, composed of the county executives of Blount, Loudon, and Monroe Counties and two people selected from each county by their respective county commissions.

12. For an appraisal of the land transferred to TRDA, see Hop Bailey, Jr., and B. Glenn Weaver to H. Wade Cowan, Nov. 3, 1982, DRPF. See also James Pounders to Cowan, Oct. 20, 1982, DRPF.

13. Interview with Billy Wolfe (local Tennessee State Game and Fish official), Mar. 16, 1983.

14. Interview with Ben Snider, Mar. 27, 1983.

A NOTE ON SOURCES

In the approximately fifty years of its existence, TVA has created an enormous records archive which can be of great use to historians. Indeed, the amount of material is truly staggering. As of September 30, 1985, the material in TVA file rooms, offices, and records centers located in Knoxville, Norris, Chattanooga, and Muscle Shoals totaled 143,429 cubic feet. This did not include numerous records stored at the National Underground Storage in Pennsylvania, the National Archives in Washington, or the Federal Records Centers in St. Louis, Mo. (mostly government personnel records), and East Point, Ga. (historical material). Of the 143,429 cubic feet, the majority consists of active or working files held in the file rooms and offices of TVA (101,168 cubic feet). The remainder (42,261 cubic feet) is retained in the three records centers at Chattanooga, Knoxville, and Muscle Shoals. To date there is no comprehensive bibliography of TVA material. A start, however, has been made in Michael J. McDonald, "Tennessee Valley Authority Records," *Agricultural History* 58 (Apr. 1984), 127–37, and in McDonald, "New Sources for Rural History: Tennessee Valley Authority Records," research monograph (Oct. 1982) in TVA Technical Library. See also the bibliographical essay in McDonald and John Muldowny, *TVA and the Dispossessed: The Resettlement of Population in the Norris Dam Area* (Knoxville, 1982), 309–24.

Although historians working on TVA encounter a staff that is uniformly knowledgeable and helpful (see "Acknowledgments"), they must expect to encounter certain problems when they attempt to use TVA records. For one thing, some TVA material has been reclassified from active to inactive and moved. For example, the General Manager's Files for 1933–58 were moved first to TVA's records staging area and then, recently, to the Federal Records Center at East Point. Former director S. David Freeman's 55 cubic feet of material have been removed from the working files to records staging, preparatory to final deposit at East Point.

Another obstacle historians confront is agency reorganization. Each time a major reorganization or minor reshuffling of the agency occurs, files are moved and their titles are changed. For example, the Project Planning Branch Files (8.3 cubic feet), extremely useful to those interested in Tellico, were housed in the Division of Water Control Planning. While the Tellico Project was going on, that division's name was changed to Division of Water Management, then to Division of Water Resources, and again to Division of Air and Water Resources. Another group of records valuable for Tellico research is that of the Office of Economic and Community Development (OECD). In October 1983, however, OECD merged with the Office of Natural Resources, to become the Division of Economic and Community Development. Within two years, that division was coupled with the Division of Land and Forest Resources and is now called the Division of Land and Economic Resources. Researchers intent on finding a particular records group might have to trace it through a labyrinth of administrative reorganization.

A minor obstacle is encountered when one seeks to do research in the "working files." By their very nature, these materials are in constant use. Although the staff is courteous, there is little doubt that research historians are "in the way." Moreover, the "working files" are arranged in reverse chronological order.

A final obstacle to doing research at TVA is the Office of General Counsel (OGC), the agency's legal arm. As the only TVA office which reports directly to the Board of Directors, over the years the OGC has acquired a great deal of power. All "working files" must be examined by a member of the OGC staff before researchers are allowed to examine them. Some material is extracted, and researchers are not told the nature or the amount withheld. It is our opinion, however, that in our case none of the withheld material would change our analysis in any significant way.

None of this is to suggest that TVA is a difficult place in which to do research. Indeed, quite the opposite is the case. Given sufficient notice, the agency endeavors to set aside work space for researchers, and hard-to-find material is located with dispatch. The system of arrangement of TVA records, however, almost inevitably poses some problems for historians.

The early history of TVA and the Tellico Project can best be traced by examining the Old Central Files (OCENTF, microfilm), Old Administrative Files (OAF), and Old Commerce Department Files

(OCOMF), all at East Point, Ga. Especially useful in delineating the ideas of Lilienthal, Clapp, and Vogel as they related to events prior to Tellico were the General Correspondence Files, especially the older Chairman's Office Files (COF), also at East Point.

Excellent material on the Tellico region, its people, and TVA plans for the area can be found in the active files of the Office of Economic and Community Development (OECD) in Knoxville. Like all of the agency's active files, the OECD Files are arranged by project, so the Tellico material is grouped together, in reverse chronological order. Within the OECD Files are the Navigation Development and Regional Studies Files (NDRSF), the Timberlake Correspondence File, the Timberlake Volume File, the Tellico Correspondence and Volume Files, and the Office of Tributary Area Development (OTAD) Files. This amounts to 40 cubic feet of extremely valuable material. Good records of the Boeing Corporation's relations with TVA are to be found here.

A day-to-day record of the planning and execution of the Tellico Project can be found in the Project Planning Branch Files (PPBF), which form part of the Division of Water Control Planning Files (WCPF) in Knoxville. That material alone comprises 8.3 cubic feet. WCPF also contains material on the stormy Fort Loudon issue, records of the archaeological projects conducted at Tellico, and benefit-cost ratio material.

The backbone of much modern research on TVA is the General Manager's Files (GMF) and the Chairman's Office Files (COF). The active files of both are in Knoxville and the inactive files at East Point. Both files contain excellent material on Tellico planning, changing philosophy within TVA, benefit-cost ratio, public opinion, and relations with state and federal governments. Altogether, we examined over 15 cubic feet of this material.

While Knoxville and East Point contain the vast majority of Tellico material, TVA records in Chattanooga and Muscle Shoals were not overlooked. In Chattanooga, the Division of Property and Services (DPS) contains roughly 16 cubic feet of material involving land transactions and condemnations for Tellico. Some people within TVA were leery of our examining those transactions, since they contain information on prices paid, mortgages, liens, appeals, and condemnations. As there is still a good deal of talk in the Tellico area about some landowners' filing suit against TVA, some agency people, probably in the OGC, were not delighted to have researchers examining land purchase and condemnation material. Nearly three months elapsed between our initial request and the grant of permission, a decision which was made by the chairman of the TVA Board of Directors. Also housed in Chattanooga were the cemetery and grave-removal records.

In Muscle Shoals, approximately 3 cubic feet of material was consulted, the major portion in the Office of Agricultural and Chemical Development (OACD). Though not voluminous, this material was significant in the Tellico story because only Agriculture provided a visible counterweight to the general TVA enthusiasm about the Tellico Project. Also in Muscle Shoals is the Office of Agricultural Relations, where we found the only complete file of the Future Dams and Reservoirs (FUDAR) Committee's deliberations.

TVA's Technical Library in Knoxville is a great boon to historians. The Technical Library has attempted to collect everything published about TVA, as well as all published material on the Tennessee Valley and its various states and counties. Housed there are some original TVA reports on the region (including the excellent county social and economic studies done in the 1930s), and an excellent newspaper and magazine clipping file, arranged by TVA project. Over a hundred newspapers and numerous magazines are combed regularly for material. Public opinion poll results are also kept there. The staff is first-rate, and there is sufficient work space. Physically close to the Technical Library are the agency's excellent photographic and mapping units, both extremely helpful to us.

References to TVA and Tellico which appeared on television news programs have not been kept by the agency. For those we consulted the Vanderbilt University Television News Archives. This facility is of enormous value to contemporary historians and is a model of organizational efficiency. Its meticulous published indexes are invaluable.

The approximately 80 cubic feet of TVA documents which we examined were supplemented by economic and demographic data from the U.S. Bureau of the Census as well as by statistical material compiled by the State of Tennessee. Although Tennessee did not become part of the "modern statistical era" until the late 1920s, fragmentary material collected before that was extremely helpful to us in understanding what the Tellico area was like before the entrance of TVA. Specific collections and reports are cited in our notes, as are various published studies which we consulted.

Documentary material was supplemented by forty-five interviews with people involved in the Tellico Project. Specific interviews are cited in our notes. Unfortunately, some of the individuals we interviewed have since died.

Finally, much information was collected through informal conversations with TVA people, people who lived in the Tellico area prior to TVA, people who opposed the Tellico Project, people who favored it, and supporters and opponents of TVA from throughout the region and state. Where appropriate, we have acknowledged their observations in our notes.

ACKNOWLEDGMENTS

This book, whatever its merits or demerits, owes a great deal to those people who helped the authors find their way through the labyrinthine records of TVA. Gloria Lucas, Eastern Division records officer, remains a tower of aid to those in the academic community interested in TVA. Jeannette Zarger, records officer, Office of Economic and Community Development (OECD), and her replacement, Catherine A. Watson, were extremely helpful in locating material on the planned town of Timberlake. Dot Bowling, records officer, Office of the General Manager, and her staff, Barbara Daughtery and Gail Stooksbury, guided the authors through the crucial correspondence of the TVA director and general manager. Kay Lemons and Sarah Stooksbury of the Division of Water Resources helped us plumb the great depths of benefit-cost analysis material; and Bernyce Thigpen, records officer, Division of Property and Services, helped with some land purchase and relocation data. Mary Jane Rhodes, records officer, Office of Agricultural and Chemical Development (OACD), Muscle Shoals, provided invaluable assistance in sifting through agricultural files to send to Knoxville and in arranging oral interviews with OACD personnel. Anthony L. Hardegree provided valuable grave relocation data on the Tellico project; the end of a long, arduous research trip to Chattanooga was made rewarding by his cordiality and helpfulness. Conroe Olive, chief, Land Branch, Division of Property and Services, was, along with Legal Office personnel (Office of General Counsel), less of a help than an obstacle in researching Land Purchase Records. But with those exceptions, the people we worked with at TVA were more than just helpful archivists and records officers. Conversations with them were enjoyable and informative and provided an early insight into the workings of an enormous organization. They bear no blame for this book, but it could not have been written without them.

The staff of the TVA Technical Library—Jesse Mills, chief librarian; Margaret Bull, assistant librarian; Bill Whitehead, reference librarian; Coleen Siler, reference librarian, and Ed Best, supervisor of services—have been unfailingly supportive. Jesse and Margaret in particular have been both warm friends and knowledgeable sources of aid, while remaining staunch supporters of the agency. Jim Pilkington of the Vanderbilt University Television News Archive was of enormous help, as well as a witty lunch companion. Joseph P. Distretti, state historian at the Fort Loudoun State Park, was immensely helpful in arranging interviews of former landowners and local residents. C. J. Mellor, a UT colleague who was active among the anti-Tellico environmentalists, made available the archives of the Society for the Preservation of the Little Tennessee River.

All the persons interviewed, too numerous to name here but all cited in the notes, deserve special thanks. Our colleagues at the University of Tennessee, J. B. Finger and Charles Johnson; Guido and Kris Ruggerio; Keith Curtis, a former TVA flood control specialist; Howard Segal, a Mellon Fellow at Harvard University; Bill Jordy and Jack Thomas of Brown University; Alan Pulsipher and Danny Schaffer of TVA; Thomas McCraw of Harvard; Bill Droze of Georgia Agrirama; and Dick Lowitt of Iowa State University—all have been much more helpful and informative than they probably think.

The UTK History Department secretarial staff deserves mention for its typing, and special thanks goes to Sandra Binkley for the final typing of the manuscript.

Carol Orr, director, and Mavis Bryant, acquisitions editor, of the University of Tennessee Press have encouraged and supported our efforts, and we are grateful. Nancy Carden and Carol Guthrie, research assistants, helped us amass some critical materials. Ted Nelson, currently completing his doctorate in Geography at UT, helped us untangle the history of the land purchase policy of TVA.

Some obligations are too great to be covered by a mere acknowledgment. To our wives, Judy Wheeler and Jeanne McDonald, whose patience and support has been a meaningful constant in our lives, we owe more than can be told here. Collaboration itself is often like marriage, and both our collaborations and our marriages were enlivened by the fact that one of our wives, Jeanne McDonald, was an assiduous, competent, lively, and amusing critic and editor of the original manuscript.

INDEX

Ackard, R.D., 75, 130

Adkins, Withers: attacks Timberlake critics, 171

Alexander, Clifford, 208

Alexander, Gov. Lamar: and letter to Wagner on exception, 187

Allbaugh, Leland: and the 1959 Watts Bar meeting, 4

Allgood, Judge Clarence: and EDF lawsuit, 143–44

Alliman, Peter: and Plater, 190–91; and Sharon Lee, 191; and revival of Cherokee opposition, 201; and conversation with Freeman and attitude, 204

Aluminum Company of America (ALCOA): and the Chilhowee Dam, 48; and 1937 strike, 58; and McDade, 69–70; and Stewart memo, 78; and RERC Report, 176

Anderson, J.C. 53

Andrus, Cecil: and Tellico opponents, 199; cabinet division, 205; and Freeman memo, 205–206; and God Committee report, 211; requests Carter veto, 213

Anheuser-Busch, 173

Appalachian Anglers: anti-dam resolution, 132

Appalachian Regional Commission (ARC): and model city funding, 157

Appleby, Philip, 41; and capture of LTRVDA, 75–76

Ashe, Victor, 146

Association for the Preservation of the Little Tennessee River (APLTR): formation at 1964 Rose Island meeting, 75; and Stewart memo, 78; and courtship of Cherokees, 149–52

Association of Dispossessed Landowners, 212

Atlanta, Knoxville, and Northern Railroad, 56

Atomic Energy Commission (AEC): and Lilienthal's chairmanship, 9

Babcock Lumber Company, 56–57

Bagwell, Joe, 79

Bagwell, Dr. Troy, 61; and refusal to allow archaeology dig at Toqua, 151

Baird, State Senator Ray: TVA opponents "eggheads," 144

Baker, Senator Howard Jr.: recipient of Phillips Report, 140; urges Supreme Court, 198; and threats against ESA, 202; and God Committee, 208; and threat to Freeman, 209; and Freeman's response, 209; reaction to God Committee report, 211; and Exemption from Endangered Species Act, 212

Baker, Howard Sr.: and Harry Carbaugh, 13

Baker, Rep. LaMar, 147

Baker, Worthington, Crossley, Stansberry and Woolf, 181

Ballplay Creek, 49

Barker, C.T.: and Wagner's early career, 16–17; and earliest analysis of Tellico, 27

Bass, Senator Ross: and endorsement of Tellico, 107

Bat Creek, 49

Beard, Rep. Robin: attempt to exempt all public works projects from Endangered Species legislation, 198

Beech River watershed project, 33–34
Bell, Corydon: and Brown's comment on
    Playboy, 154; and opposition to those
    attempting to save Cherokee village
    sites, 186
Bell, Griffen, attorney general: and Carter
    cabinet in-fighting on Tellico, 202,
    205; and argument for reversal,
    207–208
Benson, Ezra Taft, 11
Berger, Thomas, 148
Bergland, Bob: Agriculture Secretary, 208
Berry, State Senator Fred: and attempts
    to stop Tellico and amendment to
    Scenic River Act, 136–37; and Dunn's
    opposition, 146
Big Piney, 51
Birmingham News, 197
Black, Kenneth, 196
Blanton, Governor Ray: and early
    criticism of Tellico, 146–47; and dam
    opponents, 192; and request from
    Evins, 192–93
Blount County, 55–56; crude birth rates,
    59; average family size, 60
Blount County Chamber of Commerce:
    and Stewart's accusations, 79
Blowers, C.W., 159; and ignoring of TVA's
    critics, 171; and meetings with
    Boeing, 173
Boeing Corporation: and initial Timberlake
    interest, 173; and 1971 Timberlake
    ground rules, 174; and profit, 174;
    and eighth-month joint study, 174;
    and fears about federal money, 175;
    and commercial recreation, 175;
    and distrust of TVA, 175–76; and
    FANTUS Corp., 175–76; and State
    Department of Conservation, 176;
    and State Planning Office, 176; and
    RERC Report, 176–77; and TVA
    distrust, 179; and worsening rela-
    tions with TVA, 179–80; and EIS, 180;
    and annexation, 180–81; and the New
    Community Development Act, 181;
    and HUD backing out, 182; and
    requests for TVA reimbursement,
    182; and withdrawal, 182–83; and
    TVA reaction to withdrawal, 183
Bogard, G.T.: and discouraging report on
    TVA new town, 161, 163
Boileau, O.C.: Boeing ultimatum to
    Seeber, 179
Boise Cascade Corporation, 173
Bost and Associates: criticism of
    Timberlake, 171

Bortorff, James, 164
Bowaters Southern Paper Company: and
    early opposition to Tellico, 42, 70;
    and General Herbert Vogel, 42; and
    Stewart memo, 78
Brewer, Alberta: and *Valley So Wild*, 194
Brewer, Carson: and News-Sentinel, 72;
    and Dickey, 72; and Nov. 17, 1963
    article, 73–74; and *Valley So Wild*,
    194
Brock, Rep. and Senator William
    "Bill": and concern about TVA land
    purchases, 73; and inquiry into
    land purchase, 130; and Dunn's
    opposition, 146
Brogan, Beauchamp: TVA OGC alleges
    EIS not required, 144; and Charles
    Johnson, 199–200
Broome, Harvey: president of local
    Wilderness Society and Douglas
    visit, 84
Brown, Bevan: and Hawk Littlejohn, 154
Brown, Dee, 148
Brown, Richard E., 171
Burbage, Beverly: amateur archaeologist,
    TVA lawyer and friend of Myers, 70;
    and urgency of archaeological work,
    150
Burch, Bob: early opponent, 68–69, 73;
    and Douglas's 1965 visit, 84; and
    opposition policy toward Cherokee,
    134
Bureau of the Budget: and new town
    funds, 169
Burger, Chief Justice Warren: and Plater,
    208
Burgner, Dan, 198–99
Bussell Island, 49
Byrnes, Michael: Bost Associates critique,
    171

Calloway Island, 49
Carbaugh, Harry C., 13
Cardwell, John, 76
Carson, Mrs. James G.: TVA informant
    and officer of TCDA, 76
Carson, John M., Jr.: first president
    LTRVDA, 74
Carter, President "Jimmy": and
    sympathy to environmentalists,
    199; recipient of opponents'
    appeals, 199; and Brogan's letter,
    199–200; and cabinet in-fighting,
    202, 205; recommends Freeman,
    202; and efforts to exempt from
    Endangered Species Act,
    212–13

Celebrezze, Judge Anthony: and Plater, 196–97; and hearing, 197; and finding, 196–97; and reaction to finding, 197–98; and TVA appeal, 198; and Baker, 198; Supreme Court hears appeal, 207

Chance, Charles, 69

Chapman, Duane: and float protest, 137

Chapman, Jefferson: anthropological critique of TVA's work, 200–201

Cherokee Indians, Eastern Band: and delegation to Chota for Douglas, 84; mollified by Wagner, 134; and TVA's policy, 134; and opposition's Indian policy, 134; and Overhill sites, 148; passivity, 149; and Chief Noah Powell, 149; and intra-tribal discord, 149; and APLTR, 149–50; and Worth Greene, 150; seek meeting with Dunn, 151; and Gray, Foster and Evans, 152; and decline of "Cherokee problem," 156; and Timberlake project, 177–78; and Prichard's attempts to save village sites, 186; and protest over archaeological procedures, 186; and revival of opposition, 201; and lawsuit, 214

Cherokee Nation (Oklahoma): and 1972 protests, 148; and TVA invitation, 148; and Chief W.W. Keeler, 148; and Hagerstrand, 149

Chestuee watershed project, 31–32

Chilhowee Dam, 48

Chota, 186

Citico, 51–52

Citizens for TVA (CTVA): and pro-TVA lobbying to reappoint Clapp, 13; and Governor Frank Clement, 13; and Vogel, 13

Clapp, Gordon: and postwar planning functions, 9–10; and review of land acquisition policies, 11; and Eisenhower, 12–13; and CTVA, 13; and Wagner's career, 17; and Fort Loudoun Extension, 28–30

Claussen Pete: and New Town funding, 162; and Timberlake schools, 170; consulted on Alexander letter, 187

Clebsch, Ed, "Save the Little Tennessee Float Protest," 137; and EDF hearing, 144

Clement, Governor Frank: and CTVA, 13; and support for Tellico, 72–73

Cody, Dennis: criticism of Timberlake, 171

Coker Creek, 58

Colaw, Larry: on land acquisition and Boeing, 179; and HUD withdrawal, 182; and criticism of Boeing, 182

Columbia Broadcasting System (CBS): sees good story, 151; 214–15

Committee on Merchant Marine and Fisheries, 198

Congleton, Joe: support of "Alternatives Report," 210

Corntassel, 51

Corps of Engineers. *See* U.S. Army Corps of Engineers

Costle, Douglas, 208

Council on Environmental Quality (CEQ): and Russell Train, Ottinger and opposition to TVA, 134–35; and Carter administration, 205

Coytee Springs, 157, 188

Creel Club: anti-dam resolution, 132

Criley, Walter: and Tennessee State Planning Commission, 72–73; anger at non-consultation of ERA, 177; and Evins's request of Blanton, 192–93

Cronkite, Walter, 215

Crosette, George, 85

Culver, John C., 200

Cunningham, Marvis: TVA liason in LTRVDA, 74; silenced by Hicks, 75–76; and Mrs. Carson, 76

David Rockefeller Associates, 162

Davidson, Charles, 164

Davis, Albert: member of opposition, 66; and shifts in taking line, 127

Decatur, 50

DeLozier, Harold: and early condemnation, 142; and IRS, 142

Democratic Party: internal battles, 166

Dempster, George: and Guy Smith, 71

Derryberry, Dr. O.M.: and the 1959 Watts Bar meeting, 4; and Seeber memo on Environmental Impact Statement, 125

Dickey, David Dale: early opposition, 68–69, 73; and Knoxville Journal, 71; and Ed Smith of New-Sentinel, 72; and Carson Brewer, 72; and accusations by Taylor, 79; and Greater Knoxville Chamber of Commerce, 81; and his thesis on Little Tennessee River, 81–82; and thesis critique by Taylor, 81–82; TVA attempt to discredit, 81–82; and Ray Jenkins, 83; and Yandell, 82; and Wagner's criticism, 82; and TVA Board's hostility to, 81–82; and Douglas's 1965 visit, 84; and withdrawal from struggle, 85; and opposition's policy

Dickey, David Dale (*continued*)
toward Cherokees, 134; and friendship with Phillips, 139
Dicks, Newton: and FUDAR, 35
Dimmick, Ralph: and citizen groups, 32
Dingell, Rep. John: and attack on Tellico, 109–110; and TVA's non-compliance with NEPA, 134–35
Dixon-Yates controversy, 12, 15
Dorward, R.E. "Bob": and takeover by opposition of LTRVDA, 75–76; and Vonore Lions Club opposition, 77; and allegations by Stewart, 78
Douglas, Justice William O.: and invitation to visit region, 65; and visit to Tellico, 83–84; 1969 trip and attack on TVA, 130–31; and the national environmental movement, 131–32; opposition's policy toward Cherokees, 134; and Toledano, 136; recipient of Phillips Report, 140
Douglas Dam: opposition of McKellar, 29
Downey, Joe, 75
Draper, Earle S.: advocate of TVA planning and RP1, 7
Duncan, Rep. John: and endorsement of Tellico, 107; recipient of Phillips Report, 140; and Moser's complaint of TVA favoritism, 186–87; attacks Interior Dept., 193; and HR4557, 198; straw poll, 206; and exemption from Endangered Species Act, 212
Dunn, Governor Winfield: and criticism of Tellico, 142–43; and timing of opposition and lawsuit, 143; and pleasure with injunction, 145; as focal point of Republican opposition, 146; Cherokees seek meeting, 151; impact of end of term, 156; and Thackston, Schoening, and State Planning Commission, 180; and opinion of Boeing, 180–81; and New Community Development Act, 181; and pressure on NPS, 186; mentioned, 187
Durisch, Lawrence L.: and the 1959 Watts Bar meeting, 4; and enthusiasm for New Town, 160

East Tennessee Economic Development District (ETEDD): and New Town plans, 168
East Tennessee Historical Society, 194
Eastern Band Tribal Council: opposition to Tellico, 151
Economic Research Associates: Report to Boeing, 175; and non-consultation with state offices, 177

Edgar, Bob, 78
Eisenhower, President Dwight D.: and philosophical opposition to TVA, 11–12; and Dixon-Yates, 12; and Clapp, 12; and Vogel, 13; and Harry C. Carbaugh, 13; and self-financing, 14; and "partnership principle," 15; and tributary area development, 31
Elk River project, 34, 40
Ellender, Senator Allen: biographical sketch, 87; and Wagner's testimony, 87–88; and Corps of Engineers' Tellico study, 88; and harsh criticism of Wagner and Tellico, 106–107; and response to Lackey, 108; and continued criticism of Tellico, 110; and National Wildlife Federation, 110; and Wilderness Society, 110; and debate with Wagner on Timberlake, 172–73
Ellington, Buford: and support for TVA, 73; and Walter Lambert, 180
Elliot, Reed: and the 1959 Watts Bar meeting, 4; and creation of Tributary Area Committee, 19; and 1959 meeting to pick Tellico, 23–24; and FUDAR, 35–36; and FUDAR benefit-cost problems, 39; and early benefit-cost studies, 43–44; and recreation benefits, 97; and access rights, 126–27; consulted on Alexander letter, 187; Alternatives Report opposition, 210
Elliott, Roland: and letters of TVA opposition sent to President Nixon, 147
Elrod, Kenneth: campaigns for TVA Board, 145
Endangered Species Act (1973), 192; and critical habitat, 195–96; and TVA lobby for exemption, 198; and HR4557, 198; and Baker's threats, 202; and God Committee, 208–209, 211; Baker's and Duncan's efforts to exempt, 212
Endangered Species Committee (God Committee): and Baker, 208; and ESA, 211
Endangered Species List, 191
Environmental Defense Fund (EDF): and 1971 suit filed to stop Tellico, 138; and accusation by Smith, 141; lawsuit, 143–45; and Etnier and Clebsch, 144
Environmental Impact Statement: Seeber memo, 125; and shift in opposition tactics, 125; passage of NEPA, 132;

Environmental Impact (cont.)
and Wagner's opposition to submitting, 134–35; and Dingell,
Ottinger, Franson and National
Audubon Society, 134–35; and
Seeber's and Evans's opinion,
137–38; and criticism by Interior,
143; Brogan alleges not required,
144; and Etnier, 145–46; Boeing's
demand that TVA file one, 180; and
TVA, 185–86
Environmental Protection Agency, 188
Ericson, E.P. "Phil": and benefit-cost ratios
and use of traditional benefits, 91;
and comparison of Tellico benefits
to those of other projects, 93; and
land enhancement benefits, 108; and
concern over Tellico's mounting
costs, 128–29; fear of rising Timberlake costs, 170
Etnier, David: and friendship with Phillips,
139; and EDF hearing, 144; and continued criticism, 145–46; and
discovery of snail darter, 156–57,
185; and significance of the discovery, 188–89; and Plater, 190; and
Raney's criticism, 192
Evans, Paul: and the 1959 Watts Bar
meeting, 4; power and water disputes, 9; and renaming of Fort
Loudoun Extension as Tellico, 24;
Stewart and "conspiracy" to stop
TVA, 79; and TVA internal opposition
to agricultural criticism, 80; and
defense of TVA's industrial policy,
102; and shift in opposition tactics,
124–25; biographical sketch, 124; and
emergence of environmentalist opposition, 125; and TVA's Cherokee
policy, 134; and need for environmental impact statement, 137–38;
and attack on Duane Chapman, 141;
promotes Cherokee assistance, 152;
screens Gober's New Town presentation, 168; and sunshine law,
193–94
Evins, Rep. Joe L.: and support of Tims
Ford over Tellico, 105, 108–109; and
request of Blanton, 192–93; and Bert
Lance, 199
Executive Order 11593, 186

FANTUS Corporation: Reports to Boeing,
175–76
Federal Bureau of Investigation (FBI):
and investigation of Littlejohn,
152–53

Federal Housing Administration (FHA),
158; and early TVA New Town contacts, 164; and land transfer scheme,
164–65
Feiss, Carl: as TVA New Town consultant,
162; and dislike of HUD legislation,
162; and encouragement of TVA,
162–63; critical of TVA's lack of
urban planning, 163–64; and urges
TVA to purchase more land, 164; and
opinion of Norris, 164; advocates
zoning in Monroe and Loudon
County, 164; and New Town vision,
167
Fitzgerald, Michael: interview with
Brogan, 144; 205
Fort Loudoun Association, 191
Fort Loudoun Dam, 28–30
Fort Loudoun Historic Area, 215
Fort Loudoun massacre, 49
Fort Loudoun State Park, 177
Foster, Bruce, 192
Foster, Minnard "Mike": and Stewart's
allegations, 79; and extension of
taking line, 128; and concern with
Tellico's mounting costs, 128–29;
biographical sketch, 133; Wagner's
choice to consolidate TVA against
opposition, 133; reactions to
Phillips Report, 141; and Venable,
141–42; promotes Cherokee assistance, 152; and attracting New
Town industry, 166; and size of
New Town, 167; report from Colaw
on HUD, 182; on Boeing withdrawal,
183; consulted on Alexander
letter, 187; 206; push for
congressional exemption, 210
Frances, Richard, NOAA head, 208
Franson, John: and National Audubon
Society opposition over environmental impact statement, 134–35.
Freeman, S. David: and private criticism
of Tellico by Palo, 42; and Wagner's
criticism of his managerial style,
184–85; biographical sketch and
confirmation, 202–204; and early
attitudes toward Tellico, 204–205;
and memo to Seeber, 205; break
with Wagner and memo to Andrus,
205–206; straw poll, 206; complaint
by Hall, 206; and Madisonville
meeting, 206; and his supporters,
206–207; retirement of Wagner and
resignation of Jenkins, 207; and
Baker's threat, 209; and response,
209; and Nellie McCall, 214

Future Dams and Reservoirs Committee (FUDAR): creation of, 23, 35; specific analysis of Tellico, 24; early staff, 35; disappointing results, 36–38; pressure by Wagner, 39; and benefit-cost difficulties, 91–92; and land enhancement problems, 94; and recreation benefits problems, 96–97

Gangwer, H.S., 42
Gant, George F.: and postwar memo on organization, 9
Gartrell, F.E.: and FUDAR, 35
General Electric Corporation: and Phillips's planned city testimony, 161; and TVA efforts to get GE support, 161
Gibbons, Clarence, 215
Giles, Elmer, 130
Gilleland, J.E.: and FUDAR, 35
Gladstone, Robert: consultant on additional land purchase, 127; and New Town housing, 163; and increased New Town population, 169; reaction to EDF lawsuit, 174
Gober, James: and New Town task force, 164; doubts New Town federal assistance, 165; and problems with low-cost housing, 167; and 1969 meetings with TRPC, ETEDD, TSPC, 168; and Timberlake schools, 170; and meetings with Boeing, 173; meetings with F.B. "Red" Williams (1970), 174
Goodyear Tire and Rubber Corporation, 161; developer of Litchfield Park, 167
Gore, Senator Albert: and letters opposing TVA, 41; and endorsement of Tellico, 107
Gove, Doris, 195
Gray, A.J. "Flash": and power people opposition to new mission, 14; and uncertainty of Tellico, 44; reaction to Phillips Report, 141; promotes Cherokee assistance, 152; and U.S. News and World Report article, 159; and 1965 New Town memo, 160; and urging Van Mol to move on New Town, 161; as member of New Town Task Force, 164; and distrust of private developer, 167; attacks Timberlake critics, 171; consulted on Alexander letter, 187
Greater Knoxville Chamber of Commerce: and Porter Taylor's criticism of TVA opposition and Dickey, 74; and TVA pressure to approve Tellico, 80–83;

and reasons for resistance 81; and Hammer Report, 81; and William Yandell, 81; and Dickey and criticism of Ray Jenkins, 81
Greenback High School Meeting, 64–66; and LTRVDA and failure, 74–75
Greene, Worth: and TVA–Cherokee contacts, 150; and Guthe, 150; and Noah Powell, 152
Grey, Sam, 52
Grosvenor, Gilbert, 85
Gulf Oil Corporation, 161
Guthe, Alfred "Ted": and Tellico archaeology, 150–51; and Worth Greene, 150; and TVA's defense of his work, 155; and letter on Chapman, 200–201
Gwin, Louis: and the discovery of the snail darter, 188; on Plater, 190; and snail darter, 196

Hagerstrand, Col. Martin: as TVA consultant, 149; and Cherokee recreation report, 178
Hall, Charles: and outmigration of young, 61; and philosophy of project boosters, 67; and evolution of TCDA philosophy, 76–77; and protest to True magazine against Douglas's article, 131–32; receives congratulations from McBride, 132; and Douglas article and opposition, 133; complaint to Freeman, 206; records filling, 215
Hamilton, G.W., 40
Hammer Report: and criticism of Tellico as industrial location, 81
Harrison, Mark: 152
Hart, John, 137
Hart, Senator Philip: recipient of opposition letters, 132
Hays, Samuel P.: and the conservationist ethic, ix–x
Heiskell, Brent, 211
Henderson, H.A.: and criticism of TVA's agricultural policy at Tellico, 102–104
Hetch-Hetchy controversy, ix–x
Hicks, Clayton, 76
Hicks, Judge Sue: and his demand for local referendum on Tellico, 64; and opposition leaders, 66; and formation of APLTR, 75; and capture of LTRVDA, 75–76; and Monroe County Farm Bureau opposition, 79; and Douglas visit, 84; and Plater, 191
Hill, Hank, 195
Horned, Douglas W., 206

House Hasson Hardware Company, 70
Howes, Robert: and FUDAR, 35–36; and
    FUDAR difficulties, 37–38; problems
    with land enchantment calculations
    and displeasure of Wagner and Van
    Mol, 94; and recreation benefits, 97;
    and opposition to Taylor's job
    benefits analysis, 100
Hughes, Charlotte Anderson, 53; watches
    river, 215
Humble Oil and Refining Corporation, 161
Humphrey, George, 11
Humphrey, Hubert, 172

Internal Revenue Service, 142
Island Creek, 49

Jandrey, Arthur S.: and land acquisition
    review, 11; reversal of Jandrey
    Report by Wagner, 19–20; and
    Ashford Todd Jr., 19
Javits, Senator Jacob: recipient of
    opposition letters, 132
Jeffrey's Hell, 58
Jenkins, Ray: and criticism of Dickey, 82;
    and congressional testimony in
    defense of Tellico, 107
Jenkins, William "Bill," 193–94, 207
Johnson, Charles, 199–200
Johnson, Hendon, 27
Johnson, President Lyndon Baines: and
    letter from Dickey, 82; and 1966
    budget, 83; and fiscal 1966 budget,
    Tims Ford and Tellico, 104–105; and
    1966 Tellico appropriation, 110; and
    1967 budget, 161
Jones, A.R.: and early letters of
    opposition, 41; fearful of large land
    purchase policy, 44; and opposition
    to Dickey, 81–82; opposition to land
    enchantment figures, 95

Kampmeier, R.A.: and the 1959 Watts Bar
    meeting, 4; and Wagner's career, 18
Keeler, W.W.: Chief, Cherokee Nation, 148
Kefauver, Senator Estes: and Fort Loudoun
    Extension, 30
Kennedy, President John F.: and Wagner's
    appointment, 17; and Marguerite
    Owen, 17; and Muscle Shoals 30th
    anniversary speech, 39–40; and
    letters opposing TVA, 41
Kennedy, William, 59
Kercher, Kristin, 175
Kilbourne, Richard "Dick": and the 1959
    Watts Bar meeting, 4; and creation
    of Tributory Area Committee, 19; and

Beech River, 33–34; and FUDAR, 35;
    and TVA employees as officers of
    TCDA, 76; and faith in TVA's ability
    to solve regional problems, 128
Kimmons, G.H., 171
Knetsch, Jack: FUDAR committee land
    enhancement studies, 38
Knoxville Chamber of Commerce. See
    Greater Knoxville Chamber of
    Commerce
Kribbs, Lee, 169

Lackey, John M.: opposition leader, 66;
    and formation of APLTR, 75; and
    capture of LTRVDA, 75–76; and
    Vonore Lions Club opposition, 77;
    and Monroe County Farm Bureau
    opposition, 79; and testimony before
    Ellender subcommittee, 108
Lakeside, 51
Lambert, Walter: advises TVA to
    incorporate New Town, 180
Lance, Bert: and Joe L. Evins, 199
Land purchase policy: and early dams,
    10; and Jandrey Report, 11; and shift
    in policy, 11; and Jones's fear, 44;
    and Brock's concern, 73; and Taylor,
    91; in relation to Tellico's benefit-
    cost ratio and Van Mol, 93–94; and
    Gladstone and additional purchase,
    127; and Foster and extension of
    taking line, 128; and Brock's inquiry,
    130; Taylor's ruling, 142; and addi-
    tional acres for Timberlake, 171
Laycock, George: criticism of TVA in the
    Dilligent Destroyers (1970), 136
Leber, General W.P.: and Corps of
    Engineers' Tellico study and
    Ellender, 88
Lee, Norman, 211
Lee, Sharon, 191
Lehman Brothers: and Wagner's New
    Town presentation, 162
Lenoir City, Tenn., 168; and Gober's New
    Town remarks, 168
Lesesne, Ed, 201
Lilienthal, David: opposition to Arthur
    Morgan and early philosophy of TVA,
    6; opposition to regional planning
    and RP 1, 7, 16; chairman of the
    AEC, 9; and land purchase policy
    debates, 27; and McKellar, 29; defeat
    of Arthur Morgan, 160
Litchfield Park, 167
Little Tennessee River Valley Develop-
    ment Association (LTRVDA): and
    1964 Greenback meeting, 64; and

Little Tennessee River (*continued*)
OTAD, 74; and John M. Carson Jr.,
74; and Marvis Cunningham, 74; and
failure of Greenback meeting, 74–75;
and Wagner, 74–75; and takeover by
opposition, 75–76; abandonment by
TVA, 76.

Littlejohn, Hawk: and opposition, 152;
biographical sketch, 152–53; and
TVA's request of FBI investigation,
152–53; and Griscom Morgan, 153;
and Arthur Morgan, 153; and
Powell's opposition, 154; and protests
with Mashburn, 154; and fall,
154–55; decline of "Cherokee
problem," 156

Locke, John: and emerging environmentalist opposition, 124

Long, Huey, 87

Loose, William, 42

Lotspeich, Mrs. Ethel: owner of *Knoxville
Journal*, 71

Loudon County, 55–56; crude birth rates,
59; average family size, 60; and Carl
Feiss advocacy of zoning, 164

Loudon Regional Planning Commission:
and early opposition to Tellico, 41

Lowry, Robert: and FUDAR, 35

McBride, Don: congratulates Hall on
criticism of Douglas, 132; and
reaction to injunction, 145; and mail
campaign, 155–56; and enthusiasm
for GE, 161; and characterization of
Moser, 187; TVA Board and sunshine
law, 193–94

McCall, Asa, 192

McCall, Nellie: removal, 214–15

McClung museum, UT, 186

McDade, Hugh: ALCOA executive in
opposition, 69–70; and allegations by
Stewart, 78

McDonald's restaurant: at Timberlake, 169

McGhee Island, 49

McGloghlin, Marquarite: TVA staff
investigation of Phillips, 141

McGruder, Captain Beverly, 38–39

McKellar, Kenneth: and opposition to
Douglas Dam and Lilienthal, 29

McLemore, Chambers, 52

Madisonville, 50, 58

Marquis, Robert: and FUDAR, 35

Martin, Roscoe: and TVA program
definition, 11

Maryville-Alcoa Times: support of Tellico,
142

Mashburn, Fran: and Hawk Littlejohn, 154

Mattern, Don: and 1960 projects budget,
24; FUDAR and benefit-cost
problems, 39; and relocation data,
128; and Freeman, 202; Alternatives
Report opposition, 210

Mayer, Albert: and critique of New Town,
167

Mayfield, Scott: early opposition to
Tellico, 70

Maynard, Jack, 52, 59

Mellor, C.J. "Jeff", 195

Melton Hill, 40, 109

Memphis Commercial Appeal, 197

Menhinick, Howard K.: and postwar
planning reports, 9

Miller, Harold: and Tennessee State
Planning Commission, 73

Millsaps, Alean, 52

Milton, Mrs. Alice: and Douglas's 1965
visit, 84; and Plater, 191

Monroe Citizen: straw poll on TVA's
project, 77

Monroe County, 55–56; crude birth rates,
59; average family size, 60; and
dependency rate, 60; and Carl Reiss
advocacy of zoning, 164

Monroe County Farm Bureau: and anti-
Tellico resolution, 79; and Hicks and
Lackey, 79

Morgan, Arthur E.: early philosophy of
TVA, 6; and Littlejohn, 153; and
model TVA community, 159; conflict
with Harcourt Morgan and
Lilienthal, 160

Morgan, Griscom: and Littlejohn, 153

Morgan, Harcourt: early opposition to
Arthur Morgan, 6; opposition to
regional planning and RP1, 16; and
land purchase policy debates, 27;
defeat of Arthur Morgan, 160

Moroney, Pat: and TCDA's pressure on
farm bureaus, 79

Moser, Thomas Burel: opposition leader,
66; and EDF lawsuit, 143; and
complaint to Duncan about TVA
favoritism, 186–87; removal, 215

Moser family, 61

Moses, Fred J. Jr.: and opposition to
Tellico, 108

Moynihan, Daniel P.: June, 1970 urban
planning meeting, 158

Muir, John: and the Hetch-Hetchy
controversy, x

Mulberry, 51

Murphy, John M., 198

Myers, Richard, Jr: amateur archaeologist
in opposition, 70

Nash, C.W., 108–109
Nashville Banner, 197
National Audubon Society: anti-dam resolution, 132; and TVA's non-compliance with NEPA, 134–35
National Congress of American Indians, 151
National Environmental Policy Act (NEPA): 1969 passage, requiring environmental impact statement, 132; and TVA's alleged violation and Franson, Ottinger and Dingell, 134–35; Taylor's ruling and restraining order, 144; and Taylor's compliance ruling, 188
National Geographic magazine: TVA, Wagner and Douglas visit, 84–85
National Historic Preservation Act, 188
National Parks Services, 186
National Register of Historic Places, 186
National Wildlife Federation: and Ellender and opposition to Tellico, 110
Native Americans Religious Freedom Act, 214
Nelson, L.B.: and TVA agriculturalists' opposition to Tellico, 103
New Community Development Act, 180–81
New Johnsonville, TN, 10
New Philadelphia, 50
Newton, Albert, 16
Nichols, Dwight, 179
Nixon, President Richard M.: protests against TVA stirred up by Douglas's True magazine article, 132; letters prompted by Sports Afield article, 147; presidential victory and Tellico, 156; and 1970 urban planning meeting, 158; impending victory, 1968, 166
Norris, Senator George: and the Hetch-Hetchy controversy, ix–x
Norris, Tenn.: as construction town, 159; and cost overruns, 159–60; and sale, 160; and present status, 160
North American Rockwell Corporation, 173
Notchy Creek, 49

Oak Ridge National Laboratories (ORNL): and Bill Owen, 188
Office of Tributary Area Development (OTAD); and the LTRVDA, 74; and regional development optimism, 128
Oliver, John: and Estes Kefauver, 30
Ottinger, Rep. Richard: and attack on Tellico, 109–110; and TVA's non-compliance with NEPA, 134–35

Owen, Bill: ORNL, Tellico and decision making, 187–88
Owen, Marguerite: and Wagner's career, 17; and John F. Kennedy, 17; lobbying against opposition, 73

Pack, David: and recreation benefits, 96–97
Palo, George: differences between power and water people in TVA, 8; and 1959 meeting to pick Tellico, 23–24; and FUDAR, 36; and early benefit-cost problems, 40, 42; and private criticism of Tellico, 42; and early benefit-cost studies, 43–44; biography, 89; early benefit-cost doubts, 88–89; and FUDAR analysis, 91; and post-FUDAR benefit-cost work, 94–95; and recreation benefits, 97; and concern with Tellico's mounting costs, 128–29
Parker, T.B.: and early exploratory work on Little Tennessee, 29
Penegar, Dean Kenneth: and "Zyg" Plater, 194–95
Perry, Clarence: urban planner, 171
Perry, J.R., 36.
Philippe, Gerald L: and 1966 testimony on planned cities, 161
Phillips, Keith: biographical sketch, 139; origins of Phillips Report, 138–41; and friendship with Dickey and Etnier, 139; on unconstitutionality of TVA's use of eminent domain, 191
Phillips Report: origins and findings, 138–41; sent to Baker, Douglas and Duncan, 140; reactions by Foster and Gray, 141; and TVA's EIS, 185–86
Pinchot, Gifford: and the Hetch-Hetchy controversy, ix
Plant, Ben, 206
Plater, Zygmunt: mentioned by Palo to Freeman, 42; biographical sketch, snail darter and opposition, 189–92; and attempts to stop cutting, 192; and tenure fight, 194–95; and critical habitat, 196–97; and Celebrezze, 197; and Supreme Court, 208; and support of Alternatives Report, 210; and settlement of legal fees, 215
Playboy magazine: Brown's reaction, 154
Poplar Springs, 51
Powell, Noah: Chief, Eastern Band, 149; compromise with TVA, 152; and Worth Greene, 152; and opposition to Littlejohn, 154

Prichard, Mack: and Douglas visit, 84; and new recruits to opposition, 138; and Seeber's attempts to silence, 141; and attempts to save Cherokee village sites, 186

Proffitt, Harwell, 206

Public Law 89–665, Section 106, 186

Pumpkin Center, 51

Quillen, Rep. James "Jimmy", 198–99

Rabel, Ed, 215

Raney, Dr. Edward S., 192

Rasnic, Carl, 181

Real Estate Research Corporation (RERC): Report to Boeing, 176

Rechichar, Stephen: interview with Brogan, 144; 205

Reed, Nathaniel, 193

Republican Party: revival in late 1960s, 166

Reynolds Aluminum Company, 162

Rhodes, John, 109

Riggs, Fletcher: and recreation benefits, 97

Ritchey, Jean: opposition leader, 66; removal, 215

Roarks, Robert, 164

Roll, Charles W. Jr., 147

Roosevelt, President Franklin D.: and model community, 159

Rose Island Nursery: site of 1964 opposition meeting, 75; and formation of APLTR, 75

Rountree, Jack, 191

Rozek, John: and access rights, 126–27; and New Town promotion opinion, 162; as member of New Town Task Force, 164

Rural Electrification Administration (REA), 57

Rutter, E.J., 38

Ryan, Phillip, 214

Schell, Kerry: and FUDAR land enhancement, 38; and criticism of benefit-cost ratio, 91

Schoening, Niles: and New Community Development Act, 180–81

Schultze, Charles: and God committee, 208, 211

Seeber, Lynn: memorandum from Evans on opposition, 124–25; and Smith's complaints of egg attack, 137; and need for environmental impact statement, 137–38; and attempts to silence Prichard, 141; and "reason-able balance" theory, 155; and master developer for New Town, 173; ultimatum from Boileau, 179; and Wagner's criticism of Freeman's managerial style, 184–85; and refusal to stop logging, 192; and sunshine law, 193–94; and Freeman's memo, 205

Seigworth, Kenneth, 44; and TVA comments on early opposition, 69; and rebuttal of Brewer article, 74

Self-financing of TVA power: early development, 14

Selznick, David, 101

Senate Document 97, 43

Shelley, E.A., 38–39

Sierra Club: and the Hetch-Hetchy controversy, x

Sir, Joe: and Vogel, 34

Smith, Ed: and Knoxville News-Sentinel, 71–72

Smith, Frank: opposition to David Dale Dickey, 81; and Douglas visit, 84–85; and attack on Dingell and Ottinger, 109–10; complains of egging attack, 137; accuses EDF, 141; and Elrod's campaign, 145; and enthusiasm for GE, 161; and contacts with private corporations, 161–62; and Board approval of New Town scheme, 165

Smith, Guy: and Knoxville Journal opposition to Tellico, 71

Snail darter: discovery by Etnier, 156–57, 185; significance, 188–89; and critical habitat, 195–96; and Gwin, 196; and ruling by Celebrezze, 197–98; and TVA petition to delist, 198; and Baker urging, 198; and Baker threats, 202; and cabinet division, 202; and Supreme Court, 207–208; and Judge Taylor, 214

Snider, Ben, 60; and initial business support for TVA, 67–68

Snider, Carolyn: initial business support for TVA, 67

Snyder, John: and land policy debate, 28

Southeastern Indian Antiquities Association, 138

Sports Afield: and rumors of upcoming publication, 145; article on Tellico, 147

Sproul, Judge Harvey: and Wagner and TAPC concerns over Tellico slowdown, 129; and Douglas article and opposition, 133; and initial reaction to New Town plan, 168; and fears of County tax base, 168–69

Stafford, F. "Benny," 211–12.
Stansberry, F.D., 164
Stansberry, Fred: early opposition, 69
Starnes, Wayne: and the snail darter, 188
Stephenson, C.M., 178; complaint to Gober about Boeing, 179
Stern, Peter: and TVA's efforts to lure GE, 161; as members of New Town Task Force, 164; and New Town private investors, 164
Stewart, Gilbert "Pete": on alleged connections between ALCOA, Bowaters, Vonore Lions Club and APLTR, 78–79; and uncertainty over number of relocations, 127–28
Stokely-Van Camp Company, 57
Sunset International Petroleum, 161
Swafford, Wade, 126
Sweetwater, 66

Talhassee, 52
Talmadge, Senator Herman: recipient of opposition letters, 132
Tanasi, 186
Taylor, J. Porter: and the 1959 Watts Bar meeting, 4; differences between TVA power and water people, 8; and Wagner's early career, 16–17; and creation of Tributary Area Committee, 19; and earliest analysis of Tellico, 27; and FUDAR, 35–36; and accusations against Dickey, 79; and critique of Dickey's thesis, 81–82; and TVA overpurchase policy, 91; and navigation benefits, 100; and his jobs benefits analysis and Howes's criticism, 100–101; and Venable, 141–42; urges larger TVA role in New Town, 166–67; consulted on Alexander letter, 187; and critical habitat, 196–97; comments on Freeman, 202, 204
Taylor, Jonathan Ed: continued, but weaker opposition to TVA from the Cherokees, 156
Taylor, Judge Robert: ruling on land acquisition disputes, 142; and EDF lawsuit, 143–44; and restraining order, 144; rules subject to NEPA, 144; decision appealed but upheld, 144; and TVA compliance with NEPA, 188; and issue of permanent injunction, 198; and snail darter 214
Tellico Area Planning Council (TAPC): and concern over Tellico slowdown and need to lobby, 129; and Douglas article and opposition, 133

Tellico Blockhouse, 49
Tellico Plains, 66
Tellico River, 49
Tellico Regional Planning Council: and New Town plans, 168
Tellico Reservoir Development Agency: and Charles Hall, 218; and transfer of TVA land, 218; and promotion of site, 218–19
Tennessee Conservation League: anti-dam resolution, 132
Tennessee Department of Conservation: 176; and meeting with Boeing, 176
Tennessee Farm Bureau Federation: resolution of opposition to Tellico, 79
Tennessee Game and Fish Commission: early opposition, 68; estimates of trout fishing on Little Tennessee, 98
Tennessee Office of Urban and Federal Affairs, 185
Tennessee Outdoor Writers Association: early opposition, 68
Tennessee Scenic River Act, 136–137
Tennessee State Planning Commission: and divided opinions on Tellico, 72–73; and New Town plans, 168; and Dunn and Boeing, 180–81
Tennessee Valley Public Power Association: and TVA, 13
Timberlake: and funding, 158; location, 159; and naming, 169; schools, 170; detailed plans, 170; and rising costs, 170; criticism, 171; cost estimates, 171; and additional land, 171; and congressional difficulties, 172–73; and Ellender, 172–73; and master developer, 173; and Boeing, 173–83; and EDF suit, 174–75; and report on industrial potential, 176; and Boeing and state offices, 176–77; and collapse of Boeing agreement, 180, 182; and Foster and Wagner's opinion after Boeing withdrawal, 183; and threat of annexation by Vonore, 181
Timberlake Advising Board: creation and membership, 175
Timberlake Development Corporation, 179
Timberlake Task Force: creation and membership, 164; and fiscal viability study, 165; and detailed study, 169–70; Kimmons's critique of, 171
Tims Ford project, 104–108, 109
Todd, Ashford Jr.: and Jandrey Report, 19
Toledano, Ralph de: and the Douglas article, 136

Toliver Island, 49
Toqua, 151
Toqua Creek, 49
Train, Russell: CEQ and opposition to TVA's environmental policy, 135–36
Tri-Counties Development Association (TCDA): formation by TVA as replacement for LTRVDA, 76; and Mrs. Carson, 76; and TVA employees as officers, 76; and philosophy, 76–77; as reflection of local opinion on Tellico, 77; and pressure on farm bureaus to support Tellico, 79; and Douglas article and opposition, 133
Tributary Area Committee: and creation by Wagner, 19
Trout Unlimited: anti-dam resolution, 132
True magazine: and Douglas's article, 131; and generation of anti-TVA opinion, 134
Truman, President Harry S.: and political debates over TVA, 9

Union Carbide Corporation, 176
Union Oil Corporation, 161
U.S. Army Corps of Engineers: and Leber and Ellender, 88; and recreation benefits, 96
U.S. Department of Housing and Urban Development (HUD), 158; and early TVA New Town contacts, 164; and water and sewerage system, 165; strings attached to New Town funding, 175; and withdrawal of funds, 182
U.S. Department of the Interior: and criticism of TVA's EIS, 143; attacked by Duncan, 193
U.S. Fish and Wildlife Service: and critical habitat, 195–96; and TVA petition to delist, 198
U.S. Government Accounting Office (GAO): and criticism of benefit-cost ratio, 200; and desire for a review, 200
U.S. Gypsum Corporation, 173
U.S. News and World Report, 158
U.S. Supreme Court: hears appeal, 207–208; decision to uphold, 208
United States Steel Corporation, 173
University of North Carolina, Chapel Hill: study on how to incorporate blacks into New Town, 167

Valley So Wild, 194
Van Mol, James: and creation of Tributary Area Committee, 19; and FUDAR, 38;

and Loudon Regional Planning Commission, 41; and TVA employees as officers of TCDA, 76; and opposition to Dickey, 81–82; and Palo's benefit-cost fears, 89; and large land purchase, 93–94; displeasure with Howes and FUDAR, 94; and benefit-cost problems, 94–96; and land enhancement benefits, 108; and attempts to stifle internal opposition, 133; and Durisch's "New Town" enthusiasm, 160; and Board's acquiescence in the "New Town" idea, 160–61; and urging move on New Town, 161; and New Town under Section 22 of TVA Act, 165; and fear of New Town's affect on Tellico benefit-cost ratio, 165; urges New Town as industrial attractor, 165–66; screens Gober's New Town presentation, 168
Van Mol, John, 198
Venable, Sam: and trout fishing on the Little Tennessee, 98–99; and pressure not to oppose, 141–42; call to Gwin about snail darter, 188
Vietnam War: and the slowdown in Tellico appropriation, 129; and growing national frustration, 135; siphons of federal money, 166
Vines, William III, 206
Vogel, General Herbert: and appointment by Eisenhower, 13; and CTVA, 13; and Wagner's career, 19; and Joe Sir, 34; and early benefit-cost problems, 40; and meeting with Bowaters, 42
Vonore, Tennessee: and Vonore and New Town, 168
Vonore High School: meeting between environmentalists and New York lawyers, 125
Vonore Lions Club, 77; and Stewart memo, 78
Vonore Planning Commission: and annexation of Timberlake, 181
Voorduin, W.L., 29

Wagner, Aubrey "Red": and the 1959 Watts Bar meeting, 3, 6; and biography, 16–21; and assessment of early directors' ideas, 18–19; and Tributary Area Committee, 19; and reversal of Jandrey Report, 19–20; and renaming of Fort Loudoun Extension as Tellico, 24; and Beech River, 33–34; disappointment with FUDAR, 36–38; places pressure on

Wagner, Aubrey, (continued)
FUDAR, 39; and 1964 Greenback meeting, 64; and Brock's concern about land purchases, 73; and LTRVDA and failure of Greenback meeting, 74–75; and TVA employees as officers of TCDA, 76; and criticism of Dickey, 82; and Douglas visit, 84–85; and National Geographic, 85; and Ellender, 87–88; and determination to bring Tellico to Board, 91–92; displeasure with Howes and FUDAR, 94; and recreation benefits, 96; and underestimation of navigation benefits, 99; and congressional criticism of industrial policy, 105; and defense of Tellico, 109; and Sproul and TAPC concerns over Tellico slowdown, 129; and attempts to stifle internal opposition, 133; and Foster, 133; and attempts to add generating facilities, 134; mollifies local opposition, 134; mollifies Cherokees, 134; and environmentalist critics and refusal to submit environmental impact statement, 134–135; and Dingell's accusation, 134–135; and "rocks in head" statement, 141; and the "reasonable balance" theory, 155; and Timberlake planning, 159; credit for "Model City" revival in TVA 160; and meeting with GE, 161; and presentation of New Town to Lehman Brothers, 162; and Board approval of New Town scheme, 165; and response to Mayer critique, 167; and debate with Ellender over Timberlake, 171; on Boeing withdrawal, 183; and criticism of Freeman's managerial style, 184–85; and Alexander's letter, 187; and TVA Board and sunshine law, 193–94; and refusal to do GAO review, 200; management style and Freeman's, 203–204; and break with Freeman, 205–206; and retirement, 207; visits dam site, 215

Walt Disney Productions, 173
War Production Board: and TVA, 29–31
Webster, Daniel, 185
Weeks, Bradley: and fishing on the Little Tennessee, 98
Weeks, Chris: senior hippopotamus packer, 170
Wessenauer, Gabriel O.: and the 1959 Watts Bar meeting, 4; and advocacy of coal-fired steam plants, 10; and multi-purpose projects and new mission, 16; and Wagner's career, 18; and early benefit-cost problems, 40; and early benefit-cost studies, 43–44; asked to push New Town, 162
Westinghouse Corporation, 173
Wetherholt, Robert: and concern with Tellico's mounting costs, 128–29
White, Lee C.: EDF criticized by Frank Smith, 141
White's Fort, 49
Wilderness Society: and Ellender and opposition to Tellico, 110
Wiley, William: reaction to Cherokee recreation report, 178; on Timberlake land purchases, 179; and Boeing and annexation, 181
Wilkins, Price: early opponent, 68–69, 73
Williams, F.B. "Red": 1970 meetings with Gober, 174; and salesmanship in Nashville, 180
Williams, Gerald: and intra-TVA criticism of Tellico, 79–80; and TVA survey, 80
Wilson, Charles, 11
Woodworth, Roger: and agriculture opposition to Tellico within TVA, 79–80, 102; and open opposition, 206
World War II: early projects and Fort Loudoun Extension, 29–30

Yandell, William: and Greater Knoxville Chamber of Commerce, 81

TVA and the Tellico Dam, 1936–1979 has been set into type on an Alphatype CRS digital phototypesetter in ten point Melior with two points of spacing between the lines. Sans serif Harry Obese was used for display. The book was designed by Jim Billingsley, composed by Superior Type, printed offset by Thomson-Shore, Inc., and bound by John H. Dekker & Sons. The paper on which the book is printed bears acid-free characteristics for an effective life of at least three hundred years.

THE UNIVERSITY OF TENNESSEE PRESS : KNOXVILLE